# Studies in Fuzziness and Soft Computing

## Volume 315

*Series editors*

Janusz Kacprzyk, Polish Academy of Sciences, Warsaw, Poland
e-mail: kacprzyk@ibspan.waw.pl

For further volumes:
http://www.springer.com/series/2941

*About this Series*

The series "Studies in Fuzziness and Soft Computing" contains publications on various topics in the area of soft computing, which include fuzzy sets, rough sets, neural networks, evolutionary computation, probabilistic and evidential reasoning, multivalued logic, and related fields. The publications within "Studies in Fuzziness and Soft Computing" are primarily monographs and edited volumes. They cover significant recent developments in the field, both of a foundational and applicable character. An important feature of the series is its short publication time and world-wide distribution. This permits a rapid and broad dissemination of research results.

Michael B. Gibilisco · Annie M. Gowen
Karen E. Albert · John N. Mordeson
Mark J. Wierman · Terry D. Clark

# Fuzzy Social Choice Theory

Springer

Michael B. Gibilisco
Rochester
New York
USA

Annie M. Gowen
Papillion
Nebraska
USA

Karen E. Albert
Lincoln
Nebraska
USA

John N. Mordeson
Department of Mathematics
Creighton University
Omaha
Nebraska
USA

Mark J. Wierman
Department of Computer Science
Creighton University
Omaha
Nebraska
USA

Terry D. Clark
Department of Political Science
Creighton University
Omaha
Nebraska
USA

ISSN 1434-9922          ISSN 1860-0808    (electronic)
ISBN 978-3-319-35671-6     ISBN 978-3-319-05176-5    (eBook)
DOI 10.1007/978-3-319-05176-5
Springer Cham Heidelberg New York Dordrecht London

Printed on acid-free paper

Springer is part of Springer Science+Business Media (www.springer.com)

*Michael Gibilisco dedicates this book to his parents whose moral, and, at times, financial support, made the work possible. They have always encouraged him and his research throughout school and this project, and his passion for learning began with them.*

# Preface

For almost a decade, three of the authors of this book (John N. Mordeson [Mathematics], Mark J. Wierman [Computer Science], and Terry D. Clark [Political Science]) have engaged in an extensive research agenda applying fuzzy set logic to social choice theory. That collaboration has been rewarding on a number of dimensions. Among the most rewarding aspects has been the students who have joined us in that collaboration. Michael Gibilisco, the primary author of this book, is one of those students. Like Michael, many of our students have discovered the joys of research and subsequently gone on to pursue the Ph.D. Even among those who have not, the intellectual commitment and rigor that the effort has demanded has assisted d virtually all of them in discovering their life's vocation.

Of course, the discoveries that we have made along the way have been rewarding as well. While our research agenda has its genesis in the desire to apply formal models to empirical problems, the theoretical work has necessarily consumed a substantial degree of our effort and attention. This book is in many ways a summary of what we have discovered about theory. Nonetheless, at the conclusion of each of the chapters that follow we make a conscious effort to discuss empirical applications.

The social choice issues that we address are those that one familiar with the research agenda would expect. We give consideration to the effects of applying fuzzy logic to Arrow's Impossibility Theorem, Black's Median Voter Theorem, and the Gibbard-Sattherthwaite Theorem. Along the way we consider varying definitions of key concepts in social choice theory. As the chapters demonstrate, a fuzzy approach admits of a good deal more variation in these definitions than the customary approach allows. It is therefore not surprising that many of the theorems no longer hold under certain conditions. What is even more surprising, however, is how resilient the major social choice theorems are. While they no longer hold under certain fuzzy definitions, they hold under most of them.

We admit that this is contrary to what we expected when we began our effort almost a decade ago. At that time, it seemed to us that the problems that empiricists were having with applying social choice theory to their work owed to the perverse outcomes rooted in a mathematics that assumed too much precision in human thinking. The fuzzy approach intuitively seemed to offer a possible solution by modeling

less precision and clarity in human thinking on preferences and preference orders.
While this has turned out to be the case in a number of instances, thereby permitting
a marginal decrease in the estimation error on the part of fuzzy counterparts to familiar models in the comparative politics literature, the estimated outcome are still
not what we might like them to be. But we will hold that conversation for a subse-
quence volume on our empirical applications. In this volume, we focus on mostly
on our theoretical conclusions.

The volume's primary author, Michael B. Gibilisco, is currently pursuing the
Ph.D. in political science at the University of Rochester. Michael wishes to ac-
knowledge that his work benefitted from the faculty and students in the Fuzzy
Mathematics Research Colloquium throughout the years. In particular, he is grate-
ful to Carly Goodman for her patience when reading drafts and listening to the
rough beginnings of ideas. Michael also extends his thanks to Creighton University's
Graduate School, specifically, the International Relations department, for research
support. John N. Mordeson dedicates this book to his grandparents Katherine and
John Niece and Mary Ellen and Nels Mordeson. Mark J. Wierman dedicates this
book to Mary K. Dobransky. Annie Gowen thanks her co-authors, whose guidance
and patience made her work possible. She dedicates her contribution to her dearest
friend, Matthew Cockerill, for his unfailing encouragement. Karen Albert, who in-
tends to pursue the Ph.D. in political science, would like to dedicate her work in this
book to her parents, James and Carol Albert. Terry D. Clark dedicates his work in
this book to his wife of thirty-seven years, whom he adores, Marnie.

Creighton University,                                                                    John N. Mordeson
Omaha, NE,                                                                                  Mark J. Wierman
December, 2013                                                                             Terry D. Clark

# Acknowledgements

This research grew out of the Fuzzy Spatial Modeling Colloquium. The colloquium is indebted to Professor Bridget Keegan, Interim Dean of the College of Arts and Sciences at Creighton University whose support has been invaluable in sustaining our efforts. We are also indebted to Dr. George and Mrs. Sally Haddix for their generous endowments to the Department of Mathematics at Creighton University.

# Contents

# List of Figures

# List of Tables

# Acronyms

| | |
|---|---|
| FPAR | fuzzy preference aggregation rule |
| FSCF | fuzzy social choice function |
| FWPR | fuzzy weak preference relation |
| G-S | Gibbard-Sattherthwaite Theorem |
| IIA | independence of irrelevant alternatives |
| ND | non-dominated set |
| NI | non-imposition |
| PAR | preference aggregation rule |
| PC | Pareto Condition |
| WP | weak Paretianism |

# Chapter 1
# Fuzzy Social Choice

**Abstract.** This chapter presents general concepts and definitions which will be used throughout this book. A fuzzy subset is defined as a collection of values between 0 and 1 which represents a degree of membership of each element in the set. When comparing two elements from two sets and the degree of preference for one element to the other, we call this a fuzzy relation.

## 1.1   The Purpose and Plan of the Book

There is a growing literature extending fuzzy set mathematics to traditional social choice theory. Fuzzy social choice articles have been published in a number of the best journals in both economics and fuzzy mathematics. However, this literature is marked by a wide degree of variation in assumptions and definitions. Even the non-specialist would fail to note the lack of a standard model for fuzzy social choice. Moreover, the literature fails to draw connections between the major theorems, treating them in isolation from one another. The goal of this book is to present a comprehensive analysis of fuzzy set theoretic models of social choice. We address four major areas with which scholars have concerned themselves:

- the existence of a Maximal Set,
- Arrow's Theorem,
- the Gibbard-Sattherthwaite Theorem, and
- the Median Voter Theorem.

Our aim in addressing each of these problems is to contribute to the development of fuzzy social choice theory. Toward that end, we review the past literature, its assumptions, and the relationship between these assumptions. We then extend that literature while endeavoring to illustrate the relationships between these four major research concerns. Finally, our motive in considering these issues is not only to contribute to theory but to encourage empirical research using fuzzy approaches. Toward the latter end, we present applications of fuzzy social choice that we believe will be conducive to empirical research.

M.B. Gibilisco et al., *Fuzzy Social Choice Theory*,  
Studies in Fuzziness and Soft Computing 315,  
DOI: 10.1007/978-3-319-05176-5_1, © Springer International Publishing Switzerland 2014

The book is organized as follows. In chapter two, we consider the existence of a maximal set for fuzzy preference relations. We demonstrate that a non-empty maximal set is guaranteed to exist for fuzzy preference relations under conditions that are considerably less restrictive than those required in the conventional model. Moreover, a non-empty maximal set may exist even in the absence of these conditions. In chapter three, we consider Arrow's Theorem. We find that under certain conditions a fuzzy aggregation rule will satisfy all five Arrowian conditions, to include non-dictatorship. The fact that Arrow's result no longer holds under these conditions is an important conclusion that should encourage further theoretical and empirical work. In chapter four, we turn our attention to the Gibbard-Sattherthwaite (G-S) Theorem which is closely tied in the conventional social choice literature to Arrow's Impossibility Theorem. Past considerations of the G-S Theorem have focused on fuzzy individual preferences. Chapter four argues that when both individuals and groups can choose alternatives to various degrees, social choice can be both strategy-proof and non-dictatorial. In chapter five, we consider the Median Voter Theorem. When preferences are single-peaked and fuzzy strict preferences satisfy certain properties (that they are both partial and regular), we demonstrate the conditions under which a non-empty fuzzy maximal set is guaranteed. Moreover, we find that Black's Median Voter Theorem holds when fuzzy preferences are strict, but it no longer does so when fuzzy preferences are weak. We conclude and address issues that remain to be considered in chapter six. Before proceeding, we give a brief consideration to the general concepts that provide the baseline assumptions for our treatment of fuzzy social choice.

## 1.2 General Concepts

### 1.2.1 Sets

Let $S$ be a set and let $A$ and $B$ be subsets of $S$. We use the notation $A \cup B$ and $A \cap B$ to denote the *union* and *intersection* of $A$ and $B$, respectively. We also let $B \setminus A$ denote the *relative complement* of $A$ in $B$. The (relative) complement of $A$ in $S$, $S \setminus A$, is sometimes denoted by $A^c$ when $S$ is understood. It is easily verified that

$$(A \cup B)^c = A^c \cap B^c$$

and

$$(A \cap B)^c = A^c \cup B^c.$$

These equations are known as DeMorgan's Laws.

Let $x$ be an element of $S$. If $x$ is an *element* of $A$, we write $x \in A$, otherwise we write $x \notin A$. We use the notation $A \subseteq B$ or $B \supseteq A$ to denote that $A$ is a *subset* of $B$. If $A \subseteq B$ and there exists $x \in B$ such that $x \notin A$, then we write $A \subset B$ or $B \supset A$ and we say that $A$ is a *proper subset* of $B$. The *cardinality* of $A$ is denoted by $|A|$.

The *power set* of $A$, written $\mathscr{P}(A)$, is defined to be the set of all subsets of $A$, i.e., $\mathscr{P}(A) = \{B \mid B \subseteq A\}$.

Let $A$ be a subset of a set $S$. Define $1_A : S \to \{0,1\}$ by $\forall x \in S$, $1_A(x) = 1$ if $x \in A$ and $1_A(x) = 0$ otherwise. Then $1_A$ is called the *characteristic function* of $A$ in $S$.

We shall use the following conventions to represent some standard sets:

$\mathbb{N}$  the set of positive integers,
$\mathbb{Z}$  the set of integers,
$\mathbb{Q}$  the set of rational numbers, and
$\mathbb{R}$  the set of real numbers.

Let $X$ and $Y$ be sets. If $x \in X$ and $y \in Y$, then $(x,y)$ denotes the *ordered pair* of $x$ with $y$. The Cartesian cross product of $X$ with $Y$ is defined to be the set $\{(x,y) \mid x \in X,\, y \in Y\}$ and is denoted by $X \times Y$. At times we write $X^2$ for $X \times X$. In fact, for $n \in \mathbb{N}$, we let $X^n$ denote the set of all ordered $n$-tuples of elements from $X$. A *relation* $R$ of $X$ into $Y$ is a subset of $X \times Y$. Let $R$ be such a relation. Then the domain of $R$, written $Dom(R)$, is

$$Dom(R) = \{x \in X \mid \exists y \in Y \text{ such that} (x,y) \in R\}$$

and the image of $R$, written $Im(R)$, is

$$Im(R) = \{y \in Y \mid \exists x \in X \text{ such that } (x,y) \in R\}\ .$$

If $(x,y) \in R$, we sometimes write $xRy$ or $R(x) = y$. If $R$ is a relation from $X$ into $X$, we say that $R$ is a relation on $X$.

**Definition 1.1.** A relation $R$ on $X$ is called:

  (i) *reflexive* if $\forall x \in X$, $(x,x) \in R$;
  (ii) *symmetric* if $\forall x, y \in X$, $(x,y) \in R$ implies $(y,x) \in R$;
  (iii) *transitive* if $\forall\, x,y,z \in X$, $(x,y), (y,z) \in R$ implies $(x,z) \in R$.
  (iv) *antisymmetric* If $\forall x, y \in X$, $(x,y) \in R$ and $(y,x) \in R$ implies $x = y$.
  (v) *complete* If $\forall x, y \in X$, $(x,y) \in R$ or $(y,x) \in R$ or both.

**Definition 1.2.** If $R$ is a reflexive, antisymmetric and transitive relation on $X$, then $R$ is called a *partial order* on $X$ and $X$ is said to be partially ordered by $R$.

*Example 1.3.* Let A be the set $\{a,b,c\}$. The subsets of $A$ are partially ordered under the relation "subset-of". Thus $\{a,b\} \subseteq \{a,b,c\}$ but the relation is not complete since $\{a,b\}$ and $\{b,c\}$ are not comparable. Neither is a subset of the other.

**Definition 1.4.** If $R$ is a reflexive, complete, and transitive, then $R$ is a weak order.

*Example 1.5.* Suppose we have four football teams $\alpha$, $\beta$, $\gamma$ and $\delta$ in a division and the standings are $R = (\delta, \gamma, \alpha, \beta)$. Implicitly this means $\delta R \gamma R \alpha R \beta$. Then $\delta$ leads the division, but it is possible that $\delta$ and $\alpha$ are tied. In fact they all the teams could be tied. However we cannot have that $\beta$ has a better record than $\delta$. This is a weak order.

**Definition 1.6.** If, in addition to being reflexive, antisymmetric and transitive, $R$ is complete then $R$ is a *total order* which is also called a *linear order*.

*Example 1.7.* The Natural numbers under "less than or equal" are totally ordered.

**Definition 1.8.** Let $R$ be a relation of $X$ into $Y$, and $T$ a relation of $Y$ into a set $Z$. Then the *composition* of $R$ with $T$, written $T \circ R$, is defined to be the relation $\{(x,z) \in X \times Z \mid \exists y \in Y \text{ such that } (x,y) \in R \text{ and } (y,z) \in T\}$.

If $f$ is a relation of $X$ into $Y$ such that $Dom(f) = X$ and $\forall x, x' \in X$, $x = x'$ implies $f(x) = f(x')$, then $f$ is called a function of $X$ into $Y$ and we write $f : X \to Y$. Let $f$ be a function of $X$ into $Y$. Then $f$ is sometimes called a mapping of $X$ into $Y$. If $\forall y \in Y$, $\exists x \in X$ such that $f(x) = y$, then $f$ is said to be onto $Y$ or to map $X$ onto $Y$. If $\forall x, x' \in X$, $f(x) = f(x')$ implies $x = x'$, then $f$ is said to be *one-to-one* and $f$ is called an *injection*. If $f$ is a one-to-one function of $X$ onto $Y$, then $f$ is called a *bijection*. If $g$ is a function of $Y$ into a set $Z$, then the *composition* of $f$ with $g$, $g \circ f$, is a function of $X$ into $Z$ which is one-to-one if $f$ and $g$ are one-to-one and which is onto if $f$ onto $Y$ and $g$ is onto $Z$.

**Definition 1.9.** If $Im(f)$ is finite, then $f$ is called finite-valued. We say that an infinite set $X$ is countable if there exists a one-to-one function of $N$ onto $X$, otherwise $X$ is called uncountable.

We use the notation $\bigvee$ to denote maximum or supremum and $\bigwedge$ to denote minimum or infimum. For a function $f$ whose domain is $X \times X$, we sometimes write $f\rceil_A$ for the restriction of $f$ to $A \times A$, where $A$ is a subset of $X$.

## 1.2.2   Fuzzy Subsets

Fuzzy set theory holds that many things in life are matters of degree (Zadeh, 1965; Klir and Yuan, 1995). Let $S$ be a set and let $[0,1]$ denote the closed interval $\{x \in \mathbb{R} \mid 0 \leq x \leq 1\}$. A *fuzzy subset* $\mu$ of $S$ is a function $\mu : S \to [0,1]$. We think of $\mu$ as assigning to each element $x \in S$ a *degree of membership* $0 \leq \mu(x) \leq 1$. We call $\mu$ a membership function.

*Example 1.10.* Let $S = \{a,b,c,d\}$ and define the fuzzy set $\mu$ as follows, set $\mu(a) = 1.0$, $\mu(b) = 0.7$, $\mu(c) = 0.0$, and $\mu(d) = 0.3$. Thus $a$ is completely compatible with the concept $\mu$ while $b$ is only very compatible. On the other hand $c$ is incompatible with the notion conceptualized by $\mu$ and $d$ is somewhat incompatible.

Let $\mu$ be a fuzzy subset of $S$. We let $\mu^t = \{x \in S \mid \mu(x) \geq t\}$ for all $t \in [0,1]$. The set $\mu^t$ is called a *level set* or *t-level set*. We let $\text{Supp}(\mu) = \{x \in S \mid \mu(x) > 0\}$. We call $\text{Supp}(\mu)$ the *support* of $\mu$.

The set of all fuzzy subsets of $S$ is denoted by $\mathscr{F}(S)$ and is called the *fuzzy power set* of $S$.

**Definition 1.11 (null set).** We let $\theta$ denote the fuzzy subset of $S$ defined by $\theta(x) = 0$ for all $x \in S$, this null set is the equivalent of the crsip empty set.

*Example 1.12.* Let $\mu$ be as in example 1.10 then $\mu^{0.5} = \{a,b\}$ and $\text{Supp}(\mu) = \{a,b,d\}$.

For any $a,b \in [0,1]$ we denote $a \wedge b = \min(a,b)$ and $a \vee b = \max(a,b)$ . More generally, for any $\{a_i\}_{i \in I} \subseteq [0,1]$ we denote $\bigwedge_{i \in I} a_i = \inf\{a_i \mid i \in I\}$ and $\bigvee_{i \in I} a_i = \sup\{a_i \mid i \in I\}$ .

**Definition 1.13.** Let $\mu$ and $\nu$ be fuzzy subsets of $S$. Then we write:

(i) $\mu \subseteq \nu$ if $\mu(x) \leq \nu(x)$ for all $x \in S$;
(ii) $\mu \subset \nu$ if $\mu \subseteq \nu$ and $\exists x \in S$ such that $\mu(x) < \nu(x)$;
(iii) $\mu = \nu$ if $\mu \subseteq \nu$ and $\nu \subseteq \mu$.

**Definition 1.14.** Let $\mu$ and $\nu$ be fuzzy subsets of $S$. Then;

(i) the *union* of $\mu$ and $\nu$ is defined to be the fuzzy subset $\mu \cup \nu$ of S such that $\mu \cup \nu(x) = \mu(x) \vee \nu(x)$ for all $x \in S$;
(ii) the *intersection* of $\mu$ and $\nu$ is defined to be the fuzzy subset of $\mu \cap \nu$ of S such that $\mu \cap \nu(x) = \mu(x) \wedge \nu(x)$ for all $x \in S$.

**Table 1.1** Union, Intersection and Compliments of Fuzzy Sets

| S | $\mu$ | $\nu$ | $\mu \cup \nu$ | $\mu \cap \nu$ | $\mu^c$ |
|---|---|---|---|---|---|
| $a$ | 1.0 | 0.8 | 1.0 | 0.8 | 0.0 |
| $b$ | 0.7 | 0.5 | 0.7 | 0.5 | 0.3 |
| $c$ | 0.0 | 0.1 | .01 | 0.0 | 1.0 |
| $d$ | 0.3 | 1.0 | 1.0 | 0.3 | 0.7 |

**Definition 1.15.** Let $\mu$ be a fuzzy subset of $S$. The *complement* of $\mu$ in $S$, written $\mu^c$, is defined by $\mu^c(x) = 1 - \mu(x)$ for all $x \in S$.

## 1.2.3 Fuzzy Relations

**Definition 1.16.** Let $S$ and $T$ be sets. A *fuzzy relation* $\rho$ of $S$ into $T$ is a function $\rho : S \times T \to [0,1]$. If $\rho$ is a fuzzy relation of $S$ into $S$, we say that $\rho$ is a fuzzy relation on $S$.

**Definition 1.17.** Let $\rho$ be a fuzzy relation of $S$ into $T$ and $\sigma$ a fuzzy relation of $T$ into a set $U$. Define $\rho \circ \sigma : S \times U \to [0,1]$ by $\forall (x,z) \in S \times U$,

$$\rho \circ \sigma(x,z) = \bigvee_{y \in T} \rho(x,y) \wedge \sigma(y,z) .$$

The fuzzy relation $\rho \circ \sigma$ is called the *composition* of $\rho$ with $\sigma$.

*Example 1.18.* A simple example of a binary fuzzy relation $\rho$ on $S = \{1,2,3\}$, called "approximately equal" can be defined as

$$\rho(1,1) = \rho(2,2) = \rho(3,3) = 1, \tag{1.1}$$
$$\rho(1,2) = \rho(2,1) = \rho(2,3) = \rho(3,2) = 0.6, \tag{1.2}$$
$$\rho(1,3) = \rho(3,1) = 0.2. \tag{1.3}$$

In other words, $\rho(x,y) = 1$ if $x = y$, 0.6 if $|x-y| = 1$, 0.2 if $|x-y| = 2$. In matrix notation the relation $\rho$ can be represented as

| $\rho$ | 1 | 2 | 3 |
|---|---|---|---|
| 1 | 1.0 | 0.7 | 0.4 |
| 2 | 0.7 | 1.0 | 0.7 |
| 3 | 0.4 | 0.7 | 1.0 |

**Definition 1.19.** Let $\rho$ be a fuzzy relation on $S$. Then $\rho$ is called:

(i) *reflexive* if $\rho(x,x) = 1$ for all $x \in S$;
(ii) *symmetric* if $\rho(x,y) = \rho(y,x)$ for all $x,y \in S$;
(iii) *max-min transitive* if $\rho(x,z) \geq \bigvee_{y \in S} \rho(x,y) \wedge \rho(y,z)$ for all $x,z \in S$.

*Example 1.20.* Let $S = \{a,b\}$, $T = \{1,2,3\}$, $U = \{g,h\}$, and let $\rho : S \times T \to [0,1]$ and $\sigma : T \times U \to [0,1]$ be given by the following tables

| $\rho$ | 1 | 2 | 3 |
|---|---|---|---|
| $a$ | 0.8 | 0.1 | 0.3 |
| $b$ | 1.0 | 0.7 | 0.5 |

| $\sigma$ | g | h |
|---|---|---|
| 1 | 0.2 | 0.3 |
| 2 | 1.0 | 1.0 |
| 3 | 0.5 | 0.1 |

$$\tag{1.4}$$

then the sup–min composition of $\rho$ and $\sigma$, $\rho \circ \sigma$, is given by the table

| $\rho \circ \sigma$ | g | h |
|---|---|---|
| $a$ | 0.3 | 0.3 |
| $b$ | 0.7 | 0.7 |

### 1.2.4   Fuzzy Intersection and Union

Occasionally we will need types of fuzzy intersections and fuzzy unions other than minimum and maximum, respectively. These can be expressed as binary operations on the unit interval $[0,1]$.

**Definition 1.21.** A *t-norm i* is a binary operation on the unit interval $[0,1]$ that satisfies the following conditions: $\forall a,b,c \in [0,1]$,

(1) $i(a,1) = a$ (boundary condition);
(2) $b \leq c$ implies $i(a,b) \leq i(a,c)$ (monotonicity);

(3) $i(a,b) = i(b,a)$ (commutativity);
(4) $i(a, i(b, c)) = i(i(a,b), c)$ (associativity).

The following are examples of some $t$-norms that are frequently used as fuzzy intersections. . In each case, we have for $\forall a,b \in [0,1]$:

Standard intersection: $i(a,b) = a \wedge b$.
Algebraic product: $i(a,b) = ab$.
Bounded difference: $i(a,b) = 0 \vee (a+b-1)$.
Drastic intersection:

$$i(a,b) = \begin{cases} a & \text{if } b = 1, \\ b & \text{if } a = 1, \\ 0 & \text{otherwise.} \end{cases}$$

**Definition 1.22.** A $t$-conorm is a binary operation $u$ on the unit interval $[0,1]$ that satisfies the following conditions: $\forall a,b,c \in [0,1]$,

(1) $u(a,0) = a$ (boundary condition)
(2) $b \le c$ implies $u(a,b) \le u(a,c)$ (monotonicity)
(3) $u(a,b) = u(b,a)$ (commutativity)
(4) $u(a, u(b, c)) = u(u(a,b), c)$ (associativity)

The following are examples of some $t$-conorms that are frequently used as fuzzy unions. In each case, we have for $\forall a,b \in [0,1]$:

Standard union: $u(a,b) = a \vee b$.
Algebraic sum: $u(a,b) = a+b-ab$.
Bounded sum: $u(a,b) = 1 \wedge (a+b)$.
Drastic union:

$$u(a,b) = \begin{cases} a & \text{if } b = 0, \\ b & \text{if } a = 0, \\ 1 & \text{otherwise.} \end{cases}$$

## 1.2.5 Residuum

We now define a new binary operation $\rightarrow$ on $[0,1]$, called implication or residuation.

**Definition 1.23 (standard residuum).** For all $a,b \in [0,1]$ the standard residuum is defined as
$$a \rightarrow b = \bigvee \{t \in [0,1] \mid a \wedge t \le b\} \quad .$$

For any $t$-norm $i$ we can define a corresponding residuum operator via the following definition.

[residuum] For any continuous $t$-norm $i$ and for all $a,b \in [0,1]$ the residuum is defined as

$$a \rightarrow_i b = \bigvee \{t \in [0,1] \mid i(a,t) \leq b\} \ .$$

The following are examples of residuum operators corresponding to some $t$-norms that are frequently used as fuzzy intersections. In each case, we have for $\forall a,b \in [0,1]$:

Standard intersection residuum: $i(a,b) = a \wedge b$ and a simple calculation yields the standard residuum

$$a \rightarrow b = \begin{cases} 1 & \text{if } a \leq b \\ b & \text{if } a > b \end{cases} .$$

Algebraic product residuum: $i(a,b) = ab$ yields

$$a \rightarrow b = \begin{cases} 1 & \text{if } a \leq b \\ b & \text{if } a > b \end{cases} .$$

Bounded difference residuum: $i(a,b) = 0 \vee (a+b-1)$

$$a \rightarrow b = \begin{cases} 1 & \text{if } a \leq b \\ b & \text{if } a > b \end{cases} .$$

The following two lemmas collect some basic properties of the residuum and its interaction with the max and min operators.

**Lemma 1.24.** *Klement et al. (2000); Turunen (1999) For any $a,b,c \in [0,1]$ the following properties hold:*

*(1) $i(a,b) \leq c$ iff $a \leq b \rightarrow c$;*
*(2) $i(a,a \rightarrow b) = i(a,b)$;*
*(3) $a \leq b$ iff $a \rightarrow b = 1$;*
*(4) $a = 1 \rightarrow a$;*
*(5) $1 = a \rightarrow a$;*
*(6) $a \rightarrow (b \rightarrow c) = i(a,b) \rightarrow c = b \rightarrow (a \rightarrow c)$*

**Lemma 1.25.** *Klement et al. (2000); Turunen (1999) For any $\{a_i\}_{i \in} \subseteq [0,1]$ and $a \in [0,1]$ the following properties hold:*

*(1) $(\bigvee_{i \in I} a_i) \wedge a = \bigvee_{i \in I} (a_i \wedge a)$*
*(2) $a \rightarrow (\bigvee_{i \in I} a_i) = \bigvee_{i \in I} (a \rightarrow a_i)$*
*(3) $(\bigvee_{i \in I} a_i) \rightarrow a = \bigvee_{i \in I} (a_i \rightarrow a)$*

# References

Klement, E., Mesiar, R., Pap, E.: Triangular Norms. Trends in logic, Studia logica library. Springer (2000)

Klir, G.J., Yuan, B.: Fuzzy Sets and Fuzzy Logic. Theory and Applications. Prentice Hall, Upper Saddle River (1995)

Turunen, E.: Mathematics behind fuzzy logic. Advances in soft computing. Physica-Verlag (1999)

Zadeh, L.A.: Fuzzy sets. Information and Control 8(3), 338–353 (1965)

# Chapter 2
# Classical Social Choice Theorems

with John F. Zimmer V.

**Abstract.** This chapter presents the classical versions of Arrow's Theorem, the Gibbard-Sattherthwaite Theorem, the Median Voter Theorem and the maximal set. By presenting the classical versions of these theorems, this chapter sets up the fuzzy versions presented in later chapters.

## Introduction

Here we present informal proofs of the classical versions of three theorems that are the focus of the rest of this book.

## 2.1 Arrows Theorem

Let $X$ be a set of alternatives and $N = \{1, 2, \ldots, n\}$ a set of actors. We will always assume that $|X| = m$ and that $m \geq 3$.

Each actor $i$ has a preference profile $R_i \in \mathscr{P}(X \times X)$ and if $(x, y) \in R_i$ we write $x R_i y$ and say that actor $i$ prefers $x$ to $y$. We assume that the relation $R_i$ is reflexive, complete, and transitive, i.e., it is a weak order.

**Definition 2.1.** Given the binary relation, $R$, $\forall x, y \in X$, $P$ is the asymmetric derivation of $R$, where $x P y \Leftrightarrow (x R y$ and $\neg(x R y)$ and $I$ is the symmetric derivation of $R$, where $x I y \Leftrightarrow (x R y$ and $y R x)$.

**Definition 2.2 (strict preference).** A preference relation that is asymetric is called a strict preference relation.

We note that the asymmetric derivation of a preference relation $R$ is always a strict preference $P$.

A preference profile $\bar{R}$ is an $n$–tuple of preference relations so that

$$\bar{R} = (R_1, R_2, \ldots, R_n).$$

Let $\mathscr{R}$ designate the set of all weak orders on $X$ and $\mathscr{R}^n$ the set of all preferece profiles.

M.B. Gibilisco et al., *Fuzzy Social Choice Theory*, 11
Studies in Fuzziness and Soft Computing 315,
DOI: 10.1007/978-3-319-05176-5_2, © Springer International Publishing Switzerland 2014

**Definition 2.3 (restriction).** Let S be a non-empty subset of $X$ and $R$ an element of $\mathscr{R}$. We write $R]_S$ as shorthand for $R]_{S \times S}$, i.e., for $\bar{R} = (R_1, R_2, \ldots, R_n) \in \mathscr{R}^n$ by $\bar{R}]_S$ we mean $\bar{R}]_S = (R_1]_{S \times S}, \ldots, R_n]_{S \times S})$. Thus, $\bar{R}]_S$ is the *restriction* of the preference profile $\bar{R}$ to the subset $S \times S$.

**Definition 2.4 (preference aggregation rule).** A function $f : \mathscr{R}^n \to \mathscr{R}$ is called a *preference aggregation rule.*

Hence, a preference aggregation rule (PAR) relates a $\bar{R} \in \mathscr{R}^n$ to a social preference relation $f(\bar{R}) \in \mathscr{R}$. For $x, y \in X$, $x f(\bar{R}) y$ represents that society, or more specifically the set $N$ of actors, views $x$ as at least as good as $y$.

A PAR, $f$, can have any of the following properties.

**Definition 2.5 (universal admissibility).** All conceivable combinations of orderings on $X$ are admissible as rankings of the choices.

**Definition 2.6 (monotonicity).** If an actor in a preference profile increases the position of $x$ in their ranking then $x$ can not decrease in the ranking produced by the PAR.

**Definition 2.7 (independence of irrelevant alternatives (IIA)).** The ranking of $x$ and $y$ by the PAR depends only on the rankings of $x$ and $y$ by the actors. Specifically, the relative position of $z$ is irrelevant. If we have two preference profiles $\bar{R}$ and $\bar{R}'$ then $\bar{R}]_{\{x,y\}} = \bar{R}']_{\{x,y\}}$ implies $f(\bar{R})]_{\{x,y\}} = f(\bar{R}')]_{\{x,y\}}$.

**Definition 2.8 (non-imposition).** The PAR profile is not independent of the actors profiles. That is, we cannot have $xRy$ for some distinct $x, y \in X$ for all preference profiles $\bar{R}$ where $R = f(\bar{R})$.

**Definition 2.9 (non-dictatorship).** There does not exist a $k \in N$ such that $\forall x, y \in X$ $xR_k y \Rightarrow xRy$ or, equivalently, $R_k = f(\bar{R})$.

**Definition 2.10 (weakly Paretian).** If everyone agrees on $x$ over $y$, then it must be so for the PAR, $\forall i \in N$ $xR_i y \Rightarrow xRy$. This is also called unanimity.

Each preference profile $R_i$ corresponds to an ordering of the elements of $X$ such that if $x$ appears in the ordered $m$–tuple $L_i = (x_{i1}, x_{i2}, \ldots, x_{im})$ before $y$ then $xR_i y$. We will use the interchangeability of the $m$–tuple and the relation in the proofs below. When we say that $(a, b, c)$ is a preference profile we mean that $aRbRc$ (note that $aRc$ is also true by transitivity).

The following Lemma will make the proof clearer.

**Lemma 2.11 (Extremal).** *Choose an arbitrary element b from X. Let $\bar{R}$ be a preference profile where for every actor alternative b is either the first or the last element of the weak order. Let f be an aggregation rule and let R be $f(\bar{R})$. Suppose f satisfies IIA and unanimity. Then b must be the first or last element of the weak order induced by R.*

**Table 2.1** The Stairstep Method

| Actor Profile | 1 | 2 | 3 | $\cdots$ | n-1 | n | PAR |
|---|---|---|---|---|---|---|---|
| $\bar{Q}_0 = \bar{R}$ | $R_1$ | $R_2$ | $R_3$ | $\cdots$ | $R_{n-1}$ | $R_n$ | $f(\bar{Q}_0) = b$ |
| $\bar{Q}_1$ | $R'_1$ | $R_2$ | $R_3$ | $\cdots$ | $R_{n-1}$ | $R_n$ | $f(\bar{Q}_1) =?$ |
| $\bar{Q}_2$ | $R'_1$ | $R'_2$ | $R_3$ | $\cdots$ | $R_{n-1}$ | $R_n$ | $f(\bar{Q}_2) =?$ |
| $\bar{Q}_3$ | $R'_1$ | $R'_2$ | $R'_3$ | $\cdots$ | $R_{n-1}$ | $R_n$ | $f(\bar{Q}_3) =?$ |
| $\vdots$ | $\vdots$ | $\vdots$ | $\vdots$ | $\ddots$ | $\vdots$ | $\vdots$ | |
| $\bar{Q}_{n-1}$ | $R'_1$ | $R'_2$ | $R'_3$ | $\cdots$ | $R'_{n-1}$ | $R_n$ | $f(\bar{Q}_{n-1}) =?$ |
| $\bar{Q}_n = \bar{R}'$ | $R'_1$ | $R'_2$ | $R'_3$ | $\cdots$ | $R'_{n-1}$ | $R'_n$ | $f(\bar{Q}_n) = a$ |

*Proof.* For each $R_i$ in $\bar{R}$ let its associated weak order produce the $m$–tuple $L_i = (x_{i1}, x_{i2}, \ldots, x_{im})$. What happens if $b$ is not at the extremes of $R$? Then there must be alternatives $a$ and $c$ such that in $R$ $aRb$ and $bRc$. For each $R_i$ in $\bar{R}$ if $a$ comes before $c$ in the corresponding $m$–tuple then switch their positions. For this mutated preference profile $\bar{R}'$ we still have that aggregation will produce $aR'b$ and $bR'c$ because IIA says we can ignore the rankings of $a$ versus $c$. Transitivity of $R'$ says that $aR'c$ since $aR'b$ and $bR'c$; but by unanimity we would also have $cR'a$ a contradiction, proving the lemma. □

**Theorem 2.12 (Arrow).** *A PAR that satisfies unanimity, IIA, and non-dictatorship is impossible.*

*Proof.* Let $a, b, c \in X$ be three distinct elements of $X$. First we show that there is an actor that dictates the choice between $a$ and $b$. Second we show this actor Dictates the choice between $a$ and $c$. Finally we show this actor Dictates the choice between $b$ and $c$.

(1) Let $\bar{R}$ be a preference profile where every actor detest $b$. Thus in $\bar{R}$ we have for every $i \in N$ that actor $i$ ranks $b$ as the last element of its associated $m$–tuple. Let $\bar{R}'$ be a preference profile where every actor adores $b$. Thus in $\bar{R}'$ we have for every $i \in N$ that actor $i$ ranks $b$ as the last element of its associated $m$–tuple. For $x \neq b$ both $\bar{R}$ and $\bar{R}'$ are identical. .

Construct the sequence of profiles $\bar{R}_0 = \bar{R}$, $\bar{R}_1$, $\bar{R}_2$, ..., $\bar{R}_n = \bar{R}'$ where in each profile $\bar{R}_i$ actor $i$ switches from detesting $b$ to adoring $b$ so that b is now the first element of the associated $m$–tuple. This procedure is illustrated in Table 2.1.

By unanimity $R_0 = f(\bar{R}_0)$ has $b$ last and $R_n = f(\bar{R}_n)$ has $b$ first. Note also that $aR_0b$ since $b$ is last in $R_0$ and that that $bR_na$ since $b$ is first in $R_n$. By the Extremal Lemma $f(\bar{R}_i)$ always puts $b$ first or last. Let k be the first actor in the sequence of preference profiles where $b$ switches from last to first. We note that, by monotonicity, once $b$ is put first it will stay first.

actor $k$ is called the Pivotal actor.

For the next step actors $1, \ldots, k - 1$, will constitute Block I and actors $k + 1, \ldots n$ will constitute Block II.

We note that by IIA, as long as all the voters in Block I and the Pivot have $b$ before $a$ and all the actors in block II have $a$ before $b$ then the aggregation will have $a$ over $b$. When Blocks I or II are empty, the Pivot gets their preference between options $a$ and $b$ even though all other actors disagree.

(2) Construct $\bar{Q}$ as follows
  (i) Block I ranks $b$ above $c$ and $c$ above $a$.
  (ii) Pivot actor ranks $a$ above $b$ and $b$ above $c$.
  (iii) Block II ranks $a$ above $b$ and $b$ above $c$.
  and let $Q = f(\bar{Q})$. Since all players have kept the relative positions of $a$ and $b$ the same as in profile $\bar{R}_{k-1}$ of step (1), by IIA $aQb$. By unanimity $bQc$. Finally, by transitivity $aQc$. If Block II is empty this is still true even though the pivot is the only one who prefers $a$ over $c$.

(3) Now construct a new $\bar{Q}'$ as follows
  (i) Block I ranks $c$ above $b$ and $b$ above $a$.
  (ii) Pivot actor ranks $b$ above $a$ and $a$ above $c$.
  (iii) Block II ranks $a$ above $c$ and $c$ above $b$.
  and let $Q' = f(\bar{Q}')$. Since all players have kept the relative positions of $a$ and $b$ the same as in profile $\bar{R}_k$ of step (1), by IIA $bQa$. By unanimity $cQb$. Finally, by transitivity $cQa$. If Block I is empty this is still true even though the Pivot is the only one who prefers $a$ over $c$.

  The Pivot thus always gets its choice when $a$ and $c$ are considered.                    □

## 2.2  Discussion

Arrow's original presentation introduced five contradictory assumptions. It turned out that not all of the assumptions were neccesary (Inada, 1955; Blau, 1972).

**Proposition 2.13.** *Monotonicity and non-imposition imply weak Paretianism (Unanimity).*

*Proof.* Assume monotonicity and non-imposition. By non-imposition there must be a preference profile $\bar{R}$ such that $xRy$. Use the stairstep method to produce profiles $\bar{R}_i$ where at each step $R_{ii}$ switches (if neccesary) to preferring $x$ to $y$. By monotonicity we have for all $i \in N$ that $R_i = f(\bar{R}_i)$ has $xR_iy$. But $\bar{R}_n$ is unanimous prefernce for for $x$ over $y$ and we have weak Paretianism.                    □

Following the simplification of Arrow's conditions (Blau, 1972), scholars generally attempted a different approach to the theorem. It was Arrow who first suggested relaxing the restrictions of the theorem in order to circumvent its unpleasant implications for social choice (Arrow, 1951). Subsequently, scholars have attempted to weaken individual conditions of the theorem, with varying levels of success. We examine each of these in turn. Many arguments regarding the admissibility of individual preferences are preempted by Arrow's discussion of welfare economics. Attempts to reduce the minimally required domain from the universe to some finite number have resulted in only a stronger theorem (Border, 1984, 2002; Quesada, 2002).

Other endeavors regarding universal admissibility do not directly relate to the relaxation of this restriction. Campbell and Kelly, for instance, experiment with the domain of social welfare functions in order to explore the logical relation between various impossibility theorems (Campbell and Kelly, 2000, 2003). The attempts to relax the condition of universal admissibility fail in circumventing the paradox of Arrow's Theorem.

Independence of irrelevant alternatives (henceforth IIA) was actually the condition first targeted in the literature as being too restrictive. Hansson argues that the independence condition is completely inconsistent with democratic methods of aggregation (Hansson, 1969). His findings, however, lack empirical support and are based on theoretical concepts completely exogenous to Arrow's Theorem. These arguments have not had lasting impacts. A more interesting development in regard to the independence condition was introduced by Murakami (1968), and was formalized by Blau prior to her reformation of the General Possibility Theorem. Blau argues for a weakened independence condition by presenting the concept of $m$–ary independence, as opposed to binary independence (IIA). The concept of $m$-ary independence increases the subset of alternatives on which a preference must be dependent from two to some finite number. Just as IIA calls for an individual's preference between two alternatives to be based solely upon those two alternatives, Blau's concept calls for a preference relation between three (or some larger finite number) alternatives to be based solely upon those three (or larger finite number) alternatives. For some numbers of alternatives, Blau finds that m-ary independence implies IIA. For others, m-ary independence does not imply IIA, but the conclusion of Arrow's Theorem holds.

Campbell and Kelly (2007) introduce the concept of independence of some alternatives (ISA), weaker version of m-ary independence. This concept requires that a social preference over two alternatives be based solely upon the preference relation over a larger subset of the universe of alternatives. This number can vary, but must be larger than two. Coban builds upon this work, exploring the relationship between weakening unanimity and IIA (2009). The results of these works are interesting, as they have varying levels of success. The General Theorem, however, remains largely intact. Unanimity, also referred to by the literature as weak Paretianism, has been shown to have a natural tradeoff with other Arrovian conditions. A version of Arrow's Theorem which excludes unanimity has been proved by several scholars (Wilson, 1972; Kelsey, 1984; Campbell and Kelly, 1997). In general, these results are not promising. Any attempt to relax or remove unanimity as an Arrovian condition results in a weakened transitivity over preference relations. In addition to relaxing each Arrovian condition, scholars have attempted to relax the axiomatic assumptions of the theorem. An approach by contemporary scholars has been to relax the degree to which preferences are rational. Quasi-transitivity was introduced shortly following the introduction of Arrow's Theorem (Sen, 1969; Salles, 1976), but was found to have no substantial impact on the results of the theorem.

Acyclicity, an even weaker assumption of rationality, was then introduced (Blair and Pollak, 1982; Kelsey, 1984; Austen-Smith and Banks, 1999). The weakening of rationality has resulted in some intriguing findings. As scholars weaken the rationality requirement, the condition of dictatorship is generally weakened, as well. As rationality is weakened from transitivity to quasi-transitivity and acyclicity, however, a dictatorship, of sorts, still exists. Scholars have consistently shown that a dictatorship, oligarchy, or collegium will result, depending upon the level to which rationality is relaxed. (Blair and Pollak, 1982; Weymark, 1984; Kelsey, 1984; Austen-Smith and Banks, 1999). The literature has succeeded in providing academia with a simpler, generally accepted form of the theorem. The findings of the theorem, however, have continued to remain intact, despite nearly sixty years of scholarly criticism. The continual failure of conventional approaches has caused some to look for new solutions, namely fuzzy mathematical approaches.

## 2.3   Gibbard-Sattherthwaite Theorem

Let $X$ be a set of alternatives and $N = \{1, 2, \ldots, n\}$ a set of actors. We will always assume that $|X| = m$ and that $m \geq 3$.

Most proofs of the Gibbard-Sattherthwaite Theorem rely upon utility functions and the minus sign notation. We will also assume in this section that each Actor's preference relation is strict.

A utility function $u$ maps each element of $X$ to a real number, thus $u : X \rightarrow \mathbb{R}$. Now each strict preference profile $P$ corresponds to a utility function $u$ in the following way: given a utility function $u$ we create a preference $P$ by having $uPy$ if $u(x) > u(y)$. In terms of the ordered $m$–tuple $L = (x_1, x_2, \ldots, x_m)$ that corresponds to the preference relation $R$ we have that $u(x_1) > u(x_2) > \ldots > u(x_m)$. Thus we can present an actor $i$'s preference simply as the result of a utility function $u_i$.

We can thus think of a preference profile as an element in $\mathscr{U} = U^n$, so that a preference profile is $\bar{u} = (u_1, u_2, \ldots, u_n)$. The minus sign notation just divides a preference profile into $us$, $u_i$, and them, $u_{-i}$, which is all of the other $n - i$ utility/preferences. Thus $\bar{u} = (u_i, \bar{u}_{-i})$.

**Table 2.2** The Stairstep Method with Utility Notation

| Profile \ Actor | 1 | 2 | 3 | $\cdots$ | n-1 | n | PAR |
|---|---|---|---|---|---|---|---|
| $\bar{w}_0 = \bar{u}$ | $u_1$ | $u_2$ | $u_3$ | $\cdots$ | $u_{n-1}$ | $u_n$ | $f(\bar{w}_0) = a$ |
| $\bar{w}_1$ | $v_1$ | $u_2$ | $u_3$ | $\cdots$ | $u_{n-1}$ | $u_n$ | $f(\bar{w}_1) = ?$ |
| $\bar{w}_2$ | $v_1$ | $v_2$ | $u_3$ | $\cdots$ | $u_{n-1}$ | $u_n$ | $f(\bar{w}_2) = ?$ |
| $\bar{w}_3$ | $v_1$ | $v_2$ | $v_3$ | $\cdots$ | $u_{n-1}$ | $u_n$ | $f(\bar{w}_3) = ?$ |
| $\vdots$ | $\vdots$ | $\vdots$ | $\vdots$ | $\ddots$ | $\vdots$ | $\vdots$ | |
| $\bar{w}_{n-1}$ | $v_1$ | $v_2$ | $v_3$ | $\cdots$ | $v_{n-1}$ | $u_n$ | $f(\bar{w}_{n-1}) = ?$ |
| $\bar{w}_n = \bar{v}$ | $v_1$ | $v_2$ | $v_3$ | $\cdots$ | $v_{n-1}$ | $v_n$ | $f(\bar{w}_n) = b$ |

**Definition 2.14 (voting rule).** A voting rule is a mapping $f$ from $\mathcal{U}$ the set of all utility functions to $X$.

In this situation we assume that $f$ produces a unique best choice. Like presidential elections, someone must eventually be chosen.

A voting rule is manipulatable if there is some actor $i$ who can get an outcome he prefers by altering his utility function from $u_i$ to $u_i'$. Mathematically this is expressed by $u_i(f((u_i', \bar{u}_{-i}))) > u_i(f(\bar{u}))$. If a voting rule is not manipulatable, it is called strategy-proof.

In terms of a voting rule, a dictator should get a result that is among his highest utility choices. This can be expressed as $u_i(f(\bar{u})) \geq u_i(x)$ for $\forall x \in X$.

**Lemma 2.15 (Strategy proof => Monotonicity).** *Suppose that a voting rule $f$ is strategy-proof, that $\bar{u} \in \mathcal{U}$ and that $f(\bar{u}) = a$. Then $f(\bar{v}) = a$ for all utility profiles $\bar{v} \in \mathcal{U}$ where $\forall x \in X$ and $i \in N$ we have that*

$$v_i(a) \geq v_i(x) \text{ if } u_i(a) \geq u_i(x). \tag{2.1}$$

*Proof.* Suppose that there is a utility profile $\bar{v}$ that satisfies Eq. 2.1. Remember that $\bar{u} = (u_1, u_2, \ldots, u_n)$ and $\bar{u} = (v_1, v_2, \ldots, v_n)$.

Let $\bar{w}_0 = \bar{u}$ and construct $\bar{w}_1$ by changing the utility of the first actor from $u_1$ to $v_1$ so that the new utility profile looks like $\bar{w}_1 = (v_1, \bar{u}_{-1})$, see Table 2.2. If $f(\bar{w}_1) = b$ then, by strategy proofness, it follows that $v_1(a) \leq v_1(b)$, or else actor one could strategically promote a favored $a$ by substituting preference $v_1$ for $u_1$ in $\bar{u}$. Strategy-proofness also dictates that $v_1(b) \leq v_1(a)$, or else actor one could strategically promote a favored $b$ by substituting preference $u_1$ for $v_1$ in $\bar{w}_1$. Since we are dealing with strict preference $a = b$.

We now construct $\bar{w}_2$ from $\bar{w}_1$ by changing the utility of the second actor from $u_2$ to $v_2$ so that the new utility profile looks like $\bar{w}_2 = (v_2, \bar{w}_{1-1})$. If $f(\bar{w}_2) = b$ then, we repeat the argument for actor one above for actor two to show that $b = a$.

Continuing this argument until we reach $w_n$ we arrive at the truth of the lemma. □

**Lemma 2.16 (Strategy proof => Unanimity).** *Suppose that a voting rule $f$ is strategy-proof and onto. If $\bar{u} \in \mathcal{U}$ and there are distinct $a, b \in X$ such that $u_i(a) > u_i(b)$ for all actors $i \in N$, then $f(\bar{u}) \neq b$.*

*Proof.* Suppose that $f(\bar{u}) = b$. Since $f$ is onto there is some profile that produces $a$, let us call it $\bar{v}$. Create the preference profile $\bar{w}$ where for every actor $i \in N$ we have $w_i(x) = u_i(x)$ for $x \in X \setminus \{a, b\}$ and assign $a$ and $b$ large utility so that for each $x \in X \setminus \{a, b\}$

$$w_i(a) > w_i(b) > w_i(x).$$

We again build a sequence of profiles starting from $\bar{w}_0 = \bar{w}$ where actor $i$ switches from $u_i(x)$ to $v_i(x)$ for $x \in X \setminus \{a, b\}$. By the Monotonicity Lemma 2.15 it follows that $b = f(\bar{u}) = f(\bar{w}_0)$ and that $a = f(\bar{v}) = f(\bar{w}_n)$. The Pivot actor $k$, where $f$ changes from $b$ to $a$ or some $x$ distinct from $b$ can now apply strategy since $w_k(a) > w_k(b) > w_k(x)$. □

Independence of irrelevant alternatives can be restated for voting rules thusly. If $f(\bar{u}) = a$ and the utility of only alternative $b$ is changed in $\bar{v}$, that is $\bar{u}]_{X\setminus\{b\}} = \bar{v}]_{X\setminus\{b\}}$, then $f(\bar{v})$ must be either $a$ or $b$. That is, $c \neq a, b$ is irrelevant in the change from $\bar{u}$ to $\bar{v}$ and cannot be the choice.

**Lemma 2.17 (Strategy proof => IIA).** *Suppose that a voting rule $f$ is strategy-proof and onto. Then $f$ is independent of irrelevant alternatives.*

*Proof.* Suppose that $f(\bar{u}) = a$. Let some actor $i$, decreases its utility for choice $z$. By monotonicity, $a$ must remain the result of the voting rule. On the other hand, suppose actor $i$ increases its utility for $z$ from $u_i(z)$ to $v_i(z)$? Suppose that the winning candidate is $b$ when actor $i$ uses preference $v$. Let $w_i(b) = \min[u_i(b), v_i(b)]$. Under the $\bar{u}$ preference scheme $a$ is chosen, so by monotonicity $a$ is still selected when the profile is $(w_i, \bar{u}_{-i})$. However, under the $(v_i, \bar{u}_{-i})$ system $b$ is elected, and by monotonicity $b$ is elected in $(w_i, \bar{u}_{-i})$. This is a contradiction. Thus, both system must elect the same candidate. $\square$

**Theorem 2.18 (Gibbard-Satterthwaite Theorem).** *[Gibbard (1973); Satterth-waite (1975); Austen-Smith and Banks II (2005)] A voting rule that is onto, and strategy proof is dictatorial.*

*Proof.* We have shown that such a rule is IIA, monotonic and satisfies unanimity. Thus Arrow's theorem applies. $\square$

## 2.4  The Median Voter Theorem

For the median voter theorem we must move to spatial models. Let $X = \mathbb{R}$ be the real number line and assume that each actor $i \in N$ has a preference profile that is single peaked. This means that there is a utility function $u_i : X \to \mathbb{R}$ which is maximal at some ideal point $x_i \in X$ and monotonically non-increasing as we move away from the ideal point.

A majority voting rule chooses $a$ over $b$ if over half the actors prefer $a$ to $b$ or, $f(\bar{u}]_{\{a,b\}}) = a$ if $|\{i \in N \mid u_i(a) > u_i(b)\}| > \frac{n}{2}$.

**Definition 2.19 (Condorcet winner).** A Condorcet winner is a policy point $x \in X$ that is the winner versus any other feasible policy point in a pairwise vote.

**Theorem 2.20 (The Median Voter Theorem).** *Suppose that $|N|$ is an odd number, that individuals vote "truthfully" rather than strategically and that we use a majority voting rule.*

*Then, a Condorcet winner always exists and coincides with the median-ranked ideal point, $x_m$.*

*Proof.* Order the set $N$ according to the ideal points $x_i$ of each actor and label the median-ranked ideal point by $x_m$. Here the assumption that $N$ has an odd number of actors ensures $p_m$ is unique and well defined. Let $y$ be a point in $X$ such that $y < x_m$. By the assumption of single peakedness every actor to the *right* of the median voter has $u_i(y) < u_i(p_m)$ and if we add the median voter to the voters to his *right* then they will choose $x_m$ over $y$ by majority vote. If instead we pick a $z > x_m$ then the voters to *left* of the median voter together with the median voter will form a majority the will prefer $x_m$ to $z$. □

## 2.5 The Maximal Set

The *maximal set* under exact preferences $R$ is

$$M(R,X) = \{x \in X \mid xRy, \ \forall y \in X\},$$

Imagine a group consisting of three individuals, $N = \{1,2,3\}$, and three choices $X = \{a,b,c\}$.

Suppose that the preferences are $R_1 = (a,b,c)$, $R_2 = (b,c,a)$ and $R_3 = (c,a,b)$. If $f$ is majority rule then $R = f(\bar{R})$ gives $aRb$ since the subset of individuals who prefer $a$ to $b$, $\{1,3\}$, consists of two out of three actors, a majority. We also have the $f(\bar{R})$ giving $bRc$ since the subset of individuals who prefer $b$ to $c$, $\{1,2\}$, consists of two out of three actors. Finally we have $f(\bar{R})$ giving $cRa$ since the subset of individuals who prefer $c$ to $a$, $\{2,3\}$. consists of two out of three actors. This introduces a cycle and there is no maximal set in this case. This is the Condorcet paradox.

The maximal set represent the optimal choices produced by a preference rule. We see here that this set can be empty. In this case there is no best choice.

## References

Arrow, K.: Social Choice and Individual Values. Wiley, New York (1951)

Austen-Smith, D., Banks, J.S.: Positive Political Theory I: Collective Preference. University of Michigan Press, Ann Arbor (1999)

Austen-Smith, D., Banks, J.S.: Positive Political Theory II: Strategy and Structure. University of Michigan Press, Ann Arbor (2005)

Blair, D.H., Pollak, R.A.: Acyclic collective choice rules. Econometrica 50(4), 931–943 (1982), http://www.jstor.org/stable/1912770

Blau, J.H.: A direct proof of Arrow's theorem. Econometrica 40(1), 61–67 (1972), http://EconPapers.repec.org/RePEc:ecm:emetrp:v:40:y:1972:i: 1:p:61-67

Border, K.C.: Arrow's general (im)possibility theorem. Tech. rep., Division of the Humanities and Social Sciences (2002), http://www.hss.caltech.edu/~kcb/Notes/Arrow.pdf

Border, K.: An impossibility theorem for spatial models. Public Choice 43(3), 293–305 (1984), http://dx.doi.org/10.1007/BF00118938

Campbell, D.E., Kelly, J.S.: The possibility-impossibility boundary in social choice. In: Arrow, K.J., Sen, A.K., Suzumura, K. (eds.) Social Choice Re-examined (Proceedings of the 1994 International Economic Association Conference at Schloss Hernstein, Austria), pp. 179–204. Macmillan, London (1997)

Campbell, D.E., Kelly, J.S.: A simple characterization of majority rule. Economic Theory 15(3), 689–700 (2000),
http://dx.doi.org/10.1007/s001990050318

Campbell, D.E., Kelly, J.S.: A strategy-proofness characterization of majority rule. Economic Theory 22(3), 557–568 (2003),
http://dx.doi.org/10.1007/s00199-002-0344-1

Campbell, D., Kelly, J.S.: Pareto, anonymity, and independence: four alternatives. Social Choice and Welfare 29(1), 83–104 (2007),
http://dx.doi.org/10.1007/s00355-006-0194-z

Coban, C., Sanver, R.: Social choice without the pareto principle under weak independence. In: Working Papers 201005, Murat Sertel Center for Advanced Economic Studies. Istanbul Bilgi University (August 2009),
http://ideas.repec.org/p/msc/wpaper/201005.html

Gibbard, A.: Manipulation of voting schemes: A general result. Econometrica 41(4), 587–601 (1973)

Hansson, B.: Voting and group decision functions. Synthese 20(4), 526–537 (1969)

Inada, K.I.: Alternative incompatible conditions for a social welfare function. Econometrica 23(4), 396–399 (1955)

Kelsey, D.: Acyclic choice without the pareto principle. The Review of Economic Studies 51(4), 693–699 (1984)

Murakami, Y.: Logic and social choice. Monographs in modern logic. Routledge & K. Paul (1968), http://books.google.com/books?id=r-Rg0LcstqQC

Quesada, A.: More on independent decisiveness and arrow's theorem. Social Choice and Welfare 19(2), 449–454 (2002)

Salles, M.: Characterization of transitive individual preferences for quasi-transitive collective preference under simple games. International Economic Review 17(2), 308–318 (1976),
http://EconPapers.repec.org/RePEc:ier:iecrev:v:17:y:1976:i:2:p:308-18

Satterthwaite, M.A.: Strategy-proofness and Arrow's conditions: Existence and correspondence theorems for voting procedures and social welfare functions. Journal of Economic Theory 10(2), 187–217 (1975)

Sen, A.K.: Quasi-transitivity, rational choice and collective decisions. Review of Economic Studies 36 (1969)

Weymark, J.A.: Arrow's theorem with social quasi-orderings. Public Choice 42(3), 235–246 (1984)

Wilson, R.: Social choice theory without the pareto principle. Journal of Economic Theory 5(3), 478–486 (1972),
http://ideas.repec.org/a/eee/jetheo/v5y1972i3p478-486.html

# Chapter 3
# Rationality of Fuzzy Preferences

**Abstract.** Abstract Social choice theory is built upon the presupposition of rationality. At an individual level, rationality requires completeness and transitivity. Completeness refers to a preference where an individual either prefers x to y, y to x or is indifferent between the two options. Transitivity means that if there are three options and an individual prefers x to y and y to z, then they must also prefer x to z. This chapter considers how these two conditions work under fuzzy preferences to present a unified approach to the rationalization of fuzzy preferences. Specifically, fuzzy weak preference relations are shown to provide social scientists with greater flexibility when applying fuzzy social choice to empirical examples.

## Introduction

Social choice theory presupposes a basic idea of rationality. At a fundamental level, two criteria define a rational individual. First, when presented with two options, $x$ and $y$, an individual will either prefer $x$ to $y$, prefer $y$ to $x$, or be indifferent between the two. This first criterion is called *completeness*. Second, if an individual happens to prefer $x$ to $y$ and prefers $y$ to another alternative $z$, then the same individual must prefer $x$ to $z$. The second criterion is called *transitivity*. With the assumptions of completeness and transitivity, it is a fairly straightforward process to construct a binary preference relation $R$, in which $xRy$ means $x$ is at least as good as $y$, and to demarcate $R$ into two components, a strict preference relation ($P$) and an indifference relation ($I$). Using these constructions, social choice scholars restrict preferences in such a way as to replicate the supposed thought process and behavior of a rational individual.[1]

This parsimony is not retained when generalizing $R$ with a fuzzy binary relation, which allows individuals to possess varying degrees of preferences. When the relations $R$, $P$ and $I$ are fuzzy subsets and their elements possess degrees of

---

[1] For a review of the traditional conceptualizations of rational preference relations, McCarty and Meirowitz's (2007) chapter one provides an introductory text and Austen-Smith and Banks (1999) offer a more technical approach.

membership, there are no immediate fuzzy counterparts to the previous definitions of completeness and transitivity. Furthermore, there does not exist a single necessary relationship between the relations. The purpose of this chapter is to flush out these complexities of fuzzy preferences and present a unified approach to the rationalization of fuzzy preferences. To do so, we divide the the chapter into three sections. The first section considers how individuals compare one alternative to another. We propose several restrictions on fuzzy preferences, illustrate the consequences of these restrictions, and define fuzzy binary preferences used in the rest of the book. The second section then details how fuzzy preferences behave across pairwise comparisons. The section then reviews consistency conditions and demonstrates how the conditions apply to the existence of the maximal set. The third section concludes the chapter with a discussion of a fuzzy utility that is empirically applicable and its relation to the fuzzy preferences described in previous sections.

## 3.1 The Structure of Fuzzy Preference Relations

Let $X$ be a finite set of alternatives, where $|X| \geq 2$. These are the only restrictions placed on $X$ in this section. Later in the book, we will show how our results apply to cases in which $|X|$ is infinite.

**Definition 3.1 (FWPR).** A *fuzzy weak preference relation* (FWPR) is a function $\rho : X \times X \to [0,1]$.

For a FWPR, $\rho(x,y)$ is the degree to which $x$ is as good as $y$. We can interpret $\rho(x,y) = 1$ as expressing that $x$ is definitely at least as good as $y$ and $\rho(x,y) = 0$ as expressing that $x$ is definitely not at least as good as $y$. Let $\mathscr{F}(X^2)$ denote the set of all fuzzy relations on $X$. We call $\rho$ *exact* if $\rho : X \times X \to \{0,1\}$. In the exact case, $\rho$ represents the characteristic function of a traditional binary preference relation. Finally, we use a concept called the support of $\rho$, written $\text{Supp}(\rho)$ which picks out all the elements of $\rho$ with non-zero membership grades:

$$\text{Supp}(\rho) = \{(x,y) \in X \times X \mid \rho(x,y) > 0\}.$$

We wish to begin restricting $\rho$ in such a way as to faithfully replicate the rationality criterion of completeness. By doing so, we can begin to see necessary relationships between the relations. When considering how individuals compare two alternatives, we need to define and discuss two ideas: reflexivity and completeness. Reflexivity of $\rho$ has a uniform definition across the fuzzy social choice literature.[2] It is given in Definition 1.19and repeated for sake of clarity in Definition 3.2.

**Definition 3.2.** [reflexive] An FWPR $\rho$ is *reflexive* if for all $x \in X$, $\rho(x,x) = 1$.

---

[2] Of course, significant exceptions exist (see Billot (1992) and Ponsard (1990) for the most prominent examples). However, in these cases $\rho(x,y)$ refers to the preference for $x$ over $y$, and $\rho(x,x)$ refers to the intrinsic value of $x$ and varies between 0 and 1. This interpretation makes it quite difficult to dissemble $\rho$ into indifference and strict preference relations.

Reflexivity of $\rho$ allows us to say that every alternative is definitely equally as good as itself. Furthermore, with the assumption, we can now dissemble $\rho$ into its symmetric and asymmetric components.

**Definition 3.3 (symmetric).** The fuzzy binary relation $\iota$ is *symmetric* if, for all $x, y \in X$, $\iota(x,y) = \iota(y,x)$; and the fuzzy binary strict preference relation $\pi(x,y)$ is *asymmetric* if, for all $x, y \in X$, $\pi(x,y) > 0 \implies \pi(y,x) = 0$.

In Chapter 1 we introduce t–conorms which model the union operation of fuzzy sets. We denoted an arbitrary t–conorm as $u(a,b)$ but in this Chapter we will denote an arbitrary t–conorm as $\cup$ and write $\mu \cup \nu$ as the result of applying the t–conorm component-wise to the fuzzy sets $\mu$ and $\nu$.

**Definition 3.4 (components).** We say $\iota, \pi \in \mathscr{F}(X^2)$ are *components* of an FWPR $\rho$ if there exists some t–conorm $\cup: [0,1] \times [0,1] \to [0,1]$ such that $\rho = \iota \cup \pi$. In theory, $\iota$ and $\pi$ respectively represent the indifference and strict preference relations of $\rho$.

In contrast, the fuzzy counterpart of completeness varies across the social choice literature and the following definitions are more or less traditional representations of completeness.

**Definition 3.5 (connected).** A FWPR $\rho$ is

(1) **strongly connected** if, for all $x, y \in X$, $\max\{\rho(x,y), \rho(y,x)\} = 1$;
(2) **connected** if, for all $x, y \in X$, $\rho(x,y) + \rho(y,x) \geq 1$;
(3) **complete** if, for all $x, y \in X$, $\max\{\rho(x,y), \rho(y,x)\} > 0$.

Dutta (1987) implicitly assumes strong connectedness; Richardson (1998) and Fono and Andjiga (2005) use connectedness; and Barrett et al. (1986) proposes completeness in the context of fuzzy strict preference relations. A more specific form of connectedness is *reciprocalness*, which requires $\rho(x,y) + \rho(y,x) = 1$. All definitions are equivalent when $\rho$ is exact. Obviously, 3.5(1) implies 3.5(2) implies 3.5(3), but the converses are not necessarily true as the next example illustrates.

*Example 3.6.* Let $X = \{a,b,c\}$. Let $\rho_1, \rho_2, \rho_3$ be reflexive FWPRs and defined as follows:

(1) $\rho_1(a,b) = \rho_1(a,c) = \rho_1(c,b) = 1$,
(2) $\rho_2(a,b) = \rho_2(a,c) = \rho_2(c,b) = .8$,
(3) $\rho_3(a,b) = \rho_3(a,c) = \rho_3(c,b) = .4$,
(4) $\rho_k(b,a) = \rho_k(c,a) = \rho_k(b,c) = .3$ for all $k \in \{1,2,3\}$, and
(5) $\rho_k(a,a) = \rho_k(b,b) = \rho_k(c,c) = 1$ for all $k \in \{1,2,3\}$.

Then $\rho_1$ is strongly connected, connected and complete, compared to $\rho_2$, which is connected and complete but not strongly connected. Finally, $\rho_3$ is complete but not strongly connected and not connected.

Connectedness places more restrictions on FWPRs than completeness. Under connectedness, if $\rho(x,y) = 0$, then it must be true that $\rho(y,x) = 1$. The fuzzy completeness condition represents the strongest break away from traditional social choice approach, with minimal requirements on fuzzy preferences.

Throughout the subsequent sections, we will make use of all three definitions to better detail the structure of fuzzy preferences. However, for the time being, we assume that $\rho$ is complete and reflexive. Without putting any other restrictions on $\rho$, $\iota$ and $\pi$, we observe several relationships among the fuzzy relations.

**Proposition 3.7.** *Let $\rho$, $\iota$ and $\pi$ be fuzzy relations such that*

*(i)* $\text{Supp}(\rho) = \text{Supp}(\iota) \cup \text{Supp}(\pi)$,
*(ii) $\iota$ is symmetric, and*
*(iii) $\pi$ is asymmetric.*

*Then the following properties hold:*

*(1) For all $x,y \in X$, $\pi(x,y) > 0$ and $\iota(x,y) = 0$ if and only if $\rho(x,y) > 0$ and $\rho(y,x) = 0$.*
*(2) For all $x,y$, $\iota(x,y) > 0$ if and only if $\rho(x,y) > 0$ and $\rho(y,x) > 0$.*

*Proof*

(1) Suppose $\pi(x,y) > 0$ and $\iota(x,y) = \iota(y,x) = 0$. Then $(x,y) \in \text{Supp}(\rho)$ by (i). Since (iii) implies $\pi(y,x) = 0$, $\rho(y,x) = 0$ by (i). Conversely, suppose $\rho(x,y) > 0$ and $\rho(y,x) = 0$. Then $(x,y) \in \text{Supp}(\iota) \cup \text{Supp}(\pi)$ and $(y,x) \notin \text{Supp}(\iota) \cup \text{Supp}(\pi)$. Thus, $\iota(x,y) = \iota(y,x) = 0$ and so $\pi(x,y) > 0$.
(2) Suppose $\iota(x,y) > 0$, then $\iota(x,y) = \iota(y,x) > 0$, by (ii). Then, by (i), $(x,y), (y,x) \in \text{Supp}(\rho)$. Conversely, suppose $\rho(x,y) > 0$ and $\rho(y,x) > 0$. Then $(x,y), (y,x) \in \text{Supp}(\iota) \cup \text{Supp}(\pi)$. Because (iii) implies either $(x,y) \notin \text{Supp}(\pi)$ or $(y,x) \notin \text{Supp}(\pi)$, it follows that $(x,y), (y,x) \in \text{Supp}(\iota)$. □

Proposition 3.7 represents the "bare essentials" of fuzzy preferences that must be true with very minimal assumptions. Condition (1) of the proposition tells us that $x$ is strictly preferred to $y$ to some degree and the two alternatives possess no degree of indifference to each other only when $x$ is at least as good as $y$ to some degree but $y$ is definitely not at least as good to $x$. Condition (2) then stipulates $x$ and $y$ share some degree of indifference if and only if $x$ is at least as good as $y$, and vice versa, to some degree.

**Proposition 3.8.** *Let $\rho$, $\iota$ and $\pi$ be fuzzy relations such that*

*(i)* $\text{Supp}(\rho) = \text{Supp}(\iota) \cup \text{Supp}(\pi)$,
*(ii) $\iota$ is symmetric, and*
*(iii) $\pi$ is asymmetric.*

*Then (1) if and only if (2) and (2) implies (3).*

*(1) $\text{Supp}(\iota) \cap \text{Supp}(\pi) = \emptyset$.*
*(2) For all $x,y \in X$, $\pi(x,y) > 0$ implies $\rho(x,y) > 0$ and $\rho(y,x) = 0$.*
*(3) For all $x,y \in X$, $\rho(x,y) = \rho(y,x)$ implies $\pi(x,y) = \pi(y,x) = 0$.*

*Proof*

(1) $\Longrightarrow$ (2): Let $x,y \in X$. Suppose $\pi(x,y) > 0$. Since $\text{Supp}(\iota) \cap \text{Supp}(\pi) = \emptyset$, $\iota(x,y) = \iota(y,x) = 0$. By (i), $\rho(x,y) > 0$. Suppose $\rho(y,x) > 0$. Then by (i), $\pi(y,x) > 0$, but by (iii), this is impossible. Thus, $\rho(x,y) > 0$ and $\rho(y,x) = 0$.

(2) $\Longrightarrow$ (1): Contrary to the hypothesis, suppose there exists $(x,y) \in \text{Supp}(\iota) \cap \text{Supp}(\pi)$. Then $\pi(x,y) > 0$ and $\iota(x,y) = \iota(y,x) > 0$. By (2), $\pi(x,y) > 0$ implies $\rho(x,y) > 0$ and $\rho(y,x) = 0$. However, $\rho(y,x) = 0$ and $\iota(y,x) > 0$ is impossible by (1). Thus, there does not exist $(x,y) \in \text{Supp}(\iota) \cap \text{Supp}(\pi)$, and accordingly, $\text{Supp}(\iota) \cap \text{Supp}(\pi) = \emptyset$.

(2) $\Longrightarrow$ (3): Suppose $\rho(x,y) = \rho(y,x)$. If either $\pi(x,y) > 0$ or $\pi(y,x) > 0$, then by (2), $\rho(x,y) \neq \rho(y,x)$. Hence, $\pi(x,y) = \pi(y,x) = 0$. $\qquad\square$

Proposition 3.8 demonstrates the implications of a fairly inconspicuous assumption of the empty intersection between the indifference and strict preference relation, which naturally holds for exact preference relations. If the intersection is indeed empty (Proposition 3.8, condition 1), then $x$ is strictly preferred to $y$ to some degree when $y$ is definitely not as least as good as $x$. In addition, the proposition shows that there can be no strict preference of $x$ over $y$ or $y$ over $x$ if condition 1 holds and $\rho(x,y) = \rho(y,x)$.

More importantly, Proposition 3.8, condition 2 illustrates the mutual exclusiveness of $\iota$ and $\pi$ under the assumption that $\text{Supp}(\iota) \cap \text{Supp}(\pi) = \emptyset$. Formally, if (1) holds, then it is easily verified that $\iota(x,y) = \iota(y,x) > 0$ implies $\pi(x,y) = \pi(y,x) = 0$. The condition of $\text{Supp}(\iota) \cap \text{Supp}(\pi) = \emptyset$ has significant consequences for the mentality of a political actor. If he possess even the slightest degree of strict preference of $x$ over $y$, then he does not feel any degree of indifference between the two. The problem occurs when the actor's preference is extremely minimal, as in the case when $\iota(x,y) = \iota(y,x) = 0$ even though $\pi(x,y) = .001$. Likewise, if he possesses the slightest degree of indifference between the two alternatives, there is no possibility that he will also possess strict preference over the two as well. Thus, while the first and third relationships in Proposition 3.8, seem like fairly natural fuzzifications of the traditional preference relations, the fact that $\iota$ and $\pi$ are mutually exclusive under $\text{Supp}(\iota) \cap \text{Supp}(\pi) = \emptyset$, regardless of their specific values, suggests that the transformation is too rigid.

Before we can consider properties of $\rho$, $\iota$ and $\pi$, we need to assume further structure on $\rho$.

**Definition 3.9 (union axioms).** Let $A$, $B$, $C$, and $D$ be fuzzy subsets of $S$. Let $\cup : [0,1] \times [0,1] \to [0,1]$. Then, $\cup$ satisfies the *Union Axioms* if the following hold:

**Boundary Condition**

$$B(x) = 0 \implies (A \cup B)(x) = A(x);$$

**Monotonicity**

$$A(x) \geq B(x) \,\&\, C(x) \geq D(x) \implies (A \cup C)(x) \geq (B \cup D)(x).$$

The Union Axioms are quite general and represent the basics of any type of fuzzy union Klir and Yuan (1995). For instance, if we also assume that $\cup$ is associative $((A \cup B) \cup C = A \cup (B \cup C))$ and commutative $(A \cup B = B \cup A)$ then $\cup$ can be any triangular co-norm.[3] In addition, Fono and Andjiga (2005) define *quasi-subtraction* of $\cup$, denoted $\overline{\cup}$, as

$$A(x) \overline{\cup} B(x) = \bigwedge \{t \in [0,1] \mid A(x) \cup t \geq B(x)\}.$$

For the time being, we do not place any further assumptions on $\cup$ besides conditions 1 and 2 in Definition 3.9. Doing so allows Richardson (1998) and Fono and Andjiga (2005) to make the following observations.

**Proposition 3.10.** *Let $\rho$ be an FWPR on X such that*

*(i) $\rho = \iota \cup \pi$,*
*(ii) $\iota$ is symmetric, and*
*(iii) $\pi$ is asymmetric.*

*Then $\iota(x,y) = \min\{\rho(x,y), \rho(y,x)\}$.*

*Proof.* Let $x, y \in X$. Using the Union Axioms (Definition 3.9) and (i), we know $\rho(x,y) \geq \iota(x,y)$ and $\rho(y,x) \geq \iota(y,x)$. Since $\iota$ is symmetric we know that

$$\iota(x,y) = \iota(y,x) \leq \min\{\rho(x,y), \rho(y,x)\}.$$

Since $\pi$ is asymmetric either $\pi(x,y) = 0$ or $\pi(y,x) = 0$. Suppose $\pi(y,x) = 0$. Then $\iota(y,x) = \rho(y,x)$. Hence, $\iota(x,y) = \min\{\rho(x,y), \rho(y,x)\}$. The argument is symmetric when $\pi(x,y) = 0$.                                                                                                   $\square$

It is important that we are able to obtain a unique operator of $\iota(x,y)$ under very minimal assumptions. Indeed, if there is a theoretical reason not to have $\iota(x,y) = \min\{\rho(x,y), \rho(y,x)\}$, one would need to violate either the monotonicty of $\cup$ or one of the (i-iii) criteria in Proposition 3.10, which are essential axioms of traditional social choice theory. In addition, if we would like to begin constructing a unique function for $\pi$, we must adopt another characteristic of $\pi$, which is commonly assumed in the literature Banerjee (1994); Dutta (1987); Ponsard (1990).

**Definition 3.11 (simple).** Let $\rho$ be an FWPR. Then its asymmetric counter part $\pi$ is *simple* if, for all $x, y \in X$, $\rho(x,y) = \rho(y,x)$ implies $\pi(x,y) = \pi(y,x)$.

It is easy to extrapolate that, because $\pi$ is asymmetric, simplicity implies $\pi(x,y) = \pi(y,x) = 0$ when $\rho(x,y) = \rho(y,x)$. Note that simplicity is identical to condition (3) in Proposition 3.10. Simplicity merely allows us to retain this property without the more stringent assumption of $\text{Supp}(\iota) \cap \text{Supp}(\pi) = \emptyset$. In addition, the simplicity assumption creates the following relationship between $\rho$ and $\pi$.

---

[3] For a thorough discussion on triangular norms and co-norms see Klement, Mesiar and Pap (2000).

**Proposition 3.12.** *Let $\rho$ be an FWPR on $X$ such that*

(i) $\rho = \iota \cup \pi$,
(ii) $\iota$ *is symmetric, and*
(iii) $\pi$ *is asymmetric.*

*If $\pi$ is simple, then for all $x,y \in X$, the following hold:*

(1) $\rho(y,x) \leq \rho(x,y) \iff \pi(y,x) = 0$.[4]
(2) $\rho(x,y) > \rho(y,x) \iff \pi(x,y) > 0$.

*Proof*

(1) Suppose there exist $x,y \in X$ such that $\rho(y,x) \leq \rho(x,y)$ and contrary to the hypothesis, $\pi(y,x) > 0$. Then $\pi(x,y) = 0$ by (iii). Since $\pi(y,x) > \pi(x,y)$ and $\iota(x,y) = \iota(y,x)$, $\rho(y,x) \geq \rho(x,y)$ by the Union Axioms. Thus, $\rho(x,y) = \rho(y,x)$. Because $\pi$ is simple, $\pi(x,y) = \pi(y,x)$, which establishes a contradiction. Hence, $\pi(y,x) = 0$. Conversely, suppose $\pi(y,x) = 0$. Then by the Union Axioms and Proposition 3.10, $\rho(y,x) = \iota(y,x) = \min\{\rho(x,y), \rho(y,x)\}$ and so $\rho(y,x) \leq \rho(x,y)$.
(2) The desired result follows from the proof of (1) because $\rho(x,y) > \rho(y,x) \iff \pi(x,y) > 0$ is the contrapositive of $\rho(y,x) \leq \rho(x,y) \iff \pi(y,x) = 0$. □

The relationships derived in Proposition 3.12 create the basis for a regular strict preference relation.

**Definition 3.13 (regular).** Let $\rho$ be an FWPR. Then its asymmetric counterpart $\pi$ is *regular* if $\pi(x,y) > 0$ if and only if $\rho(x,y) > \rho(y,x)$.

In fact, we can combine Propositions 3.10 and 3.12 to obtain the converse relationships between the structure on $\rho$ and the relationships among the fuzzy relations.

**Corollary 3.14.** *Suppose*

(i) $\iota(x,y) = \min\{\rho(x,y), \rho(y,x)\}$,
(ii) $\pi$ *is regular and*
(iii) $\rho(x,y) > \rho(y,x)$ *implies* $\rho(x,y) = \iota(x,y) \cup \pi(x,y)$ *where $\cup$ satisfies the Union Axioms.*

*Then the following hold:*

(1) $\rho(x,y) = \iota(x,y) \cup \pi(x,y)$,
(2) $\iota$ *is symmetric,*
(3) $\pi$ *is asymmetric,*
(4) $\pi$ *is simple.*

---

[4] Richardson (1998) demonstrates the necessity of this condition. Here the proof is strengthened to show sufficiency as well.

*Proof.* By (i), $\iota$ is symmetric. By (ii), $\pi$ is asymmetric. Suppose $\rho(x,y) = \rho(y,x)$, then $\pi(x,y) = 0 = \pi(y,x)$ by (ii). Thus, $\pi$ is simple.

Suppose $\rho(y,x) \geq \rho(x,y)$. Then $\iota(x,y) \cup \pi(x,y) = \rho(x,y) \cup 0 = \rho(x,y)$ by (i) and (ii). Now suppose $\rho(x,y) > \rho(y,x)$. Then, $\rho(x,y) = \iota(x,y) \cup \pi(x,y)$ by (iii).     $\square$

We are unable to establish a unique function for $\pi(x,y)$ even when assuming regularity. The problem derives from the lack of structure on $\cup$, even while assuming the Union Axioms. For instance, we assume that $\rho(x,y) = \iota(x,y) \cup \pi(x,y)$ and that $\pi$ is regular. By Corollary 3.14, we know $\rho(x,y) = \min\{\rho(x,y), \rho(y,x)\} \cup \pi(x,y)$. If $\rho(x,y) \leq \rho(y,x)$, then we have a trivial case where $\rho(x,y) = \iota(x,y) = \rho(x,y)$ because $\pi(x,y) = 0$ by regularity. When $\rho(x,y) > \rho(y,x)$, we have $\rho(x,y) = \rho(y,x) \cup \pi(x,y)$. To solve for $\pi(x,y)$, we must consider specific operations of the set union of two fuzzy sets.

**Definition 3.15 (unions 1-4).** Let $A$ and $B$ be fuzzy subsets. Then, $\cup_1$, $\cup_2$, $\cup_3$ and $\cup_4$ are defined as follows:

(1) Gödel union $\cup_1$
$$(A \cup_1 B)(x) = \max\{A(x), B(x)\},$$

(2) Łukasiewicz union $\cup_2$
$$(A \cup_2 B)(x) = \min\{1, A(x) + B(x)\},$$

(3) Strict union $\cup_3$
$$(A \cup_3 B)(x) = \begin{cases} B(x) & \text{if } A(x) = 0 \\ A(x) & \text{if } B(x) = 0 \\ 1 & \text{otherwise,} \end{cases}$$

(4) Algebraic union $\cup_4$
$$(A \cup_4 B)(x) = A(x) + B(x) - A(x)B(x).$$

$\cup_1$, $\cup_2$, $\cup_3$ and $\cup_4$ are three t-conorms, among many, for taking the set union of two fuzzy sets. In fact, Fono and Andjiga (2005) use the Intermediate Value Theorem to show that we can derive an operation for $\pi$ when $\cup$ is any continuous t-conorm. We consider $\cup_1$ and $\cup_2$, the Gödel and Łukasiewicz t-conorms, respectively, because they are common to the literature (Banerjee, 1993; Barrett et al., 1986; Richardson, 1998) and are the most intuitive. We consider $\cup_3$ because, for any t-conorm $\cup$,

$$(A \cup_1 B)(x) \leq (A \cup B)(x) \leq (A \cup_3 B)(x)$$

(Fodor and Roubens, 1994). Finally, $\cup_4$ is the algebraic sum t-conorm and expresses the union of two independent events in probability theory. Using Definition 3.15, we can now derive a unique derivation of $\pi$.

**Proposition 3.16.** *Let* $\rho$ *be an FWPR on X such that*

*(i)* $\rho = \iota \cup_1 \pi$,
*(ii)* $\iota$ *is symmetric, and*
*(iii)* $\pi$ *is asymmetric.*

*Then* $\pi$ *is simple if and only if the following equation holds:*[5]

$$\pi(x,y) = \begin{cases} \rho(x,y) & \text{if } \rho(x,y) > \rho(y,x), \\ 0 & \text{otherwise.} \end{cases}$$

*Proof.* Assume $\pi$ is simple. Let $x,y \in X$. By (i), $\rho(x,y) \geq \pi(x,y)$. We have two cases to consider. First, suppose $\rho(x,y) > \rho(y,x)$. Assume $\rho(x,y) > \pi(x,y)$. Then $\rho(x,y) = \iota(x,y) = \iota(y,x) = \rho(y,x)$ by Definition 3.15 and Proposition 3.10. However, this contradicts $\rho(x,y) > \rho(y,x)$, and thus, $\rho(x,y) = \pi(x,y)$. Second, suppose $\rho(x,y) \leq \rho(y,x)$. If $\rho(x,y) = \rho(y,x)$, then $\pi(x,y) = \pi(y,x) = 0$, by simplicity and (iii). If $\rho(x,y) < \rho(y,x)$, then, by the previous argument, $\pi(y,x) = \rho(y,x) > 0$. Hence, $\pi(x,y) = 0$, by (iii).

Conversely, suppose

$$\pi(x,y) = \begin{cases} \rho(x,y) & \text{if } \rho(x,y) > \rho(y,x), \\ 0 & \text{otherwise.} \end{cases}$$

Then $\rho(x,y) = \rho(y,x)$ implies $\pi(x,y) = \pi(y,x) = 0$. Hence, $\pi$ is simple. $\square$

The unique $\pi$ derived in Proposition 3.16 is not a new derivation. In fact, Ovchinnikov (1981) first proposed the definition of $\pi$, and Dutta (1987) used Proposition 3.16 to describe the dictatorial results of fuzzy aggregation rules, a subject considered in detail in Chapter 4. However, the definition was subsequently rejected in Banerjee (1994); Richardson (1998) as an illogical conception because it ignores the value of $\rho(y,x)$. The following example demonstrates the concern.

*Example 3.17.* Let $X = \{a,b\}$ and let $\rho_1$ and $\rho_2$ be reflexive FWPRs. Suppose the following values for $\rho_1$ and $\rho_2$.

(1) $\rho_1(a,b) = 1$ and $\rho_1(b,a) = .99$;
(2) $\rho_2(a,b) = 1$ and $\rho_2(b,a) = .01$.

Then $\pi_1(a,b) = \pi_2(a,b) = 1$, where $\pi_1$ and $\pi_2$ are the asymmetric components of $\rho_1$ and $\rho_2$, respectively.

Thus, if we use the definition of fuzzy strict preference derived in Proposition 3.16, then the difference between $\rho(x,y)$ and $\rho(y,x)$ is irrelevant to the value of $\pi(x,y)$.

In response, Banerjee (1994) suggests using the definition of $\cup_2$; however, $\cup_2$ is insufficient to derive a unique $\pi$ operation with the conditions presented in Proposition 3.16. Consequently, he suggests two more conditions for consideration:

---

[5] Necessity was first shown in Dutta (1987), but we strengthen the relationship here to include sufficiency.

(1) $\iota(x,y) + \pi(x,y) \leq 1$, for all $x, y \in X$, and
(2) $\rho(x,y) < 1$ implies $\pi(y,x) > 0$.

In effect, condition (1) constrains the degree of strict preference and indifference. Thus, if an individual possesses some degree of strict preference of $x$ over $y$, then this degree of strict preference allows the same individual to possess a limited amount of indifference. For example, if $\pi(x,y) = .6$, then $\iota(x,y) \leq .4$. The second condition is less intuitive. It requires that if an individual is not completely confident that $x$ is at least as good as $y$, then the individual *must* possess some degree of strict preference for alternative $y$ over $x$. In doing so, the second condition imposes strong connectedness on $\rho$ because of the asymmetry of $\pi$. Under this condition, it is easily verified that $\rho(x,y) = 1$ or $\rho(y,x) = 1$ because $\pi(x,y)$ or $\pi(y,x)$ must equal zero. Under these conditions, Banerjee (1994) proves the following proposition:

**Proposition 3.18.** *Let* $x, y \in X$. *Let* $\rho$ *be an FWPR on X such that*

*(i)* $\rho = \iota \cup_2 \pi$,
*(ii)* $\iota$ *is symmetric,*
*(iii)* $\pi$ *is asymmetric,*
*(iv)* $\iota(x,y) + \pi(x,y) \leq 1$, *and*
*(v)* $\rho(x,y) < 1$ *implies* $\pi(y,x) > 0$.

*Then the following equation holds:*

$$\pi(x,y) = 1 - \rho(y,x).$$

*Proof.* By (i) and (iv), $\rho(x,y) = \pi(x,y) + \iota(x,y)$. There are two cases to consider. First, suppose, $\rho(y,x) = 1$. Then $\rho(x,y) \leq \rho(y,x)$. By Proposition 3.16, $\pi(x,y) = 0$ and $\rho(x,y) = \iota(x,y)$. Thus, $\rho(x,y) = \iota(x,y)$ and $\pi(x,y) = 1 - \rho(y,x)$.

Second, suppose $\rho(y,x) < 1$. Then by (v), $\pi(x,y) > 0$, and $\pi(y,x) = 0$ by (iii). By (v), $\rho(x,y) = 1$. Because $\rho(x,y) > \rho(y,x)$, $\iota(x,y) = \rho(y,x)$. Thus,

$$1 = \rho(x,y)$$
$$= \pi(x,y) + \iota(x,y)$$
$$= \pi(x,y) + \rho(y,x).$$

Hence, $\pi(x,y) = 1 - \rho(y,x)$.                                                    □

Like Proposition 3.16, the derivation of $\pi$ in Proposition 3.18 has its own logical inconsistencies for two reasons (Richardson, 1998). First, $\pi(x,y)$ is completely independent of $\rho(x,y)$. However, this reason is minor because under the conditions of Proposition 3.18, $\pi(x,y) > 0$ implies $\rho(x,y) = 1$ across all alternatives. Second, this definition of strict preference only applies to strongly connected fuzzy preferences. Due to reasons discussed earlier in the section, Richardson (1998) drops condition (v) from this proposition and still obtains a unique operation for $\pi$.[6]

---

[6] Richardson (1998) demonstrates necessity. Here Proposition 2.16 is strengthened to show sufficiency as well.

**Proposition 3.19.** *Let $\rho$ be an FWPR on $X$ such that*

*(i)* $\rho = \iota \cup_2 \pi$,
*(ii)* $\iota$ *is symmetric and*
*(iii)* $\pi$ *is asymmetric.*

*Then* $\iota(x,y) + \pi(x,y) \leq 1$ *if and only if the following equation holds:*

$$\pi(x,y) = max\{0, \rho(x,y) - \rho(y,x)\}.$$

*Proof.* Suppose $\iota(x,y) + \pi(x,y) \leq 1$. Then, by (i),

$$\rho(x,y) = (\pi \cup_2 \iota)(x,y)$$
$$= min\{1, \iota(x,y) + \pi(x,y)\}.$$

By hypothesis, we obtain

$$\pi(x,y) = \rho(x,y) - \iota(x,y)$$
$$= \rho(x,y) - min\{\rho(x,y), \rho(y,x)\}$$
$$= max\{(\rho(x,y) - \rho(x,y)), (\rho(x,y) - \rho(y,x)\}$$
$$= max\{0, \rho(x,y) - \rho(y,x)\}.$$

Conversely, suppose $\pi(x,y) = max\{0, \rho(x,y) - \rho(y,x)\}$. By Proposition 3.10,

$$\pi(x,y) + \iota(x,y) = max\{0, \rho(x,y) - \rho(y,x)\} + min\{\rho(x,y), \rho(y,x)\}$$
$$= \begin{cases} \rho(x,y) - \rho(y,x) + \rho(y,x) & \text{if } \rho(x,y) > \rho(y,x), \\ 0 + \rho(x,y) & \text{if } \rho(x,y) \leq \rho(y,x). \end{cases}$$
$$= \begin{cases} \rho(x,y) & \text{if } \rho(x,y) > \rho(y,x), \\ \rho(x,y) & \text{if } \rho(x,y) \leq \rho(y,x). \end{cases}$$

Hence, $\iota(x,y) + \pi(x,y) \leq 1$. □

Richardson (1998) settles for the $\pi$ operation in Proposition 3.19 because it incorporates both $\rho(x,y)$ and $\rho(y,x)$ and does not require preferences to be connected or strongly connected. However, this conceptualization of strict preference is completely relative and tells us little about the values of $\rho$ across cases, as the next example illustrates.

*Example 3.20.* Let $X = \{a,b\}$, and let $\rho_1$, $\rho_2$, and $\rho_3$ be reflexive FWPRs. Suppose the following values for $\rho_1$, $\rho_2$, and $\rho_3$.

(1) $\rho_1(a,b) = .1$ and $\rho_1(b,a) = 0$,
(2) $\rho_2(a,b) = .55$ and $\rho_1(b,a) = .45$, and
(3) $\rho_3(a,b) = 1$ and $\rho_3(b,a) = .9$.

Then $\pi_1(a,b) = \pi_2(a,b) = \pi_3(a,b) = .1$, where $\pi_1$, $\pi_2$ and $\pi_3$ are the asymmetric components of $\rho_1$, $\rho_2$ and $\rho_3$, respectively.

Even though the derivations of $\pi$ presented in Propositions 3.16, 3.18 and 3.19 are the definitions most common in the fuzzy social choice literature, still other operations emerge for $\pi$ when we use the fuzzy set unions $\cup_3$ and $\cup_4$.

**Proposition 3.21.** *Let $\rho$ be an FWPR on X such that*

*(i) $\rho = \iota \cup_3 \pi$,*
*(ii) $\iota$ is symmetric and*
*(iii) $\pi$ is asymmetric.*

*Then the following holds for all $x, y \in X$:*

$$\pi(x,y) = \begin{cases} \rho(x,y) & \text{if } \rho(y,x) = 0, \\ 0 & \text{otherwise.} \end{cases}$$

*Proof.* Suppose $\rho(x,y) > \rho(y,x)$ and $\rho(y,x) > 0$. Then

$$\bigwedge \{t \in [0,1] \mid \rho(y,x) \cup_3 t \geq \rho(x,y)\}$$

does not exist since for all $t > 0$, $\rho(y,x) \cup_3 t = 1$ and when $t = 0$, $\rho(y,x) \cup_3 t = \rho(y,x) < \rho(x,y)$. Thus,

$$\bigwedge \{t \in [0,1] \mid \rho(y,x) \cup_3 t \geq \rho(x,y)\}$$

exists and equals $\rho(x,y)$ if and only if $\rho(y,x) = 0$. Hence,

$$\pi(x,y) = \begin{cases} \rho(x,y) & \text{if } \rho(y,x) = 0, \\ 0 & \text{otherwise.} \end{cases}$$

as desired.                                                                                          □

The $\pi$ derived in Proposition 3.21 has been used in fuzzy revealed preference theory (Georgescu, 2007a). When $\rho = \iota \cup_4 \pi$, we have the following.

**Proposition 3.22.** *Let $\rho$ be an FWPR on X such that*

*(i) $\rho = \iota \cup_4 \pi$,*
*(ii) $\iota$ is symmetric and*
*(iii) $\pi$ is asymmetric.*

*Then the following holds for all $x, y \in X$:*

$$\pi(x,y) = \begin{cases} \frac{\rho(x,y) - \rho(y,x)}{1 - \rho(y,x)} & \text{if } \rho(x,y) > \rho(y,x), \\ 0 & \text{otherwise.} \end{cases}$$

*Proof.* Suppose $\rho(x,y) > \rho(y,x)$. Then, by Proposition 3.16 and (i), $\rho(x,y) = \rho(y,x) \cup_4 \pi(x,y)$. By the definition of $\cup_4$,

$$\rho(x,y) = \rho(y,x) + \pi(x,y) - \rho(y,x)\pi(x,y).$$

Simplifying this expression yields

$$\rho(x,y) - \rho(y,x) = \pi(x,y)(1 - \rho(y,x)) \, .$$

The desired result now follows.                    □

Propositions 3.21 and 3.22 conclude our unique derivations for $\pi$ considered in the book. We summarize them here because, as later chapters will show, their constructions significantly impact the social choice implications of fuzzy sets.

**Corollary 3.23 ($\pi$ 1-5).** *Let $\rho$ be an FWPR on X such that*

*(i) $\rho = \iota \cup \pi$,*
*(ii) $\iota$ is symmetric, and*
*(iii) $\pi$ is asymmetric.*

**Proposition 3.24.** *Then we list the following results.*

*(1) If $\cup = \cup_1$, where Gödel union $\cup_1$ is given by*

$$(A \cup_1 B)(x) = max\{A(x), B(x)\} \, ,$$

*then $\pi = \pi_{(1)}$ where*

$$\pi_{(1)}(x,y) = \begin{cases} \rho(x,y) & \text{if } \rho(x,y) > \rho(y,x), \\ 0 & \text{otherwise.} \end{cases}$$

*(2) If $\cup = \cup_2$, where Łukasiewicz union $\cup_2$ is given by*

$$(A \cup_2 B)(x) = min\{1, A(x) + B(x)\} \, ,$$

*and conditions (iv) and (v) from Proposition 3.18 hold, then $\pi = \pi_{(2)}$ where*

$$\pi_{(2)}(x,y) = 1 - \rho(y,x).$$

*(3) If $\cup = \cup_2$, then $\pi = \pi_{(3)}$ where*

$$\pi_{(3)}(x,y) = max\{0, \rho(x,y) - \rho(y,x)\}.$$

*(4) If $\cup = \cup_3$, where strict union $\cup_3$ is given by*

$$(A \cup_3 B)(x) = \begin{cases} B(x) & \text{if } A(x) = 0 \\ A(x) & \text{if } B(x) = 0 \\ 1 & \text{otherwise,} \end{cases}$$

*then $\pi = \pi_{(4)}$ where*

$$\pi_{(4)}(x,y) = \begin{cases} \rho(x,y) & \text{if } \rho(y,x) = 0, \\ 0 & \text{otherwise.} \end{cases}$$

*(5) If* $\cup = \cup_4$, *where* algebraic union $\cup_4$ *is given by*

$$(A \cup_4 B)(x) = A(x) + B(x) - A(x)B(x)$$

*then* $\pi = \pi_{(5)}$ *where*

$$\pi_{(5)}(x,y) = \begin{cases} \frac{\rho(x,y) - \rho(y,x)}{1 - \rho(y,x)} & \text{if } \rho(x,y) > \rho(y,x), \\ 0 & \text{otherwise.} \end{cases}$$

Even though this section has restricted an FWPR $\rho$ to being reflexive and complete and has derived various forms of $\pi$ to best reflect traditional understandings of strict preference, we are not able to guarantee that an individual is necessarily rational. This stems from the fact that we have not yet considered the condition of transitivity of fuzzy preference relations, which restricts the behavior of FWPRs across all pairwise comparisons of alternatives. Example 2.21 demonstrates this problem.

*Example 3.25.* Let $X = \{a,b,c\}$, let $\rho$ be a reflexive and complete FWPR and let $\pi = \pi_{(3)}$. Suppose $\rho$ is defined as follows:

(1) $\rho(a,b) = .7$ and $\rho(b,a) = .2$,
(2) $\rho(b,c) = .6$ and $\rho(c,b) = .4$, and
(3) $\rho(a,c) = .1$ and $\rho(c,a) = .7$.

Now $\pi(x,y) = \rho(x,y) - \rho(y,x)$ when $\rho(x,y) \geq \rho(y,x)$ and $\pi(x,y) = 0$, otherwise. In this case, $\pi(a,b) = .5$, $\pi(b,c) = .2$ and $\pi(c,a) = .6$. Thus, $a$ is strictly preferred to $b$ to some degree, $b$ is strictly preferred to $c$ to some degree, and $c$ is strictly preferred to $a$ to some degree. Hence, a political actor possessing $\rho$ has no clear method to vote for or nominate an alternative in $X$.

Obviously, the preferences presented in Example 3.25 do not immediately correspond to our conception of rationality. If an actor with such preferences were to be presented with a choice between alternatives $a$, $b$, and $c$, the actor *could* choose an alternative; but it remains unclear as to how this choice would relate to the actor's preferences. Because the model we have presented thus far does not provide another tool that can explain the behavior of individuals besides preferences, we not only consider what an individual best alternative set could like like under fuzzy preferences but also we also restrict $\rho$ in such a way as to ensure such a "maximal" choice set always exists. It is to this issue that we now turn.

## 3.2 Consistency of Fuzzy Preferences and the Fuzzy Maximal Set

In the preceding section, we considered how an FWPR $\rho$ can be dissected into its symmetric, $\iota$, and asymmetric, $\pi$, components. In addition, we proposed the restrictions of reflexivity and completeness on $\rho$ to model the basic tenet of rational preference given two alternatives. However, the section concluded that there is no clear method for an individual to identify an alternative as a maximal one when there are more than three alternatives. When every alternative in an individual's choice set is strictly preferred by another alternative, the individual is said to have cyclic preferences (1999).

This section is concerned with restricting how FWPRs behave across pairwise comparisons, a task involving more than two alternatives. More specifically, we wish to guarantee the existence of a maximal set. Formally, the *maximal set* under exact preferences is

$$M(R,X) = \{x \in X \mid xRy, \ \forall y \in X\},$$

where $R$ is an exact preference relation and $X$ is a set of alternatives (1999). In words, an alternative $x$ is in the maximal set if it is at least as good all other alternatives. When fuzzifying the maximal set, we must consider the fuzzy counterparts of $R$ and $X$. Section 3.1 detailed the transformation of an exact preference relation into a fuzzy one. In this section, we also allow the set of alternatives to be a fuzzy subset, $\mu : X \to [0,1]$. For our purposes, $\mu(x)$ denotes the degree to which $x$ is in a fuzzy subset of $X$, and $\mathscr{F}(X)$ denotes the set of all fuzzy subsets of $X$ not equal to the null set $\theta$ (Definition 1.11). In words, $\mu(x)$ represents the degree to which $x$ is a feasible alternative. This is an important addition to modeling political preferences. For instance, a legislator may prefer alternative $x$ over alternative $y$; but she also knows that $y$ is more feasible than $x$ in terms of state capacity, bureaucratic performance, and other factors. The function $\mu$ allows these realities to be incorporated into the legislator's decision-making calculus.

In the first attempts to fuzzify the maximal set, Orlovsky (1980) and later Kołodziejczyk (1986) and Montero and Tejada (1988) conceptualize maximal alternatives as those that are never dominated. To do so, they use fuzzy strict preference relations to obtain the degree to which $x \in X$ is not dominated: $nd(x) = 1 - \max_{y \in X}\{\pi(y,x)\}$. When $\pi(y,x) = 0$ for all $y \in X$, $nd(x) = 1$. When this occurs, the set of non-dominated solutions, $ND(\pi,X) = \{x \in X \mid nd(x) = 1\}$, is non-empty.

**Definition 3.26 (degree of non-domination).** The degree of non-domination of $x$ is

$$nd(x) = 1 - \max_{y \in X}\{\pi(y,x)\}.$$

**Definition 3.27 (non-dominated set).** The non dominated set is

$$ND(\pi,X) = \{x \in X \mid nd(x) = 1\}.$$

It is readily verified that Example 3.25 demonstrates a case where $ND(\pi,X) = \varnothing$. For $\mu$ a fuzzy subset of $X$, we define the fuzzy subset $nd(\mu)$ of $X$ by $\forall x \in X$,

$$nd(\mu)(x) = 1 - \max\{\pi(x,y) \mid y \in \text{Supp}(\mu)\} .$$

For $S$, a subset of $X$, we define

$$ND(\pi, S) = \{x \in S \mid nd(x) = 1\}.$$

Using $ND(\pi, X)$ as a maximal set raises two significant problems. First, as Section 3.1 illustrated, constructing a fuzzy strict preference relation can be excessively complicated and requires significant restrictions on FWPRs. Second, the definition does not make full use of fuzzy set theory, as it still views alternatives as either in or out of the set of alternatives *and* the set of nondominated alternatives. For these reasons, we also consider the fuzzy maximal set proposed by Georgescu (2007b), which is similar to the one found in Dasgupta and Deb (1996):

$$M(\rho, \mu)(x) = \mu(x) * \bigodot_{w \in \text{Supp}(\mu)} \bigvee \{t \in [0,1] \mid \mu(w) * \rho(w,x) * t \leq \rho(x,w)\}$$

$$= \mu(x) * \bigodot_{w \in \text{Supp}(\mu)} \mu(w) * \rho(w,x) \to_* \rho(x,w),$$

where $*$ and $\odot$ are arbitrary t-norms and $t \in [0,1]$.[7] Recall that a t-norm, i.e. triangular norm, is a function $t : [0,1] \times [0,1] \to [0,1]$ that is also associative, commutative and monotonic. The minimum function, for example, is a t-norm. For a more thorough discussion on t-norms see Klement et al. (2000). For our purposes, we do not specify a t-norm in this section. Throughout, we assume that $*$ and $\odot$ have no zero divisors; specifically, $a \neq 0$ and $b \neq 0$ imply $a * b \neq 0$. This assumption eliminates perverse cases where, for some $x \in X$ such that $\mu(x) > 0$ and some $t \in (0,1]$, $M(\rho, \mu)(x) = \mu(x) * t = 0$. Then $M(\rho, \mu)$ is the *fuzzy maximal subset* associated with $\rho$ and $\mu$, and $M(\rho, \mu)(x)$ denotes the degree to which $x$ is maximal.

*Example 3.28.* Let $X = \{a, b, c\}$. Let $\mu \in \mathscr{F}(X)$ be such that $\mu(a) = 0$, $\mu(b) = .25$ and $\mu(c) = .5$. Let $\rho$ be a reflexive FWPR defined as $\rho(a,b) = .25$, $\rho(b,c) = .1$, $\rho(a,c) = .75$, and $\rho(\_,\_) = 0$ otherwise.

Since $\mu(a) = 0$, it is immediate that $M(\rho, \mu)(a) = 0$. Now

$$M(\rho, \mu)(b) = \mu(b) * \left[ \left( \bigvee \{t \in [0,1] \mid \mu(b) * \rho(b,b) * t \leq \rho(b,b)\} \right) \right.$$

$$\left. \odot \left( \bigvee \{t \in [0,1] \mid \mu(c) * \rho(c,b) * t \leq \rho(b,c)\} \right) \right]$$

$$= .25 * (1 \odot 1)$$

$$= .25$$

---

[7] The purpose of the t-norm differentiation between $*$ and $\odot$ is a generalization of the fuzzy maximal set considered by Georgescu (2007a) in which $\odot$ is assumed to be the minimum function.

and

$$M(\rho,\mu)(c) = \mu(c) * \left[\left(\bigvee\{t \in [0,1] \mid \mu(c) * \rho(c,c) * t \le \rho(c,c)\}\right)\right.$$
$$\left. \odot \left(\bigvee\{t \in [0,1] \mid \mu(b) * \rho(b,c) * t \le \rho(c,b)\}\right)\right]$$
$$= .5 * (1 \odot 0)$$
$$= 0$$

The fuzzy maximal set presented here enables a much more nuanced consideration of solution sets than those under exact preference relations. Not only does the fuzzy maximal set $M(\rho,\mu)$ utilize an FWPR and a fuzzy set of alternatives; for each pairwise comparison between $x$ and $y$, the equation accounts for $\mu(x)$, $\mu(y)$, $\rho(x,y)$ and $\rho(y,x)$. Furthermore, unlike $ND(\pi,X)$, $M(\rho,\mu)$ considers the weak preference relation, which incorporates the fuzzy indifference and strict preference relations, instead of just the fuzzy strict preference relations. Hence, $M(\rho,\mu)$ accounts for much more information. Later in the section, we show how this implies $x$ is in $\mathrm{Supp}(M(\rho,\mu))$ if it is in $ND(\pi,X)$ (see Proposition 3.39), but the converse is not necessarily true. The latter claim is demonstrated by the following example.

*Example 3.29.* Let $X = \{a,b,c\}$, and let $\rho$ be a reflexive and complete FWPR. Assume $\mu(x) = 1$ for all $x \in X$. Suppose $\pi$ is regular and suppose $\rho$ is defined in the same manner as Example 3.28. By Proposition 3.21 we can calculate $nd(x)$ for all $x \in X$ as follows:

$nd(a) = 1 - \pi(c,a)$ because $\pi(b,a) = 0$,
$nd(b) = 1 - \pi(a,b)$ because $\pi(c,b) = 0$,
$nd(c) = 1 - \pi(b,c)$ because $\pi(a,c) = 0$.
Hence, $ND(\pi,X) = \varnothing$ because $nd(x) < 1$ for all $x \in X$. To see that $M(\rho,\mu)$ is nonempty, consider

$$M(\rho,\mu)(a) = \mu(a) * \left[\left(\bigvee\{t \in [0,1] \mid \mu(a) * \rho(a,a) * t \le \rho(a,a)\}\right)\right.$$
$$\odot \left(\bigvee\{t \in [0,1] \mid \mu(b) * \rho(b,a) * t \le \rho(a,b)\}\right)$$
$$\left. \odot \left(\bigvee\{t \in [0,1] \mid \mu(c) * \rho(c,a) * t \le \rho(a,c)\}\right)\right]$$
$$= 1 * \left[\left(\bigvee\{t \in [0,1] \mid 1 * 1 * t \le 1\}\right)\right.$$
$$\odot \left(\bigvee\{t \in [0,1] \mid 1 * .2 * t \le .7\}\right)$$
$$\left. \odot \left(\bigvee\{t \in [0,1] \mid 1 * .7 * t \le .1\}\right)\right]$$

This implies

$$M(\rho,\mu)(a) > 0$$

since $*$ and $\odot$ have no zero divisors.

The difference between $ND(\pi,X)$ and $M(\rho,\mu)$ stems from the fuzzification of $M(R,S)$ and the different criteria that determine whether $x \in X$ is maximal. Proposition 3.30 demonstrates the logic behind $M(\rho,\mu)$.

**Proposition 3.30.** *Let* $\mu \in \mathcal{F}(X)$, $x \in Supp(\mu)$ *and* $\rho$ *be a reflexive and complete FWPR. Then* $M(\rho,\mu)(x) = 0$ *if and only if there exits a* $w \in Supp(\mu)$ *such that* $\rho(x,w) = 0$.

*Proof.* Suppose $x \in Supp(\mu)$ and $M(\rho,\mu)(x) = 0$. Then there exists $w \in Supp(\mu)$ such that

$$\bigvee \{t \in [0,1] \mid \mu(w) * \rho(w,x) * t \leq \rho(x,w)\} = 0$$

because $\mu(x) > 0$. This implies $\rho(x,w) = 0$.

Conversely, suppose $x \in Supp(\mu)$ and there exists a $w \in Supp(\mu)$ such that $\rho(x,w) = 0$. Then we have the following:

$$M(\rho,\mu)(x) = \mu(x) * \left[ \bigvee \{t \in [0,1] \mid \mu(w) * \rho(w,x) * t \leq \rho(x,w)\} \right.$$

$$\left. \odot \bigvee \{t \in [0,1] \mid \mu(v) * \rho(v,x) * t \leq \rho(x,v), \forall v \in Supp(\mu) \neq w\} \right]$$

Substituting, we have

$$M(\rho,\mu)(x) = \mu(x) * \left[ \bigvee \{t \in [0,1] \mid \mu(w) * \rho(w,x) * t \leq 0\} \right.$$

$$\left. \odot \bigvee \{t \in [0,1] \mid \mu(v) * \rho(v,x) * t \leq \rho(x,v), \forall v \in Supp(\mu) \neq w\} \right]$$

$$= \mu(x) * 0 \odot t_2 \odot ... \odot t_n$$

$$= 0.$$

Hence, $M(\rho,\mu)(x) = 0$ if and only if there exists a $w \in Supp(\mu)$ such that $\rho(x,w) = 0$.                                                                        $\square$

Thus, an alternative $x \in X$ is in $Supp(M(\rho,\mu))$ if $x$ is at least as good as all other alternatives in $Supp(\mu)$ to some degree.[8] In contrast, $x \in X$ is in $ND(\pi,X)$ if there does not exist another alternative $y \in X$ that is strictly preferred to $x$ to any degree. In the exact case, these definitions would be equivalent because $xRy$ implies not $yPx$. However, as Example 3.29 and Proposition 3.30 illustrate, this relationship no longer holds under fuzzy preferences. Obviously, if $x$ is in $ND(\pi,X)$ then $x$ is in $Supp(M(\rho,\mu))$ to some degree, but the converse is not necessarily true.[9]

Before we consider what restrictions on $\rho$ guarantee $M(\rho,\mu) \neq \theta$, we need to verify that the fuzzy maximal set includes the exact case, a task which Billot Billot (1992) calls the "credo" of fuzzy social choice scholars.

---

[8] The maximal set proposed by Dasgupta and Deb (1991), $M'(\rho,\mu)(x) = \min\limits_{y \in X} \{\rho(x,y)\}$, also shares this logic.

[9] This finding reflects those in the revealed preference literature where, for an exact preference relation $R$, there is a difference between $R$ − maximality ($x \in ND(\pi,X)$) and $R$ − greatness ($x \in Supp(M(\rho,\mu))$).

**Proposition 3.31.** *Suppose $R$ is a reflexive relation on $X$ and let $S$ be a nonempty subset of $X$. Then the following holds if and only if $R$ is complete:*

$$M(1_R, 1_S)(x) = \begin{cases} 1 & \text{if } x \in S \text{ and } (x,w) \in R, \ \forall w \in S, \\ 0 & \text{otherwise.} \end{cases}$$

*Proof.* First, suppose $R$ is complete. Clearly, $M(1_R, 1_S)(x) = 1$ if $x \in S$ and $(x,w) \in R$ for all $w \in S$. Now suppose $x \notin S$. Then, $M(1_R, 1_S)(x) = 0$. Now suppose $(x,w) \notin R$ for some $w \in S$. Then, $1_R(x,w) = 0$. In this case, $1_S(w) * 1_R(w,x) \neq 0$ because $R$ is complete and $*$ has no zero divisors. Thus, $\bigvee \{t \in [0,1] \mid 1_S(w) * 1_R(w,x) * t \leq 1_R(x,w)\} = 0$. Thus, $M(1_R, 1_S)(x) = 0$.

To prove the converse, suppose $R$ is not complete. Then there exists $x, y \in X$ such that $(x,y) \notin R$ and $(y,x) \notin R$. Let $S = \{x,y\}$. Since $1_R(y,x) = 0$ and $1_S(z) = 0$ for all $z \in X \setminus \{x,y\}$,

$$M(1_S, 1_R)(x) = 1_S(x) * \Big[ \bigvee \{t \in [0,1] \mid 1_S(x) * 1_R(x,x) * t \leq 1_R(x,x)\}$$
$$\odot \bigvee \{t \in [0,1] \mid 1_S(y) * 1_R(y,x) * t \leq 1_R(x,y)\} \Big]$$
$$= 1 * (1 \odot 1)$$
$$= 1.$$

It is not the case that $(x,w) \in R$ for all $w \in S$. □

Once we know that $M(\rho, \mu)$ extends the logic of the exact maximal set to fuzzy preferences, we can now restrict $\rho$ in such a ways as to guarantee a non-empty fuzzy maximal set, i.e. $M(\rho, \mu) \neq \theta$. This ensures that given any set of alternatives, an individual will be able to identify at least one alternative that is a best choice to some degree. To do so requires considering assumptions that designate the behavior of $\rho$ across pairwise comparisons.

**Definition 3.32.** A FWPR $\rho$ on $X$ is

(1) **max-star transitive** if, for all $x,y,z \in X$, $\rho(x,z) \geq \rho(x,y) * \rho(y,z)$;
(2) **weakly transitive** if, for all $x,y,z \in X$, $\rho(x,y) \geq \rho(y,x)$ and $\rho(y,z) \geq \rho(z,y)$ imply $\rho(x,z) \geq \rho(z,x)$;
(3) **acyclic** if, for all $\{x_1, x_2, x_3, ..., x_{n-1}, x_n\} \subseteq X$, $\pi(x_1, x_2) * \pi(x_2, x_3) * ... * \pi(x_{n-1}, x_n) > 0$ implies $\pi(x_n, x_1) = 0$;
(4) **partially acyclic** if, for all $\{x_1, x_2, x_3, ..., x_{n-1}, x_n\} \subseteq X$, $\pi(x_1, x_2) * \pi(x_2, x_3) * ... * \pi(x_{n-1}, x_n) > 0$ implies $\rho(x_1, x_n) > 0$.

Definition 3.32(1) was first proposed by Zadeh (1971) and Orlovsky (1978) and is now standard among fuzzy social choice papers (Banerjee, 1993, 1994; Dutta, 1987; Kołodziejczyk, 1986; Orlovsky, 1980; Richardson, 1998). Throughout, we use max-star and max-* interchangeably. Weak transitivity has appeared in Luo (1986), Banerjee (1994), Georgescu (2007a) under different names and in various forms. Billot (1992) also selects Definition 3.32(2) as the basis of fuzzy rationality. The form of acyclicity in Definition 3.32(3) is similar to Richardson's (1998)

concept of "negative transitivity" and is almost identical to acyclicity of exact preference relations (Austen-Smith and Banks, 1999). In addition when $\pi$ is regular,
acyclicity corresponds to Wang's (1997) $\omega$-transitivity. Finally, partial acyclicity,
the weakest of the four conditions, is our own creation; but the definition is quite
important to guaranteeing the existence of the fuzzy maximal set using specific types
of fuzzy strict preference. Certainly there exist numerous other definitions that place
consistency restrictions on $\rho$; however, we highlight the four in Definition 3.32 because either they are standards in the literature (Definitions 3.32(1) and (2)) or they
guarantee a non-empty maximal set (Definitions 3.32(3) and (4)). Proposition 3.33
lays out the relationships between the consistency requirements in Definition 3.32
when $\pi$ is regular. For a more complete review, see Wang (1997) and Dasgupta and
Deb (1996).

**Proposition 3.33.** *Let $\rho$ be a reflexive and complete FWPR. Suppose $\pi$ is regular.
Then the following relationships hold:*

*(1) weak transitivity implies acyclicity and*
*(2) acyclicity implies partial acyclicity.*

*Proof*

(1) Suppose $\rho$ is weakly transitive. Further, suppose there exists an order-
    ing of alternatives $\{x_1, x_2, x_3, ..., x_{k-1}, x_k\} \subseteq X$, where $3 \leq k \leq n$, such
    that $\pi(x_1,x_2) * \pi(x_2,x_3) * ... * \pi(x_{k-1},x_k) > 0$. First, by the hypothesis,
    $\pi(x_i,x_{i+1}) > 0$ for all $x_i, x_{i+1} \in \{x_1, x_2, x_3, ..., x_{k-1}, x_k\}$. Second, by the regular-
    ity assumption, $\rho(x_i,x_{i+1}) > \rho(x_{i+1},x_i) \geq 0$.
    Now by weak transitivity, $\rho(x_1,x_3) \geq \rho(x_3,x_1)$ because $\rho(x_1,x_2) > \rho(x_2,x_1)$
    and $\rho(x_2,x_3) > \rho(x_3,x_2)$. Similarly, $\rho(x_1,x_4) \geq \rho(x_4,x_1)$ because $\rho(x_1,x_3) \geq$
    $\rho(x_3,x_1)$ and $\rho(x_3,x_4) > \rho(x_4,x_3)$. It can now be shown that $\rho(x_1,x_i) \geq$
    $\rho(x_i,x_1)$ by repeating the previous argument $i - 2$ times. Thus,
    $\rho(x_1,x_k) \geq \rho(x_k,x_1)$. Hence $\pi(x_k,x_1) = 0$, and $\rho$ is acyclic.
(2) Suppose now that $\rho$ is acylic and that there exists an ordering of alternatives
    $\{x_1,x_2,x_3,...,x_{k-1},x_k\} \subseteq X$, where $3 \leq k \leq n$, such that $\pi(x_1,x_2) * \pi(x_2,x_3) *$
    $... * \pi(x_{k-1},x_k) > 0$. Then we know $\pi(x_k,x_1) = 0$ by hypothesis. Then regularity
    of $\pi$ implies $\rho(x_1,x_k) \geq \rho(x_k,x_1)$. By completeness, $\rho(x_1,x_k) > 0$. Hence, $\rho$ is
    partially acyclic.                                                                    $\square$

For parsimony, it would be beneficial to assert some type of relationship between
max-$*$ transitivity and weak transitivity; however, the following example demonstrates that the former does not imply the latter.

**Definition 3.34 (quasi–transitive).** Let $\rho$ be a FWPR on $X$ and $\pi$ its asymmetric
component. Then $\rho$ is said to be quasi–transitive with respect to a t–norm $*$ if

$$\pi(x,y) * \pi(y,z) \leq \pi(x,z)$$

for $\forall x,y,z \in X$.

*Example 3.35.* Let $X = \{x_1, x_2, x_3\}$. Define the fuzzy preference relation $\rho$ on $X$ as follows:

$$\rho(x_1, x_1) = \rho(x_2, x_2) = \rho(x_3, x_3) = 1,$$
$$\rho(x_1, x_3) = \rho(x_1, x_2) = \rho(x_2, x_3) = 1/2,$$
$$\rho(x_3, x_1) = 5/8,$$
$$\rho(x_2, x_1) = 3/8,$$
$$\rho(x_3, x_2) = 3/8.$$

We first show that $\rho$ is max-$*$ transitive, where $*$ denotes product. We have

$$\rho(x_3, x_2) = 3/8 \geq 5/8 \cdot 1/2$$
$$= \rho(x_3, x_1) * \rho(x_3, x_2),$$
$$\rho(x_2, x_1) = 3/8 \geq 5/8 \cdot 1/2$$
$$= \rho(x_2, x_3) * \rho(x_3, x_1).$$

The remaining inequalities are easily established. For $\pi$ regular, we have $\pi(x_1, x_2) > 0$, $\pi(x_2, x_3) > 0$, $\pi(x_1, x_3) = 0$, and $\pi(x_3, x_1) > 0$. Hence we see that $\rho$ is not max-product quasi-transitive and not acyclic. If also follows that $\rho$ is not weakly transitive.

*Example 3.36.* Let $X = \{a, b, c\}$ and let $\rho$ be a reflexive FWPR over $X$ defined as follows.

$$\rho(a, b) = \rho(b, a) = \rho(b, c) = \rho(c, b) = .3,$$
$$\rho(a, c) = .4,$$
$$\rho(c, a) = .6.$$

It is easily verified that $\rho$ satisfies max $- *$ transitivity for any $*$. To see that $\rho$ violates weak transitivity, consider that $\rho(a, b) \geq \rho(b, a)$ and $\rho(b, c) \geq \rho(c, b)$. However, $\rho(a, c) \not\geq \rho(c, a)$.

Once the relationships between the consistency definitions are clear, we can present our main result on the existence of a non-empty fuzzy maximal set.

**Theorem 3.37.** *Let $\rho$ be a reflexive and complete FWPR and let $\pi$ be regular. If $\rho$ is partially acyclic, then*

$$M(\rho, \mu) \neq \theta$$

*for all $\mu \in \mathscr{F}(X)$.*

*Proof.* Suppose $\rho$ is partially acyclic. Let $\mu \in \mathscr{F}(X)$. Then $\mathrm{Supp}(\mu) \neq \varnothing$. Let $x_1 \in \mathrm{Supp}(\mu)$. If $\rho(x_1, w) > 0$ for all $w \in \mathrm{Supp}(\mu)$, then, by Proposition 3.33 $M(\rho, \mu)(x_1) > 0$, and so $M(\rho, \mu) \neq \theta$. Suppose there exists an $x_2 \in \mathrm{Supp}(\mu) \setminus \{x_1\}$ such that $\rho(x_1, x_2) = 0$. By completeness, $\rho(x_2, x_1) > 0$. Thus, $\pi(x_2, x_1) > 0$ because $\rho(x_2, x_1) > \rho(x_1, x_2)$ and $\pi$ is regular. If $\rho(x_2, w) > 0 \forall w \in \mathrm{Supp}(\mu)$, then

$M(\rho,\mu)(x_2) > 0$. Suppose there exists $x_1,...,x_k \in \text{Supp}(\mu)$ such that $\pi(x_i,x_{i-1}) > 0$ for all $i = 2,...,k$. If $\rho(x_k,w) > 0$ for all $w \in \text{Supp}(\mu)$, then $M(\rho,\mu)(x_k) \neq 0$. Suppose this is not the case and there exists $x_{k+1} \in \text{Supp}(\mu)\backslash\{x_1,...,x_k\}$ such that $\rho(x_k,x_{k+1}) = 0$. Then $\rho(x_{k+1},x_k) > 0$ by completeness.

By induction, either there exists an $x \in \text{Supp}(\mu)$ such that $\rho(x,w) > 0$ for all $w \in \text{Supp}(\mu)$, in which case $M(\rho,\mu) \neq \theta$, or since $\text{Supp}(\mu)$ is finite, $\text{Supp}(\mu) = \{x_1,...,x_n\}$ is such that $\pi(x_i,x_{i-1}) > 0$ for $i = 2,...,n$. Since $\rho$ is partially acyclic and $\rho$ is reflexive, $\rho(x_n,x_i) > 0$ for $i = 1,...,n$. Hence, $M(\rho,\mu)(x_n) > 0$ and $M(\rho,\mu) \neq \theta$.                                                                                    $\square$

Theorem 3.37 demonstrates that under the very weak condition of partial acyclicity, $M(\rho,\mu)$ will not equal $\theta$. Unlike the exact case (see Austen-Smith and Banks (1999) for reference), cyclic behavior in fuzzy strict preference relation does not render the maximal set empty. Indeed, when $\pi(x_1,x_2) > 0$, $\pi(x_2,x_3) > 0$, and $\pi(x_3,x_1) > 0$, $x_1$ can still be in the maximal set if $\rho(x_1,x_3) > 0$. A further departure from the exact case occurs because the relationship in Theorem 3.37 cannot be strengthened to include the converse. A non-empty maximal set does not necessarily imply partial acyclicity of $\rho$; the following proposition lays this out in greater detail.

*Example 3.38.* Let $X = \{x_1,x_2,x_3\}$. Define the FWPR $\rho$ as follows: $\rho(x,x) = 1 \, \forall x \in X$,

$$\rho(x_1,x_2) > 0,$$
$$\rho(x_2,x_3) > 0,$$
$$\rho(x_3,x_1) > \rho(x_1,x_3) > 0.$$

Let $\pi$ be regular. Then $\rho$ is not acyclic, yet $M(\rho,1_{\{x\}})(x_1) > 0$.

As Example 3.38 demonstrates, we know very little about the consistency of $\rho$ when $M(\rho,\mu) \neq \theta$; it may not even be partially acyclic. $M(\rho,\mu) \neq \theta$ merely implies there exists a $x \in \text{Supp}(\mu)$ where $\rho(x,w) > 0$ for all $w$ in $X$. Because of this, it may be useful to consider the consistency property that guarantees a non-empty $ND(\pi,X)$, which imposes more structure on $\rho$. In an attempt to fuzzify $ND(\pi,X)$ we write it as

$$ND(\pi,\mu)(x) = \begin{cases} \mu(x) & \text{if } x \in ND(\pi,\text{Supp}(\mu)), \\ 0 & \text{otherwise.} \end{cases}$$

As the following proposition explains, this is an intuitive fuzzification of $ND(\pi,X)$ because if $x$ is undominated then $M(\rho,\mu)(x) = \mu(x)$.

**Proposition 3.39.** *Suppose $\rho$ is a reflexive and complete FWPR and $\pi$ is regular. If $x \in ND(\pi,X)$, then $M(\rho,\mu)(x) = \mu(x)$ for all $\mu \in \mathscr{F}(X)$.*

*Proof.* Let $\mu \in \mathscr{F}(X)$. Then there exists $x \in \text{Supp}(\mu)$. Now suppose $x \in ND(\pi,X)$. By definition, $nd(x) = 1$. Therefore, $\max_{y \in X}\{\pi(y,x)\} = 0$. It follows that $\pi(y,x) = 0$ for all $y \in X$.

By regularity, we know $\pi(y,x) > 0$ if and only if $\rho(y,x) > \rho(x,y)$. Thus, $\rho(x,y) \geq \rho(y,x)$ for all $y \in X$. Hence, $\rho(x,y) > 0$ for all $y \in X$. Because there does not exist a $w \in X$ such that $\rho(x,w) = 0$, $x \in \mathrm{Supp}(M(\rho,\mu))$. Further, $\bigvee\{t \in [0,1] \mid \mu(y) * \rho(y,x) * t \leq \rho(x,y)\} = 1$ for all $y \in X$ because $\rho(x,y) \geq \rho(y,x)$. Hence $M(\rho,\mu)(x) = \mu(x) * 1 \odot \ldots \odot 1 = \mu(x)$. □

Hence, $ND(\pi,\mu)(x) = M(\rho,\mu)(x) = \mu(x)$ if $x \in ND(\pi,X)$ and $ND(\pi,\mu)(x) = 0$ otherwise. In words, Proposition 3.39 provides support for the fuzzification of $ND(\pi,X)$. Now we can proceed with this section's second major result that guarantees the non-emptiness of $ND(\pi,\mu)$.

**Theorem 3.40.** *Let $\rho$ be a reflexive and complete FWPR and suppose $\pi$ is regular. Then $\rho$ is acyclic if and only if $ND(\pi,\mu) \neq \theta$ for all $\mu \in \mathscr{F}(X)$.*

*Proof.* Suppose $\rho$ is acyclic. Let $\mu \in \mathscr{F}(X)$. Then $\mathrm{Supp}(\mu) \neq \varnothing$. Let $x_1 \in \mathrm{Supp}(\mu)$. If $\rho(x_1,w) \geq \rho(w,x_1)$ for all $w \in \mathrm{Supp}(\mu)$, then $\pi(w,x_1) = 0$ for all $w \in \mathrm{Supp}(\mu)$ by regularity. Hence $ND(\pi,\mu)(x_1) > 0$ and so $ND(\pi,\mu) \neq \theta$.

Suppose there exists $x_2 \in \mathrm{Supp}(\mu) \backslash \{x_1\}$ such that $\rho(x_2,x_1) > \rho(x_1,x_2)$. Then $\pi(x_2,x_1) > 0$ and $ND(\pi,\mu)(x_1) = 0$. Suppose there exist $x_1,...,x_k \in \mathrm{Supp}(\mu)$ such that $\rho(x_i,x_{i-1}) > \rho(x_{i-1},x_i)$ for $i = 2,...,k$. If $\rho(x_k,w) \geq \rho(w,x_k)$ for all $w \in \mathrm{Supp}(\mu)$, then $ND(\pi,\mu) \neq \theta$ as above. If not, there exists $x_{k+1} \in \mathrm{Supp}(\mu) \backslash \{x_1,...,x_k\}$ such that $\rho(x_{k+1},x_k) > \rho(x_k,x_{k+1})$. Then $\pi(x_{k+1},x_k) > 0$ by regularity.

By induction, either there exists $x \in \mathrm{Supp}(\mu)$ such that $\rho(x,w) \geq \rho(w,x)$ for all $w \in \mathrm{Supp}(\mu)$, in which case $ND(\pi,\mu) \neq \theta$, or since $\mathrm{Supp}(\mu)$ is finite, $\mathrm{Supp}(\mu) = \{x_1,...,x_n\}$ is such that $\pi(x_i,x_{i-1}) > 0$ for $i = 2,...,n$. Since $\rho$ is partially acyclic and reflexive, $\pi(x_i,x_n) = 0$ for $i = 2,...,n$. Thus, $\rho(x_n,x_i) \geq \rho(x_i,x_n)$ and $\rho(x_n,x_i) > 0$ for $i = 1,...,n$. By Proposition 3.39, $ND(\pi,\mu)(x_n) > 0$ and accordingly, $ND(\pi,\mu) \neq \theta$.

Conversely, suppose $ND(\pi,\mu) \neq \theta$ for all $\mu \in \mathscr{F}(X)$. Let $\mu \in \mathscr{F}(X)$. Suppose $x_1,...,x_n \in \mathrm{Supp}(\mu)$ are such that $\pi(x_i,x_{i+1}) > 0$ for all $i = 1,...,n-1$. We must show that $\pi(x_n,x_1) = 0$. Let $S = \{x_1,...,x_n\}$. Then $ND(\pi,1_S) \neq \emptyset$. Since $\pi(x_i,x_{i+1}) > 0$ for all $i = 1,...,n-1$, $ND(\pi,1_S)(x_i) = 0$ for all $i = 2,...,n$. Thus, $ND(\pi,1_S)(x_1) > 0$ and so $nd(x_1) = 1$. Therefore, $\pi(y,x_1) = 0 \forall y \in X$. Thus $\pi(x_n,y_1) = 0$. Hence, $\rho$ is acyclic. □

Theorem 3.40 can be found in its original form in Montero and Tejada (1988), in which the crisp set $ND(\pi,X)$ is considered. In contrast to Theorem 3.37, $ND(\pi,\mu)$ allows us to infer a considerable amount of information about the consistency of $\rho$. More specifically, $\rho$ is acyclic if and only if $ND(\pi,\mu)$ is non-empty. Generally, $\mathrm{Supp}(ND(\pi,\mu)) \subseteq \mathrm{Supp}(M(\rho,\mu))$, but under certain conditions, $\mathrm{Supp}(M(\rho,\mu)) = \mathrm{Supp}(ND(\pi,\mu))$ for all $\rho \in \mathscr{F}(X^2)$. This occurs when using a specific derivation for strict preference as the following proposition demonstrates.

**Proposition 3.41.** *Suppose $\rho$ is a reflexive and complete FWPR. Then*

$$\mathrm{Supp}(M(\rho,\mu)) = \mathrm{Supp}(ND(\pi,\mu))$$

*for all $\mu \in \mathscr{F}(X)$ if and only if $\pi$ is of type $\pi_{(4)}$.*

*Proof.* Suppose $\pi$ is of type $\pi_{(4)}$. Let $\mu \in \mathcal{F}(X)$. Let $x \in \text{Supp}\ (M(\rho,\mu))$. Then $\rho(x,w) > 0$ for all $w \in \text{Supp}(\mu)$ by Proposition 3.30. Thus $\pi(w,x) = 0\ \forall w \in \text{Supp}\ (\mu)$ since $\pi$ is of type $\pi_{(4)}$. Hence $nd(\mu)(x) = 1$ and so $x \in ND(\pi,\text{Supp}\ (\mu)) = \text{Supp}\ (ND(\pi,\mu))$. Thus, $\text{Supp}(M(\rho,\mu)) \subseteq \text{Supp}(ND(\pi,u))$.

Conversely, suppose $\pi$ is not of type $\pi_{(4)}$. Then there exist $x,y \in X$ such that $\pi(x,y) > 0$ and $\rho(y,x) > 0$. Since $\pi(x,y) > 0$, $nd(y) < 1$. Let $\mu^* \in \mathcal{F}(X)$ be such that $\text{Supp}(\mu^*) = \{x,y\}$. Since $\rho(y,x) > 0$, $M(\rho,\mu^*)(y) > 0$. Thus, $\text{Supp}(ND(\pi,\mu^*)) \subset \text{Supp}(M(\rho,\mu^*))$ since $ND(\pi,\mu^*)(y) = 0$.

Because the various consistency definitions in Definition 3.32 represent more or less rational behavior in $\rho$, the least restrictive condition of partial acyclicity ensures a non-empty $M(\rho,\mu)$, and the slightly stronger condition of acyclicity ensures a non-empty $ND(\pi,\mu)$, hence the reason for the relationship in Proposition 3.39. Thus, when considering other social choice problems such as the aggregation of individual preferences or the specification of a collective choice, we should, at the very least, consider consistency conditions that guarantee a non-empty $M(\rho,\mu)$. This standard eliminates some transitivity definitions found in other fuzzy social choice studies, as the following example illustrates.

*Example 3.42.* Let $X = \{a,b,c\}$ and $\mu(x) = 1$ for all $x \in X$. Let $\rho$ be a reflexive and complete FWPR and $\pi$, $\rho$'s asymmetric component, be regular. Suppose $\rho$ defined as follows:

$$\rho(a,b) = \rho(b,c) = \rho(c,a) = 0,$$
$$\rho(b,a) = \rho(c,b) = \rho(a,c) = .5.$$

In this case, it is easily verified that $\rho$ is what fuzzy social choice scholars call *T2–transitive*, i.e. $\rho(x,z) \geq \rho(x,y) + \rho(y,z) - 1$ for all $x,y,x \in X$. T2-transitivity is widely used concerning the aggregation of fuzzy preferences (Dutta, 1987; Nurmi, 1981; Richardson, 1998), where authors relax max-$*$ transitivity to accommodate an aggregation rule that does not follow the reverse conclusions of Arrow's (1951) theorem. The definition can also be found under various names in hierarchical studies on fuzzy consistency relations (Bezdek and Harris, 1978; Dasgupta and Deb, 1996). In this previous research, $\rho$ is often assumed to be connected or strongly connected. However, when relaxing the assumption to completeness, T2-transitivity no longer guarantees that a political actor will be able to identify a maximal alternative. To see this, first consider $ND(\pi,\mu)$ of the FWPR defined above. Because $\pi$ is regular, we know for every $x \in X$ there exists a $w \in X$ such that $\pi(w,x) > 0$. Accordingly, $nd(x) < 1$ for all $x \in X$, and $ND(\pi,\mu) = \theta$.

Second, consider the the implications of Proposition 3.31. In this example, for every $x \in X$ there exists a $w \in X$ such that $\rho(x,w) = 0$. Likewise, we can conclude that $M(\rho,\mu)(x) = 0$ for all $x \in X$. Thus, $M(\rho,\mu)(x) = \theta$. Also, $\rho$ is not partially acyclic because $\pi(b,a) \wedge \pi(a,c) > 0$ does not imply that $\rho(b,c) > 0$. Hence, under the conditions presented in this chapter, we may not want to consider T2-transitivity as a consistency condition on $\rho$ because it does not meet the minimal conditions of rationality when assuming completeness.

As Example 3.42 demonstrates, $ND(\pi,\mu)$ and $M(\rho,\mu)$ provide the social choice scholar with tools to assess the rationality of an FWPR under various conditions. Of course, as stated previously in this section, a given FWPR may have an empty maximal set. In which case, it becomes exceedingly difficult to establish what an individual or a collective choice outcome would be without applying another sort of assumption that "forces" a choice to emerge from $\rho$. Such assumptions include a time horizon in which a choice must be made or an enforcement of consequences if such a choice is not made. Thus, we can now proceed with a very basic idea of the structure of $\rho$ when the FWPR is broken down into its symmetric and antisymmetric components and restricted in such a way as to guarantee a non-empty choice set.

## 3.3   Empirical Application I: Deriving an FWPR from a Fuzzy Preference Function

The previous two sections have made use of FWPRs requiring an actor to specify the degree to which an alternative $x$ is at least as good as alternative $y$, and vice versa. In theory, this seems like an intuitive application of fuzzy sets because traditional preference relations, where $xRy$ or $yRx$, appear unnecessarily restrictive. However, in practice, social scientists will have great difficultly in collecting data where every individual must assign values to $\rho(x,y)$ and $\rho(y,x)$ for all alternatives. This problem is exasperated by Euclidean space or other infinite sets of alternatives. Even in the exact case, social scientists are concerned when a preference can be summarized or "represented" by a simpler function (see Debreu (1954) for the seminal work). For this reason, the purpose of this section is two fold. First, we present a conceptualization of a fuzzy preference function that incorporates some properties of the ambiguity discussed above. Second, we propose two methods for extracting an underlying FWPR—one corresponds to max-∗ transitivity and the other to weak transitivity.

In response to this potential problem in the application of fuzzy social choice theory, Clark, Larson, Mordeson, Potter, and Wierman (2008) propose a new method for modeling individual preferences with fuzzy sets. They apply Nurmi's (1981) construct and assume that a political actor possess vague assessments as to whether an alternative is more or less ideal. Formally, this is modeled with a fuzzy subset when $\sigma \in \mathscr{F}(X)$ represents the actors *preference function*, where for $x \in X$, $\sigma(x) = 1$ signifies alternatives that are fully ideal and $\sigma(x) = 0$ signifies those that are completely unacceptable. Further, it is assumed that $\sigma \in \mathscr{F}(X)$ is *normal*, i.e. there exists $x \in X$ such that $\sigma(x) = 1$. This assumption guarantees that the political actor possesses an ideal alternative.

One obvious benefit from conceptualizing individual preferences in this way comes from the flexibility of $\sigma$, which, thus far, only has one assumption of an ideal alternative. This allows $\sigma$ to easily fit the data or the demands of the modeler. The flexibility is best displayed when the set of alternatives is represented by Euclidean space. Figure 1(a) demonstrates a natural first extension of $\sigma$ to a one-dimensional space, with fuzzy preference profile with an ideal position at the peak of the function, where $\sigma(x) = 1$. While this depiction may seem identical to the traditional single-peaked utility functions, the fuzzy profile introduces the concept of a policy horizon, past which, an actor is unwilling to accept any policy. Essentially, the policy horizon is Supp($\sigma$), and the actor will consider all alternatives outside of this range as equally unacceptable.

Figures 3.1a and 3.1b demonstrate further departures from the traditional concept of spatial preferences. Figure 3.1b shows a $\sigma$ with significant areas of indifference over discrete areas in space. Not only does the preference function incorporate single-plateau preferences, where actors have an ideal policy range rather than a ideal point (see Ching and Serizawa (1998) and Massó and Neme (2001) for formal applications); but it also includes discrete areas of indifference throughout the policy space, which are similar to the preferences modeled by Sloss (1973), Tovey (2010) and Koehler (2001). In this case, Im($\sigma$) $= \{0,.25,.5,.75,1\}$, and the actor's preferences are similar to those of a Likert scale. Finally, Figure 3.2 shows a $\sigma \in \mathscr{F}(X)$ where $X$ is two-dimensional Euclidean space. Obviously, fuzzy sets allow preferences to manifest very irregular shapes, suggesting a complex relationship between the two policy dimensions. This can be done without specifying a Euclidean-distance utility function, i.e $u(y) = -(y-x)A(y-x)^T$, where $x$ is an ideal point, $y$ is a point in two-dimensional space, $T$ denotes transpose, and $A$ is a $2 \times 2$ matrix.[10] In addition, there is no assumption requiring $\sigma$ to be convex or pseudo-convex; and fuzzy profiles can model actors with two ideal regions. Such situations can arise when a collective actor, a political party for example, contains polarized factions, or another dimension, irrelevant to the policy space, induces a trade-off between two ideal alternatives.

Applications of fuzzy sets in this manner have only used $\sigma$ to derive exact preferences relations (Clark et al., 2008; Nurmi, 1981; Mordeson et al., 2011). Hence, $\rho(x,y) = 1$ if and only if $\sigma(x) \geq \sigma(y)$ and $\rho(x,y) = 0$ otherwise. However, this need not be the case. Drawing on revealed preference theory, Georgescu (2005; 2007a) proposes an operation for fuzzy revealed preferences that can be applied to the context discussed here. For any $x,y \in X$ and $\sigma \in \mathscr{F}(X)$, $\rho_1$ can be defined as follows:

$$\rho_1(x,y) = \bigvee \{t \in [0,1] \mid \sigma(y) * t \leq \sigma(x)\}.$$

Further, when $*$ is the minimum t-norm, $\rho_1$ can be written as

$$\rho_1(x,y) = \begin{cases} 1 & \text{if } \sigma(x) \geq \sigma(y), \\ \sigma(x) & \text{otherwise.} \end{cases}$$

Obviously, because $\sigma(x) \geq \sigma(y)$ or $\sigma(y) \geq \sigma(x)$, $\rho_1$ is not only complete and reflexive but strongly connected as well. In addition, the following proposition reveals that $\rho_1$ is max-$*$ transitive.

**Proposition 3.43.** *Let* $\sigma \in \mathscr{F}(X)$. *Suppose* $\rho_1(x,y) = \begin{cases} 1 & \text{if } \sigma(x) \geq \sigma(y), \\ \sigma(x) & \text{otherwise.} \end{cases}$
*Then* $\rho_1$ *is max-$*$ transitive over X for any $*$.*

---

[10] In $k$-dimensional Euclidean space, $A$ would be a $k \times k$ matrix.

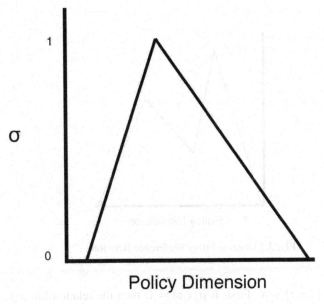

(a) Simple Discrete Fuzzy Preference

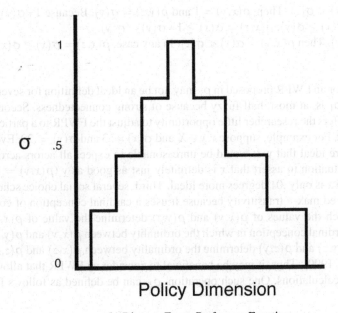

(b) Discrete Fuzzy Preference Function

**Fig. 3.1** Examples of Fuzzy Preference Functions

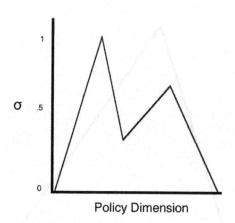

**Fig. 3.2** Discrete Fuzzy Preference Relation

*Proof.* Suppose $\{x,y,z\} \in X$. First, if $\rho(x,z) = 1$, then the relationship $\rho(x,z) \geq \rho(x,y) * \rho(y,z)$ holds. If $\rho(x,z) < 1$, then $\sigma(x) < \sigma(z)$ and $\rho(x,z) = \sigma(x)$. There are two cases to consider.

(1) $\sigma(y) \leq \sigma(x) < \sigma(z)$. Then, $\rho(x,y) = 1$ and $\rho(y,z) = \sigma(y)$. Because $1 * \sigma(y) = \sigma(y)$ and $\sigma(x) \geq \sigma(y)$, $\rho(x,z) = \sigma(x) \geq 1 * \sigma(y) = \sigma(y)$.
(2) $\sigma(x) < \sigma(y)$. Then $\rho(x,y) = \sigma(x) < \sigma(z)$. In any case, $\rho(x,z) = \sigma(x) \geq \sigma(x) * \rho(y,z)$.
                                                                                              □

The operation for an FWPR proposed in $\rho_1$ may not be an ideal definition for several reasons. First, $\rho_1$ is, at most, half fuzzy because of strong connectedness. Second, the operation offers the researcher little opportunity to adjust the FWPR to a particular type of actor. For example, suppose $x,y \in X$ and $\sigma(x) = .3$ and $\sigma(y) = .29$. Even though $x$ is more ideal that $y$, it would be unreasonable to expect all actors across a varieties of situation to assert that $x$ is definitely just as good as $y$ ($\rho(x,y) = 1$), especially when $x$ is only .01 degrees more ideal. Third, several social choice scholars have criticized max-* transitivity because it uses a cardinal conception of consistency in which the values of $\rho(x,y)$ and $\rho(y,z)$ determine the value of $\rho(x,z)$ rather than an ordinal conception in which the ordinality between $\rho(x,y)$ and $\rho(y,x)$ and between $\rho(y,z)$ and $\rho(z,y)$ determine the ordinality between $\rho(x,z)$ and $\rho(z,x)$ (Ponsard, 1988, 1990). Thus, it may be beneficial to consider an FWPR that allows variation in its calculations. One such operation, $\rho_2$, can be defined as follows for all $x,y \in X$:

$$\rho_2(x,y) = \begin{cases} 1 & \text{if } x = y, \\ (\sigma(x) - \sigma(y) + c) \wedge 1 & \text{if } \sigma(x) \geq \sigma(y), \\ 1 - [(\sigma(y) - \sigma(x) + 1 - c) \wedge 1] & \text{otherwise,} \end{cases}$$

where $c \in [0,1]$ and $\sigma \in \mathscr{F}(X)$. Obviously, $\rho_2$ is reflexive. Further, when $c = 1$, $\rho_2$ is strongly connected; and when $c > 0$, $\rho_2$ is complete. In addition, the definition of $\rho_2$ allows for greater substantive interpretations of the FWPR. First, the variable $c$ could account for the degree of certainty in $\rho$. If $c = 0$, an individual will prefer one alternative to another less strongly than when $c = 1$. For example, if $c = 0$, $\sigma(x) = .75$ and $\sigma(y) = .25$, then $\rho(x,y) = .5$ and $\rho(y,x) = 0$. In contrast, if $c = 1$, then $\rho(x,y) = 1$ and $\rho(y,x) = .5$. While the difference between $\rho(x,y)$ and $\rho(y,x)$ remains unchanged in these two examples, the actor becomes more certain that $x$ is at least as good as $y$ as $c$ approaches one. In addition, $c$ could also be used to denote types of fuzzy weak preference. Examples include $\rho(x,y)$ when $c \in (0.5]$ denoting the degree to which $x$ is strongly at least as good as $y$ and $\rho(x,y)$ when $c \in (.5.1]$ denoting the degree to which $x$ is weakly at least as good as $y$. In addition, $\rho_2$ induces a weakly transitive FWPR rather than a max-$*$ one as the following proposition demonstrates.

**Proposition 3.44.** *Let* $\sigma \in \mathscr{F}(X)$. *Suppose* $\rho_2$ *is defined as discussed above. Then* $\rho_2$ *is weakly transitive over* $X$.

*Proof.* Suppose $\{x,y,z\} \in X$. Further, suppose $\rho_2(x,y) \geq \rho_2(y,x)$ and $\rho_2(y,z) \geq \rho_2(z,y)$. This proof will show that $\rho_2(x,z) \geq \rho_2(z,x)$.

First, $\rho_2(x,y) \geq \rho_2(y,x)$ implies $\sigma(x) \geq \sigma(y)$. To see this, suppose the contrary. Then $\rho_2(y,x) = (\sigma(y) - \sigma(x) + c) \wedge 1$ and

$$\rho_2(x,y) = 1 - [(\sigma(y) - \sigma(x) + 1 - c) \wedge 1]$$
$$= (1 - \sigma(y) + \sigma(x) - 1 + c) \vee 0$$
$$= (-\sigma(y) + \sigma(x) + c) \vee 0.$$

Because $\sigma(y) > \sigma(x)$,

$$(-\sigma(y) + \sigma(x) + c) \vee 0 < (\sigma(y) - \sigma(x) + c) \wedge 1$$

and $\rho_2(y,x) > \rho_2(x,y)$, a contradiction. Thus, $\rho_2(x,y) \geq \rho_2(y,x)$ implies $\sigma(x) \geq \sigma(y)$. Hence, $\sigma(x) \geq \sigma(y) \geq \sigma(z)$.

Second, to see that $\sigma(x) \geq \sigma(z)$ implies $\rho_2(x,z) \geq \rho_2(z,x)$, consider the case where $\sigma(x) > \sigma(z)$. Then,

$$\rho_2(x,z) = (\sigma(x) - \sigma(z) + c) \wedge 1$$

and

$$\rho_2(z,x) = (-\sigma(x) + \sigma(z) + c) \vee 0.$$

Because $\sigma(x) > \sigma(z)$,

$$(\sigma(x) - \sigma(y) + c) \wedge 1 > (-\sigma(x) + \sigma(z) + c) \vee 0.$$

Second, consider the case $\sigma(x) = \sigma(z)$. Then, $\rho_2(x,z) = \rho_2(z,x) = c$ by definition. Hence, $\rho_2(x,z) \geq \rho_2(z,x)$, and $\rho_2$ is accordingly weakly transitive.    $\square$

The two operations $\rho_1$ and $\rho_2$ are only two procedures that extract an FWPR from a preference function. Example 2.37 better illustrates their differences.

*Example 3.45.* Let $x, y, z \in X$ and suppose $\sigma(x) = .7$, $\sigma(y) = 1$ and $\sigma(z) = .2$. Obviously, $\rho_1(w, w) = 1$ and $\rho_2(w, w) = 1$ for all $w \in X$. The remainder of $\rho_1$ is defined as follows:

$$\rho_1(y, x) = \rho_1(y, z) = \rho_1(x, z) = 1$$
$$\rho_1(x, y) = .7$$
$$\rho_1(z, y) = \rho_1(z, x) = .2.$$

In contrast, $\rho_2$ is defined as follows:

$$\rho_2(x, y) = (-.3 + c) \vee 0$$
$$\rho_2(x, z) = (.5 + c) \wedge 1$$
$$\rho_2(y, x) = (.3 + c) \wedge 1$$
$$\rho_2(y, z) = (.8 + c) \wedge 1$$
$$\rho_2(z, x) = (-.5 + c) \vee 0$$
$$\rho_2(z, y) = (-.8 + c) \vee 0.$$

In sum, deriving FWPRs from individual preference functions, which denote the degree to which an alternative is ideal, provides the social scientist with greater flexibility when applying fuzzy social choice to empirical examples. Essentially, there is no need to intensively gather data on an actor's true FWPR, which requires knowledge about the degree to which an actor views $x$ at least as good as $y$, and vice versa, for all pairwise comparisons. Further, as Clark et al. (2008) suggest, if $\sigma(x) > 0$ represents all acceptable policy positions, then the analysis of group decision-making reduces down to analyzing the alternatives which are acceptable to at least one actor, thereby restricting the policy space.

# References

Arrow, K.: Social Choice and Individual Values. Wiley, New York (1951)

Austen-Smith, D., Banks, J.S.: Positive Political Theory I: Collective Preference. University of Michigan Press, Ann Arbor (1999)

Banerjee, A.: Rational choice under fuzzy preferences: The orlovsky choice function. Fuzzy Sets and Systems 53, 295–299 (1993)

Banerjee, A.: Fuzzy preferences and Arrow-type problems. Social Choice and Welfare 11, 121–130 (1994)

Barrett, C., Pattanaik, P.K., Salles, M.: On the structure of fuzzy social welfare functions. Fuzzy Sets and Systems 19(1), 1–10 (1986)

Bezdek, J.C., Harris, J.D.: Fuzzy partitions and relations; an axiomatic basis for clustering. Fuzzy Sets and Systems 1(2), 111–127 (1978),
http://www.sciencedirect.com/science/article/pii/
016501147890012X

Billot, A.: Economic theory of fuzzy equilibria: an axiomatic analysis. Lecture notes in economics and mathematical systems. Springer (1992),
http://books.google.com/books?id=ml-7AAAAIAAJ

Ching, S., Serizawa, S.: A maximal domain for the existence of strategy-proof rules. Journal of Economic Theory 78(1), 157–166 (1998),
http://www.sciencedirect.com/science/article/pii/
S0022053197923371

Clark, T.D., Larson, J.M., Mordeson, J.N., Potter, J.D., Wierman, M.J. (eds.): Applying Fuzzy Mathematics to Formal Models in Comparative Politics. STUDFUZZ, vol. 225. Springer, Heidelberg (2008)

Dasgupta, M., Deb, R.: Fuzzy choice functions. Social Choice and Welfare 8(2), 171–182 (1991), http://dx.doi.org/10.1007/BF00187373

Dasgupta, M., Deb, R.: Transitivity and fuzzy preferences. Social Choice and Welfare 13(3), 305–318 (1996), http://dx.doi.org/10.1007/BF00179234

Debreu, G.: Theory of Value, an Axiomatic Analysis of Economic Equilibria. Wiley, New York (1954)

Dutta, B.: Fuzzy preferences and social choice. Mathematical Social Sciences 13(3), 215–229 (1987)

Fodor, J., Roubens, M.: Fuzzy Preference Modelling And Multi-Criteria Decision Support. Kluwer Academic Publishers, Dordrecht (1994)

Fono, L.A., Andjiga, N.G.: Fuzzy strict preference and social choice. Fuzzy Sets Syst. 155, 372–389 (2005), http://dx.doi.org/10.1016/j.fss.2005.05.001

Georgescu, I.: Rational Choice and Revealed Preference: A Fuzzy Approach. Ph.D. thesis, Abo Akademi University Turku Centre for Computer Science, Lemminkainengatan 14B Fin-20520 Abo, Finland (2005)

Georgescu, I. (ed.): Fuzzy Choice Functions - A Revealed Preference Approach. STUDFUZZ, vol. 214. Springer, Heidelberg (2007a)

Georgescu, I.: Similarity of fuzzy choice functions. Fuzzy Sets and Systems 158(12), 1314–1326 (2007b), http://www.sciencedirect.com/science/article/pii/
S0165011407000346

Klement, E., Mesiar, R., Pap, E.: Triangular Norms. Trends in logic, Studia logica library. Springer (2000)

Klir, G.J., Yuan, B.: Fuzzy Sets and Fuzzy Logic; Theory and Applications. Prentice Hall, Upper Saddle River (1995)

Koehler, D.H.: Convergence and restricted preference maximizing under simple majority rule: Results from a computer simulation of committee choice in two-dimensional space. American Political Science Review, 155–167 (March 2001),
http://journals.cambridge.org/article_S0003055401000065

Kołodziejczyk, W.: Orlovsky's concept of decision-making with fuzzy preference relation-further results. Fuzzy Sets and Systems 19(1), 11–20 (1986),
http://www.sciencedirect.com/science/article/pii/
S0165011486800732

Zhong Luo, C.: Fuzzy relation equation on infinite sets. BUSEFAL 26, 57–66 (1986)

Massó, J., Neme, A.: Maximal domain of preferences in the division problem. Games and Economic Behavior 37(2), 367–387 (2001),
http://www.sciencedirect.com/science/article/pii/
S0899825601908504

McCarty, N., Meirowitz, A.: Political Game Theory: An Introduction, 1st edn. Cambridge University Press (2007)

Montero, F., Tejada, J.: A necessary and sufficient condition for the existence of orlovsky's choice set. Fuzzy Sets and Systems 26(1), 121–125 (1988), http://www.sciencedirect.com/science/article/pii/0165011488900103

Mordeson, J.N., Clark, T.D., Miller, N.R., Casey, P.C., Gibilisco, M.B.: The uncovered set and indifference in spatial models: A fuzzy set approach. Fuzzy Sets and Systems 168(1), 89–101 (2011), http://www.sciencedirect.com/science/article/pii/S0165011410004471, Theme: Aggregation operations

Nurmi, H.: A fuzzy solution to a majority voting game. Fuzzy Sets and Systems 5, 187–198 (1981)

Orlovsky, S.: Decision-making with a fuzzy preference relation. Fuzzy Sets and Systems 1, 155–167 (1978)

Orlovsky, S.: On formalization of a general fuzzy mathematical problem. Fuzzy Sets and Systems 3(3), 311–321 (1980), http://www.sciencedirect.com/science/article/pii/0165011480900263

Ovchinnikov, S.V.: Structure of fuzzy binary relations. Fuzzy Sets and Systems 6(2), 169–195 (1981)

Ponsard, C.: Fuzzy mathematical models in economics. Fuzzy Sets and Systems 28(3), 273–283 (1988), http://www.sciencedirect.com/science/article/pii/0165011488900346

Ponsard, C.: Some dissenting views on the transitivity of individual preference. Annals of Operations Research 23(1), 279–288 (1990), http://dx.doi.org/10.1007/BF02204852

Richardson, G.: The structure of fuzzy preferences: Social choice implications. Social Choice and Welfare 15, 359–369 (1998)

Sloss, J.: Stable outcomes in majority rule voting games. Public Choice 15(1), 19–48 (1973), http://dx.doi.org/10.1007/BF01718841

Tovey, C.A.: The instability of instability of centered distributions. Mathematical Social Sciences 59(1), 53–73 (2010)

Wang, X.: An investigation into relations between some transitivity-related concepts. Fuzzy Sets and Systems 89(2), 257–262 (1997)

Zadeh, L.A.: Similarity relations and fuzzy orderings. Information Sciences 3(2), 177–200 (1971)

# Chapter 4
# Arrow and the Aggregation of Fuzzy Preferences

**Abstract.** This chapter builds off of chapter 3 by examining the aggregation of fuzzy weak preference relations in order to determine how a social preference relation emerges. Specifically, this chapter focuses on Arrow's theorem which employs a deductive analysis of aggregation rules and establishes five necessary conditions for an ideal aggregation rule. When Arrow's theorem is applied with fuzzy preferences, not only do serious complications arise when conceiving the fuzzy definitions of an ideal aggregation rule, but there exist specific combinations of conditions that allow for a fuzzy aggregation rule to satisfy all of the fuzzy counterparts of Arrow's conditions. Moreover, this chapter shows that a fuzzy aggregation rule exists which satisfies all five Arrowian conditions including non-dictatorship.

## Introduction

Chapter 3 detailed the underlying structure of FWPRs and the complications that arise when trying to incorporate the logic of exact preferences into the fuzzy framework. Essentially, there is no obvious one-to-one procedure that fuzzifies the underlying assumptions of a rational preference relation. Among these complications, there exist several methods for extracting a fuzzy choice set, and there is little guarantee that these methods will return equivalent results. However, a proper specification of the fuzzy maximal set, along with other characteristics of an FWPR, identifies obvious best outcomes that should emerge given a preference relation of an individual or a collective body. Yet in the case of social preference relations, it is very unlikely that one will be specified *a priori,* without the use of some social welfare function relating individual preferences, i.e. those belonging to voters, committee members or legislators, to those of a social relation. Even if such an example exists, the applications of the various fuzzy maximal sets can be done without complication. Thus, it is worthwhile to consider situations where individual FWPRs are aggregated to form a fuzzy social preference relation.

The goal of this chapter is to examine aggregation of FWPRs in order to determine how a social preference relation emerges. In doing so, it focuses on a classic result in social choice theory: Arrow's Theorem (1951). Because the number

M.B. Gibilisco et al., *Fuzzy Social Choice Theory,*
Studies in Fuzziness and Soft Computing 315,
DOI: 10.1007/978-3-319-05176-5_4, © Springer International Publishing Switzerland 2014

of aggregation rules is quite large and considering each aggregation individually can become quite tedious, Arrow employs a deductive analysis of aggregation rules and establishes five requiste conditions of an ideal rule that possess inherit trade offs. More simply, if an aggregation rule posssesses four of the five coniditions, it must violate the fifth, thereby demonstating the impossibility of an ideal aggregation rule. Nonetheless, these traditional results rely on exact preferences. When the formal logic of Arrow's theorem is extended into the fuzzy framework, not only do serious complications arise when conceiving the fuzzy definitions of an ideal aggregation rule, but there exist specific combinations of conditions that allow for a fuzzy aggregation rule to satisfy all of the fuzzy counterparts of Arrow's conditions.

The chapter is organized as follows. The first section introduces the classic results of Arrow's theorem and then proposes several fuzzifications of the original five conditions. Next, Section 2 presents the formal proof of fuzzy Arrow's theorem and demonstrates under what conditions a fuzzy aggregation rule will satisfy the five criteria proposed in Section 1. Finally, Section 3 concludes the chapter with a discussion on the empirical applications of fuzzy aggregation.

## 4.1 Fuzzifying Arrow's Conditions

This section lays out the preliminary definitions used in Arrow's formal considerations of aggregation rules. To do so, we use the following notation. Let $N = \{1, \ldots, n\}$ be a finite set of individuals where $n \geq 2$. As in Chapter 3, $X$ is a finite set of alternatives such that $3 \leq |X|$. Throughout the chapter, each individual $i$ is assumed to possess an FWPR, $\rho_i \in \mathscr{F}(X^2)$, such that $\rho_i$ is reflexive and complete. In this case, we call $\rho_i$ a *fuzzy weak order*.[1]

Let $\mathscr{FR}$ denote the set of all fuzzy weak orders on $X$. Then a *preference profile* is an $n$-tuple of fuzzy weak orders, $\bar{\rho} = (\rho_1, \ldots, \rho_n) \in \mathscr{FR}^n$ and describes the fuzzy preferences of all individuals. Throughout, we manipulate the consistency conditions concerning the weak orders of individuals. When doing so, we will write "assume $\bar{\rho}$ satisfies a particular consistency condition" or "suppose $\rho_i$ is max-$*$ transitive for all $i \in N$." Finally, our definitions related to FPAR's are written generally (that is, with domain $\mathscr{FR}^n$), but our results often assume that these definitions reflect the transitivity restrictions when appropriate.

For any non-empty $S \subseteq X$, let $\bar{\rho}]_S = (\rho_1|_{S \times S}, \ldots, \rho_n|_{S \times S})$. In words, $\bar{\rho}]_S$ denotes the restriction of the preference profile to the subset $S \times S$ and, accordingly, $\bar{\rho}]_S$ describes only $\rho(x, y)$ and $\rho(y, x)$ for $x, y \in S$ and every $i \in N$. In addition, for any FWPR $\rho$ and all $\alpha \in [0, 1]$, $\rho^\alpha = \{(x, y) \in X \times X \mid \rho(x, y) \geq \alpha\}$. Often, $\rho^\alpha$ is called the $\alpha$-*cut* of $\rho$.

Finally, for all $\bar{\rho} \in \mathscr{FR}^n$ and $x, y \in X$,

$$R(x, y; \bar{\rho}) = \{i \in N \mid \rho_i(x, y) > 0\}$$

---

[1] Fuzzy weak orders usually possess some consistency or transitivity condition. However, throughout this chapter, we vary these types of assumptions. The more general definition given here permits us to do so.

and

$$P(x,y;\bar{\rho}) = \{i \in N \mid \pi_i(x,y) > 0\}.$$

In words, $R(x,y;\bar{\rho})$ denotes the collection of individuals who view $x$ as at least as good as $y$ to some degree and $P(x,y;\bar{\rho})$ the collection of individuals who strictly prefer $x$ to $y$ to some degree.

**Definition 4.1.** A function $\tilde{f} \colon \mathscr{FR}^n \to \mathscr{FR}$ is called a *fuzzy preference aggregation rule*.

Hence, a fuzzy preference aggregation rule (FPAR) relates a $\bar{\rho} \in \mathscr{FR}^n$ to a social preference relation $\tilde{f}(\bar{\rho}) \in \mathscr{FR}$. When this occurs, $\tilde{f}(\bar{\rho})(x,y)$ represents the degree to which society, or more specifically the set of $N$ actors, views $x$ as at least as good as $y$. Obviously, this encompasses the exact case where $\tilde{f}(\bar{\rho})(x,y) \in \{0,1\}$. At times, we suppress the $\tilde{f}(\bar{\rho})$ and let $\rho$ denote the social preference relation. In this manner, we can derive $\rho$'s components $\iota$ and $\pi$, which correspond to the social fuzzy indifference and social fuzzy strict preference relations, respectively. Furthermore, we will at times restrict FPAR's to particular domains of fuzzy weak orders that satisfy consistency conditions. For example, we may assume $\rho_i$ is weakly transitive for all $i \in N$. Then we analyze $\tilde{f} \colon D_w^n \to \mathscr{FR}$, where $D_w$ is the set of all weakly transitive fuzzy weak orders. While this may appear to be an unnecessary technical complication, the intent is to illustrate the consequences of various types of consistency conditions without needless notation to redefine FPAR's in every case. With this in mind, we assume that any FPAR has an *unrestricted domain*. That is, an FPAR must assign a social preference relation to every fuzzy preference profile with the consistency condition under consideration regardless of the specific combination of the indvidual $\rho_i$s. Unrestricted domain is fairly innocuous because the assumption allows individuals to choose any fuzzy weak order in $\mathscr{FR}$. In democratic terms, the aggregation rule does not require individuals to possess certain types of opinions about the possible alternatives. The understanding of an FPAR in Definition 4.1 allows for a greater variety of aggregation rules than that of exact rules.

*Example 4.2.* Let $\bar{\rho} \in \mathscr{FR}^n$. Then the following are examples of fuzzy preference aggregation rules:

(1) For all $x,y \in X$,

$$\rho(x,y) = \frac{1}{n}\sum_{i=1}^{n}\rho_i(x,y),$$

(2) For all $x,y \in X$ and any $\beta \in (0,1)$,

$$\rho(x,y) = \begin{cases} 1 & \text{if } \rho_i(x,y) \geq \rho_i(y,x), \ \forall i \in N, \\ \beta & \text{otherwise,} \end{cases}$$

(3) For all $x,y \in X$,

$$\rho(x,y) = \max_{i \in N}\{\rho_i(x,y)\}.$$

It is easily verified that $\rho$ is complete and reflexive in all three cases.[2]

Arrow's seminal work lays out five requisite and incompatible conditions for preference aggregation. The original conditions are

- universal admissibility,
- non-negative monotonicity,
- independence of irrelevant alternatives,
- non-imposition and
- non-dictatorship.

Efforts to dismiss the relevance of the theorem outright (Little, 1952) were followed by attempts to replace certain original conditions. For example, some studies eliminated non-negative monotonicity (Blau, 1972; Inada, 1955) while others replaced it with positive responsiveness (Black, 1969; Fishburn, 1974; May, 1952). The ultimate result of these reinterpreations was a simpler form of Arrow's theorm by Blau (1972) that is generally accepted by contemporary scholars (Austen-Smith and Banks, 1999). In this form, any preference aggregation rule that is transitive, weakly Paretian and independent of irrelevant alternatives must be dictatorial. In the remainder of this section we discuss these terms further and provide several definitions of their fuzzy counterparts.

### 4.1.1  Transitivity

There are several fuzzy consistency conditions that correspond to transitivity in the traditional sense of determing how FWPRs behave across pairwise comparions. In the fuzzy Arrow literature, the most pervasive approach is the use of some specific form of max-star transitivity. The definition can be used to derive an infinite number of transitivity conditions and few tudies consider the general condition of max-star transitivity (Duddy et al., 2011; Fono and Andjiga, 2005; Fono et al., 2009). However, the most common definitions make use of the Gödel (minimum) and Łukasiewicz t-norm (Banerjee, 1994; Dutta, 1987; Ovchinnikov, 1991; Richardson, 1998). In these two cases, for all $x, y, z \in X$ and $\rho \in \mathscr{F}(X^2)$, $\rho(x,z) \geq \min\{\rho(x,y), \rho(y,z)\}$ or $\rho(x,z) \geq \rho(x,y) + \rho(y,z) - 1$, respectively. As Duddy, Perote-Peña and Piggins (2007) demonstrate, designating a specific t-norm for max-star transitivity has important consequences on whether Arrow's conclusions hold in the fuzzy frame work. Hence, it is important to consider a variety of consistency defintions. In one of the first applications of fuzzy sets to Arrow's theorem, Barrett, Pattanaik and Salles (1992) propose the following for asymmetric preferences.

**Definition 4.3.** Let $\rho$ be a complete and reflexive FWPR and let $\pi$ be its asymmetric component.

---

[2] Example 4.2(1) was first proposed by Skala (1978) and is now standard in the fuzzy social choice literature, 4.2(2) comes from Dutta (1987), and 4.2(3)'s first application to Arrow's theorem can be found in Fung and Fu (1975).

(1) **(partially transitive)** $\rho$ is said to be *partially transitive* if, for all $x, y, z \in X$, $\rho(x,y) > 0$ and $\rho(y,z) > 0$ implies $\rho(x,z) > 0$,
(2) **(partially quasi-transitive)** $\rho$ is said to be *partially quasi-transitive* if, for all $x, y, z \in X$, $\pi(x,y) > 0$ and $\pi(y,z) > 0$ implies $\pi(x,z) > 0$.

The relationship in Definition 4.3(1) relates to a special case of max-$*$ transitivitiy where $*$ has no zero divisors. In this case, for three alternatives $x, y, z \in X$, $\rho(x,y) > 0$ and $\rho(y,z) > 0$ implies $\rho(x,y) * \rho(y,z) > 0$. In addition, Definition 4.3(2) strengthens the condition of acyclicity in Definition (3). Specifically, partial quasi-transitivity requires not only $\pi(z,x) = 0$, as in acyclicity, but also $\pi(x,z) > 0$ when $\pi(x,y) > 0$ and $\pi(y,z) > 0$ for all $x, y, z \in X$. An application of partial quasi-transitivity can also be found in Dasgupta and Deb (1999). In a similar manner, we can define consistency conditions of fuzzy aggregation rules, which, like FWPRs, possess more or less strictness.

**Definition 4.4.** Let $\tilde{f}$ be an FPAR.

(1) **(max-star transitive)** $\tilde{f}$ is said to be *max-$*$ transitive* if, for all $\bar{\rho} \in \mathscr{FR}^n$, $\tilde{f}(\bar{\rho})$ is max-$*$ transitive,
(2) **(weakly transitive)** $\tilde{f}$ is said to be *weakly transitive* if, for all $\bar{\rho} \in \mathscr{FR}^n$, $\tilde{f}(\bar{\rho})$ is weakly transitive,
(3) **(partially quasi-transitive)** $\tilde{f}$ is said to be *partially quasi-transitive* if, for all $\bar{\rho} \in \mathscr{FR}^n$, $\tilde{f}(\bar{\rho})$ is partially quasi-transitive,
(4) **(partially acyclic)** $\tilde{f}$ is said to be *partially acyclic* if, for all $\bar{\rho} \in \mathscr{FR}^n$, $\tilde{f}(\bar{\rho})$ is partially acyclic.

Definition 4.4 presents the consistency conditions of fuzzy aggregations rules used in this text, but there are other consistency conditions previously explored in the fuzzy Arrow literature, which we will not focus on because they have already been explicated in the existing literature. These other conditions include *minimal transitivity*, i.e. $\min\{\rho(x,y), \rho(y,x)\} = 1$ implies $\rho(x,z) = 1$ for all $x, y, z \in X$, and *negative transitivity*, i.e. $\pi(x,y) > 0$ implies $\max\{\pi(x,z), \pi(z,y)\} > 0$ for all $x, y, z \in X$, the contrapositive of which is called *positive transitivity* Fono et al. (2009); Fung and Fu (1975); Richardson (1998).

### 4.1.2 Weak Paretianism

Weak Paretianism, as the name suggests, determines how an FPAR will behave when every actor in society holds a certain preference between two alternatives. In the exact case, an aggregation rule is weakly Paretian if, for two possible alternatives $x$ and $y$, every $i \in N$ strictly prefers $x$ to $y$ then the social preference must prefer $x$ to $y$ (Austen-Smith and Banks, 1999; Blau, 1972). In this sense, weak Paretianism has little to say about the final social preferences if all actors possess the same *weak* preferences between two alternatives or if all actors in $N \setminus \{i\}$ strictly prefer $x$ to $y$, but individual $i$ is indifferent between the two. Weak Paretianism in the fuzzy context, often called the "Pareto Condition", has a fairly uniform definition across the fuzzy literature (Banerjee, 1994; Barrett et al., 1992; Dasgupta and Deb, 1999; Dutta, 1987; Fono et al., 2009; Fung and Fu, 1975; Richardson, 1998).

**Definition 4.5 (Pareto Condition).** Let $\tilde{f}$ be an FPAR. Then $\tilde{f}$ is said to satisfy the *Pareto Condition* if, for all $\bar{\rho} \in \mathscr{F}\mathscr{R}^n$ and $x, y \in X$, $\pi(x, y) \geq \min_{i \in N} \{\pi_i(x, y)\}$.

Of course, derivations from Definition 4.5 exist in the fuzzy literature. Examples include the *strict Pareto Condition* where $\min_{i \in N} \{\pi_i(x, y)\} = 1$ implies $\pi(x, y) = 1$ for all $x, y \in X$ (Ovchinnikov, 1991) and *unanimity*, which, for all $x, y \in X$ and $t \in [0, 1]$, requires $\rho(x, y) = t$ if $\rho_i(x, y) = t$ for all $i \in N$ (Duddy et al., 2011; García-Lapresta and Llamazares, 2000). In addition, when formal arguments do not require constructing a fuzzy strict preference relation, Definition 4.5 can be applied to FWPRs (Duddy et al., 2011; Perote-Peña and Piggins, 2007). To better explicate fuzzy Arrow's theorem, we also consider a weaker assumption than the Pareto Condition that was first proposed by Mordeson and Clark (2009).

**Definition 4.6 (weakly Paretian).** Let $\tilde{f}$ be an FPAR. Then $\tilde{f}$ is said to be *weakly Paretian* if, for all $\bar{\rho} \in \mathscr{F}\mathscr{R}^n$ and $x, y \in X$, $\min_{i \in N} \{\pi_i(x, y)\} > 0$ implies $\pi(x, y) > 0$.

Obviously, Definition 4.6 relaxes Definition 4.5 because Definition 4.6 no longer restricts the social strict preference between the two alternatives to a more specific alpha level. Nonetheless, both definitions correspond to weak Paretianism in the exact case because $\min\{\pi_i(x, y)\} > 0$ implies $\min\{\pi_i(x, y)\} = 1$, which, under a weakly Paretian aggregation rule, implies $\pi(x, y) = 1 \geq \min\{\pi_i(x, y)\}$, $i \in N$. It is still important to distinguish between these two definitions because, as discussed in a subsequent section, there is an important relationship between these conditions and the types of FPARs that satisfy all Arrowian conditions. Example 4.7 illustrates some basic differences between the conditions.

*Example 4.7.* Let $\tilde{f}$ be an FPAR and $X = \{a, b\}$. Suppose $\bar{\rho}$ is reflexive and defined as follows:

$$\rho_i(a, b) = .5$$
$$\rho_i(b, a) = .3$$
$$\pi_i(a, b) = .2$$

for all $i \in N$. If $\tilde{f}$ is unanimous, then the social weak preference, $\rho$, will be $\rho(a, b) = .5$ and $\rho(b, a) = .3$. If $\tilde{f}$ satisfies the Pareto Condition, the social strict preference relation, $\pi$, will be $\pi(a, b) \geq .2$. Finally, if $\tilde{f}$ is weakly Paretian, $\pi(a, b) > 0$. Notice the Pareto Condition and weak Paretianism do not guarantee any specific value of $\rho(a, b)$ or $\rho(b, a)$; however, assuming that the social strict preference relation is regular, all three conditions ensure that $\rho(a, b) > \rho(b, a)$.

### 4.1.3  Independence of Irrelevant Alternatives

Unlike some of the other Arrowian conditions, independence of irrelevant alternatives is less normatively democratic, i.e. where the FPAR responds to some conditions of the preference profile, and more technically desirable. In theory, an aggregation rule satisfies the independence of irrelevant alternatives conditions if the social preference between $x$ and $y$ is solely determined by individuals' preferences between $x$ and $y$. According to Austen-Smith and Banks (1999), the traditional independence criterion implies two requirements:

(1) the social preference between two alternatives is specifically determined by individual preferences between two alternatives and
(2) cardinal and relative information contained in indivdual preferences is unrelated to the societal preference.

In other words, these requirements stipulate that each individual can produce a ranked list of the alternatives, including ties, and that the aggregation rule only considers the ordinal relationship between $x$ and $y$ when determining the social preference. Information such as $x$ is four alternatives higher in the preference ranking than $y$ or $x$ is 2.5 times more preferred than $y$ becomes trivial. In the fuzzy framework, the literature has most frequently relied on one definition for independence of irrelevant alternatives Banerjee (1994); Barrett et al. (1992); Duddy et al. (2011); Fono and Andjiga (2005); Fono et al. (2009); García-Lapresta and Llamazares (2000); Ovchinnikov (1991); Richardson (1998).

**Definition 4.8 (IIA-1).** Let $\tilde{f}$ be an FPAR. Then $\tilde{f}$ is said to be *independent of irrelevant alternatives, type 1* (IIA-1), if for all $\bar{\rho},\bar{\rho}' \in \mathscr{FR}^n$ and all $x,y \in X$, $\rho_i(x,y) = \rho_i'(x,y)$ for all $i \in N$ implies $\tilde{f}(\bar{\rho})(x,y) = \tilde{f}(\bar{\rho}')(x,y)$.

In terms of the two previously discussed criteria, Definition 4.8 certainly satisfies the first condition where $\tilde{f}(\bar{\rho})(x,y)$ is only related to $\bar{\rho}]_{\{x,y\}}$ because the values of $\rho(w,z)$ are left undefined for all $w \neq x$ and $z \neq y$. However, IIA-1 does not faithfully reproduce the second condition of ordinality where the strength of an actor's preference for one alternative over another becomes arbitrary.

One recent effort to reconsider a fuzzy version of the independence condition appears in Mordeson and Clark (2009) where the support of fuzzy preference relations is used.

**Definition 4.9 (IIA-2).** Let $\tilde{f}$ be an FPAR. Then $\tilde{f}$ is said to be *independent of irrelevant alternatives, type 2* (IIA-2), if for all $\bar{\rho},\bar{\rho}' \in \mathscr{FR}^n$ and $x,y \in X$, $\mathrm{Supp}(\rho_i]_{\{x,y\}}) = \mathrm{Supp}(\rho_i']_{\{x,y\}})$ for all $i \in N$ imples $\mathrm{Supp}(\tilde{f}(\rho_i)]_{\{x,y\}}) = \mathrm{Supp}(\tilde{f}(\rho_i')]_{\{x,y\}})$.

Definition 4.9 certainly captures some aspects of the ordinal quality of the crisp independence condition. In words, if there exist two profiles $\bar{\rho},\bar{\rho}' \in \mathscr{FR}^n$ such that, when restricted to two alternatives $x$ and $y$, the supports of the individual fuzzy weak orders in $\bar{\rho}$ are identical to those in $\bar{\rho}'$, then the support of the two social preference relations generated by an IIA-2 FPAR should be identical as well, regardless of the relationship between the other alternatives and regardless of the specific values for $\rho_i(x,y)$ and $\rho_i(y,x)$. However, constructing the independence condition in this manner offers no guarantee that the relationship between $\tilde{f}(\bar{\rho})(x,y)$ and $\tilde{f}(\bar{\rho})(y,x)$ will be preserved in the fuzzy social preference relation generated by $\tilde{f}(\bar{\rho}')$. This can have important consequences when constructing a social strict preference relation as the following example demonstrates.

*Example 4.10.* Let $X = \{x,y\}$ and let $\bar{\rho}, \bar{\rho}' \in \mathscr{FR}^n$. Suppose the fuzzy social preference relations derived from $\bar{\rho}$ and $\bar{\rho}'$, denoted $\rho$ and $\rho'$, respectively, are derived as follows:

$$\rho(x,y) = \rho'(y,x) = .5,$$
$$\rho(y,x) = \rho'(x,y) = .2.$$

Obviously, $\text{Supp}(\rho) = \text{Supp}(\rho')$. However, if we were to construct a fuzzy social strict preference relation by assuming that social strict preference relations, $\pi$ and $\pi'$, are regular, then $\pi(x,y) > 0$ and $\pi'(y,x) > 0$.

Example 4.10 begs the question: How truly similar are two preferences relations when their supports are identical? If we are also interested in creating a social strict preference, then we may want to consider an independence condition that maintains the ordinal relationships between two FWPRs. Such a definition is proposed by Billot (1992), which has remained largely overlooked in the literature. Before proceeding, we need the following definition.

**Definition 4.11 (equivalent).** Let $\rho, \rho' \in \mathscr{F}(X^2)$ and let $Im(\rho) = \{s_1, \ldots, s_m\}$ and $Im(\rho') = \{t_1, \ldots, t_n\}$ be such that $s_1 < \ldots < s_m$ and $t_1 < \ldots < t_n$. We then say $\rho$ and $\rho'$ are *equivalent*, written $\rho \sim \rho'$, if and only if

(1) $s_1 = 0 \iff t_1 = 0$,
(2) $n = m$,
(3) $\rho^{s_i} = \rho'^{t_i}$, for all $i = 1, \ldots, m$.

Using this concept of analogous preference relations, we can model a third variant of the independence condition in the manner of Billot (1992).

**Definition 4.12 (IIA-3).** Let $\tilde{f}$ be an FPAR. Then $\tilde{f}$ is said to be *independent of irrelevant alternatives, type 3* (IIA-3), if for all $\bar{\rho}, \bar{\rho}' \in \mathscr{FR}^n$ and $x, y \in X, \rho_i]_{\{x,y\}} \sim \rho_i']_{\{x,y\}}$ for all $i \in N$ implies $\tilde{f}(\bar{\rho})]_{\{x,y\}} \sim \tilde{f}(\bar{\rho}')]_{\{x,y\}}$.

Proposition 4.13 demonstrates that the binary relation $\sim$ preserves the ordinal relationship between $\rho(x,y)$ and $\rho(y,x)$ across analogous preference relations.

**Proposition 4.13.** *Let $\rho$ and $\rho'$ be FWPRs on $X$ where $x, y \in X$. Suppose $\rho \sim \rho'$. Then $\rho(x,y) > \rho(y,x) \iff \rho'(x,y) > \rho'(y,x)$.*

*Proof.* Suppose $\rho(x,y) > \rho(y,x)$ and $\rho(x,y) = s_i$. Then $s_i > \rho(y,x)$ and $\rho(y,x) \notin \rho^{s_i}$. Thus, $\rho(y,x) \notin \rho'^{t_i}$. Now $(x,y) \in \rho^{s_i}$ implies $(x,y) \in \rho'^{t_i}$. Hence $\rho'(x,y) \geq t_i > \rho'(y,x)$.      □

Proposition 4.13 helps us to interpret IIA-3. For some fuzzy preference profile $\bar{\rho}$, suppose there exists another profile $\bar{\rho}'$ such that $\rho_i(x,y) > \rho(y,x)$ if and only if $\rho_i'(x,y) > \rho_i'(y,x)$ for all $i \in N$. Then an IIA-3 FPAR will associate equivalent social preferences over $x$ and $y$ to $\bar{\rho}$ and $\bar{\rho}'$, where $\tilde{f}(\bar{\rho})(x,y) > \tilde{f}(\bar{\rho})(y,x)$ if and only if $\tilde{f}(\bar{\rho}')(x,y) > \tilde{f}(\bar{\rho}')(y,x)$. Hence, IIA-3 preserves the ordinal relationship between the social preference over $(x,y)$ and over $(y,x)$ without considering the specific values of social preference, thereby satifying the conditions presented earlier in this subsection.

### 4.1.4 Dictatorship

In contrast to the other fuzzy Arrow conditions, dictatorship or a dictatorial aggregation rule exhibits very little variation over definitions throughout the literature (Banerjee, 1994; Barrett et al., 1992; Duddy et al., 2011; Fono and Andjiga, 2005; Fono et al., 2009; Mordeson and Clark, 2009; Richardson, 1998; Salles, 1998).

**Definition 4.14 (dictatorial).** Let $\tilde{f}$ be an FPAR. Then $\tilde{f}$ is said to be *dictatorial* if there exists an $i \in N$ such that for all $\bar{\rho} \in \mathscr{FR}^n$ and $x,y \in X$, $\pi_i(x,y) > 0$ implies $\pi(x,y) > 0$.

Definition 4.14 is standard in the literature. Obviously a dictatorship over an FPAR corresponds neatly to a dictatorship in the case of exact preferences, where society striclty prefers one alternative to another if the dictator does as well. As discussed previously, some scholars have chosen to avoid fuzzy strict preference relation and rely on another definition of dictatorship (Billot, 1992; Duddy et al., 2011).

**Definition 4.15 (strongly dictatorial).** Let $\tilde{f}$ be an FPAR. Then $\tilde{f}$ is said to be *strongly dictatorial* if there exists an $i \in N$ such that for all $\bar{\rho} \in \mathscr{FR}^n$ and $x,y \in X$

$$\rho_i(x,y) = \tilde{f}(\bar{\rho})(x,y) \, .$$

A strong dictatorship implies a dictatorship assuming that $\pi$ is regular on both the individual and social levels.

## 4.2 Making and Breaking Arrow's Theorem

The traditional proofs of Arrow's theorem use exact preference relations. This section demonstrates the conditions under which Arrow's conclusion holds in the fuzzy framework discussed in the previous section. Further, we also detail under what conditions there exists an FPAR that satisfies certain combinations of fuzzy Arrowian conditions. To prove our main results, we make use of the following definition.

**Definition 4.16.** Let $\tilde{f}$ be an FPAR, let $(x,y) \in X \times X$ and let $\lambda$ be a fuzzy subset of $N$.

(1) **(semidecisive)** $\lambda$ is called *semidecisive for $x$ against $y$*, written $x\tilde{D}_\lambda y$, if for every $\bar{\rho} \in \mathscr{FR}^n$,

$$\pi_i(x,y) > 0 \text{ for all } i \in \text{Supp}(\lambda) \text{ and } \pi_j(y,x) > 0 \text{ for all } j \notin \text{Supp}(\lambda)$$

implies $\pi(x,y) > 0$.

(2) **(decisive)** $\lambda$ is called *decisive for x against y*, written $xD_\lambda y$, if for every $\bar{\rho} \in \mathscr{FR}^n$,

$$\pi_i(x,y) > 0 \text{ for all } i \in \text{Supp}(\lambda)$$

implies $\pi(x,y) > 0$.

In words, we call $\lambda$ a *fuzzy coalition* when $|\text{Supp}(\lambda)| \geq 1$. In addition we say a coalition $\lambda$ is *semidecisive* or *decisive* if it is *semidecisive* or *decisive* for all ordered pairs of alternatives.

There are two comments worth making about Definition 4.16 before proceeding to the formal arguments of fuzzy Arrow's theorem. First, the fuzzy definition of (semi)decisiveness introduces another application of fuzzy sets to social choice theory. Here we use a fuzzy subset of the actors rather than a traditional crisp case. Such a nuance is necessary when actors possess varied levels of influences within a coalition. These situations can arise in informal committees where the preferences of a more senior member may have more influence on the group's final preferences than those of a more junior member. Second, it is important to emphasize how very little semidecisiveness implies about a specific coalition $\lambda$. Obviously, decisiveness implies semidecisiveness, but the converse does not hold because semidecisiveness incorporates the preferences of individuals not in $\text{Supp}(\lambda)$. Hence, if there exists a $j \in \text{Supp}(\lambda)$ such that $\pi_j(y,x) = 0$, we cannot conclude that $\lambda$ is semidecisive for $x$ against $y$, and we know very little about the social preference between $x$ and $y$. Given these restrictions on semidecisiveness, the following lemma is quite remarkable in the fact that additional structure on the FPAR implies a semidecisive coalition over an ordered pair is actually a decisive coalition over all pairs of alternatives.

**Lemma 4.17.** *Let* $\lambda$ *be a fuzzy subset of N. Let* $\tilde{f}$ *be a partially quasi-transitive FPAR that is weakly Paretian and IIA-3 where* $\pi$ *is regular. If* $\lambda$ *is semidecisive for x against y, then for all* $(v,w) \in X \times X$, $\lambda$ *is decisive for v against w.*

*Proof.* Suppose $\lambda$ is semidecisive for $x$ against $y$. Let $\bar{\rho}$ be a preference profile such that $\pi_i(x,z) > 0$, for all $i \in \text{Supp}(\lambda)$ and all $z \in X \setminus \{x,y\}$. Let $\bar{\rho}'$ be a fuzzy preference profile such that

$$\rho_i'(x,z) = \rho_i(x,z) \text{ and } \rho_i'(z,x) = \rho_i(z,x), \forall i \in N \tag{4.1}$$
$$\pi_i'(x,y) > 0, \forall i \in \text{Supp}(\lambda)$$
$$\pi_j'(y,x) > 0, \forall i \in N \setminus \text{Supp}(\lambda)$$
$$\pi_i'(y,z) > 0, \forall i \in N.$$

Since $\pi_i(x,z) > 0$ for all $i \in \text{Supp}(\lambda)$, $\pi_i'(x,z) > 0$ for all $i \in \text{Supp}(\lambda)$ by the definition of $\bar{\rho}'$. Since $x\tilde{D}_\lambda y$, $\pi'(x,y) > 0$ by hypothesis. Since $\tilde{f}$ is weakly Paretian, $\pi'(y,z) > 0$. Since $\tilde{f}$ is partially quasi-transitive, $\pi'(x,z) > 0$ and $\rho'(x,z) > \rho'(z,x)$. Since $\rho_i]_{\{x,z\}} = \rho_i']_{\{x,z\}}$ for all $i \in N$ and $\tilde{f}$ is IIA-3, $\rho]_{\{x,z\}} \sim \rho']_{\{x,z\}}$ implies $\rho(x,z) > \rho(z,x)$. Hence $\pi(x,z) > 0$. Since $\bar{\rho}$ is arbitrary, $xD_\lambda z$. Since $z$ was arbitrary in $X \setminus \{x,y\}$,

$$x\tilde{D}_\lambda y \implies xD_\lambda z, \forall z \in X \setminus \{x,y\}. \tag{4.2}$$

Since $\lambda$ is decisive for $x$ against $z$ implies $\lambda$ is semidecisive for $x$ against $z$, interchanging $y$ and $z$ in Eq. (4.2) implies $\lambda$ is decisive for $x$ against $y$.

Now let $\bar{\rho}^*$ be another profile such that $\pi_i^*(y,z) > 0$ for all $i \in \text{Supp}(\lambda)$ and let $\bar{\rho}^+$ be such that

$$\rho_i^+(y,z) = \rho_i^*(y,z) \text{ and } \rho_i^+(z,y) = \rho_i^*(z,y), \forall i \in N$$
$$\pi_i^+(y,x) > 0, \forall i \in N$$
$$\pi_i^+(x,z) > 0, \forall i \in \text{Supp}(\lambda)$$
$$\pi_j^+(z,x) > 0, \forall j \in N \backslash \text{Supp}(\lambda).$$

Then $\pi_i^+(y,z) > 0$ for all $i \in \text{Supp}(\lambda)$. Since $xD_\lambda z$, $\pi^+(x,z) > 0$. Since $\tilde{f}$ is weakly Paretian, $\pi^+(y,x) > 0$. Since $\tilde{f}$ is partially quasi-transitive, $\pi^+(y,z) > 0$. Since $\rho_i^*\rceil_{\{y,z\}} = \rho_i^+\rceil_{\{y,z\}}$ for all $i \in N$ and $\tilde{f}$ is IIA-3, $\rho^*\rceil_{\{y,z\}} \sim \rho^+\rceil_{\{y,z\}}$, and so $\rho^*(y,z) > \rho^*(z,y)$. Thus, $\pi^*(y,z) > 0$ and so $yD_\lambda z$ because $\bar{\rho}^*$ is arbitrary. Because $z$ is arbitrary in $X \backslash \{x,y\}$,

$$x\tilde{D}y \implies yD_\lambda z, \forall z \notin \{x,y\}. \tag{4.3}$$

Now because $\lambda$ is decisive for $y$ against $z$, $\lambda$ is semidecisive for $y$ against $z$. Thus by 4.13, $\lambda$ is decisive for $y$ against $x$. To summarize, we have, for all $(v,w) \in X \times X$,

$$x\tilde{D}_\lambda y \implies xD_\lambda v \text{ (by 4.13)} \implies x\tilde{D}v \implies vD_\lambda w$$

by Eq. (4.3). $\qquad\qquad\qquad\qquad\qquad\qquad\qquad\qquad\qquad\qquad\qquad\qquad\qquad\square$

Lemma 4.17 lays out the formal argument in the fuzzy framework of what Sen (1976) labels the "Paretian epidemic", where a coalition that is semidecisive over an ordered pair becomes globally decisive after adopting the Arrowian conditions. An important aspect of Lemma 4.17 is the generalization of strict preference to a regular fuzzy strict preference relation, which as Chapter 3 illustrated, imposes minimal assumptions on the structure of FWPRs. Nonetheless, the argument still holds for certain non-regular strict preference relations but requires a new specification of IIA. The following definition and proposition explores this relationship formally using the *cosupport* of a fuzzy subset $U$ of $X$. That is, $\text{Cosupp}(U) = \{x \in X \mid U(x) < 1\}$.

**Definition 4.18 (IIA-4).** Let $\tilde{f}$ be an FPAR. Then $\tilde{f}$ is said to be *independent of irrelevant alternatives, type 4* (IIA-4), if for all $\bar{\rho}, \bar{\rho}' \in \mathscr{F}\mathscr{R}^n$ and $x, y \in X$,

$$\text{Cosupp}(\bar{\rho}_i \rceil_{\{x,y\}}) = \text{Cosupp}(\bar{\rho}_i' \rceil_{\{x,y\}})$$

for all $i \in N$ implies

$$\text{Cosupp}(\tilde{f}(\bar{\rho}_i) \rceil_{\{x,y\}}) = \text{Cosupp}(\tilde{f}(\bar{\rho}_i') \rceil_{\{x,y\}}).$$

It is easily verified that $\pi_{(2)}(x,y) = 1 - \rho(y,x)$ and

$$\pi_{(4)}(x,y) = \begin{cases} \rho(x,y) & \text{if } \rho(y,x) = 0, \\ 0 & \text{otherwise,} \end{cases}$$

are not regular when there is no further structure placed on $\rho$ besides completeness and reflexivity.

**Lemma 4.19.** *Let $\lambda$ be a fuzzy subset of $N$. Let $\tilde{f}$ be a partially quasi-transitive FPAR that is weakly Paretian and IIA-2 when $\pi = \pi_{(4)}$ and IIA-4 when $\pi = \pi_{(2)}$. If $\lambda$ is semidecisive for $x$ against $y$, then for all $(v,w) \in X \times X$, $\lambda$ is decisive for $v$ against $w$.*

*Proof.* Suppose $\lambda$ is semidecisive for $x$ against $y$. Consider a profile $\bar{\rho} \in \mathcal{FR}^n$ such that $\pi_i(x,z) > 0$, for all $i \in \mathrm{Supp}(\lambda)$. Let $\bar{\rho}'$ be the fuzzy preference profile as defined in Lemma 4.17. By an identical argument, we know $\pi_i'(x,z) > 0$ for all $i \in \mathrm{Supp}(\lambda)$ and $\pi'(x,y) > 0$. Likewise, $\pi'(y,z) > 0$, because $\tilde{f}$ is weakly Paretian. Since $\tilde{f}$ is partially quasi-transitive, $\pi'(x,z) > 0$.

For $\pi = \pi_{(4)}$, $\pi_{(4)}'(x,z) > 0$ implies $\rho'(x,z) > 0$ and $\rho'(z,x) = 0$. Since

$$\mathrm{Supp}(\rho_i]_{\{x,z\}}) = \mathrm{Supp}(\rho_i']_{\{x,z\}})$$

for all $i \in N$ and $\tilde{f}$ is IIA-2,

$$\mathrm{Supp}(\rho]_{\{x,z\}}) = \mathrm{Supp}(\rho']_{\{x,z\}}).$$

Thus, $\rho(x,z) > 0$ and $\rho(z,x) = 0$, which implies $\pi_{(4)}(x,z) > 0$ by the definition of $\pi_{(4)}$.

For $\pi = \pi_{(2)}$, $\pi_{(2)}'(x,z) > 0$ implies $\rho'(z,x) < 1$. Since

$$\mathrm{Cosupp}(\rho_i]_{\{x,z\}}) = \mathrm{Cosupp}(\rho_i']_{\{x,z\}})$$

for all $i \in N$ and $\tilde{f}$ is IIA-4,

$$\mathrm{Cosupp}(\rho]_{\{x,z\}}) = \mathrm{Cosupp}(\rho']_{\{x,z\}}).$$

Thus, $\rho'(z,x) < 1$, which implies $\pi_{(2)}(x,z) > 0$ by definition of $\pi_{(2)}$.

Because $\pi(x,z) > 0$ and $\bar{\rho} \in \mathcal{FR}^n$ and $z \in X \setminus \{x,y\}$ are arbitrary, we obtain the following result:

$$x\tilde{D}_\lambda y \implies xD_\lambda z, \forall z \in X \setminus \{x,y\}.$$

The remainder of the proof follows easily from a similar argument using Lemma 4.17. $\qquad\square$

Before presenting the main results, we prove the following proposition.

**Proposition 4.20.** *Let $\rho$ be an FWPR on $X$. Then the following properties are equivalent:*

*(1) $\rho$ is weakly transitive.*
*(2) For all $x,y,z \in X$, $\rho(x,y) \geq \rho(y,x)$ and $\rho(y,z) \geq \rho(z,y)$ with a strict equality holding at least once, then $\rho(x,z) > \rho(z,x)$.*

*Proof.* Suppose 4.20(1). Assume that $\rho(x,y) \geq \rho(y,x)$ and $\rho(y,z) > \rho(z,y)$. Then $\rho(x,z) \geq \rho(z,x)$. Suppose $\rho(z,x) \geq \rho(x,z)$. Then $\rho(z,y) \geq \rho(y,z)$ by 4.20(1), a contradiction. Hence, $\rho(x,z) > \rho(z,x)$. A similar argument shows that $\rho(x,y) > \rho(y,x)$ and $\rho(y,z) \geq \rho(z,y)$ implies $\rho(x,z) > \rho(z,x)$.

Suppose 4.20(2). Let $x,y,z \in X$. Suppose $\rho(x,y) \geq \rho(y,x)$ and $\rho(y,z) \geq \rho(z,y)$. Suppose $\rho(z,x) > \rho(x,z)$. Then by (2), $\rho(z,x) > \rho(x,z)$ and $\rho(x,y) \geq \rho(y,x)$ imply $\rho(z,y) > \rho(y,z)$, a contradiction. Hence, $\rho(x,z) \geq \rho(z,x)$. $\qquad\square$

**Corollary 4.21.** *Let $\rho$ be an FWPR on $X$. If $\rho$ is weakly transitive, then $\rho$ is partially quasi-transitive.*

As Proposition 4.20 and Corollary 4.21 show, weak transitivity is more restrictive than partial quasi-transitivity. This added assumption, when paired with the conditions of independence and weak Paretianism, implies a dictatorial FPAR. To illustrate this formally, the results in Lemmas 4.17 and 4.19 make it sufficient to show that $\mathrm{Supp}(\lambda) = \{i\}$, where $\lambda$ is any semidecisive coalition under the Arrowian conditions. In such a case, $\pi_i(x,y) > 0$ implies $\pi(x,y) > 0$ for all $\bar{\rho} \in \mathscr{FR}^n$ and $x,y \in X$, and $\lambda$ is a dictator rather than a coalition.

**Theorem 4.22 (Fuzzy Arrow's Theorem).** *Let $\tilde{f} : D_w^n \to \mathscr{FR}$ be a fuzzy aggregation rule. Suppose $\pi$ is regular, and $\tilde{f}$ is weakly Paretian, weakly transitive and IIA-3. Then $\tilde{f}$ is dictatorial.*

*Proof.* Since $\tilde{f}$ is weakly Paretian, there exists a decisive $\lambda$ for any pair of alternatives, namely, $\mathrm{Supp}(\lambda) = N$. For all $(u,v) \in X \times X$, let $m(u,v)$ denote the size of the smallest $|\mathrm{Supp}(\lambda)|$ for a $\lambda$ semidecisive for $u$ against $v$. Let $m = \wedge\{m(u,v) \mid (u,v) \in X \times X\}$. Without loss of generality, suppose $\lambda$ is semidecisive for $x$ against $y$ where $|\mathrm{Supp}(\lambda)| = m$. If $m = 1$, the proof is complete. Suppose $m > 1$. Let $i \in \mathrm{Supp}(\lambda)$, and let $z \in X \setminus \{x,y\}$. Consider any fuzzy profile $\bar{\rho}$ such that

$$\pi_i(x,y) > 0, \pi_i(y,z) > 0 \text{ and } \pi_i(x,z) > 0$$
$$\pi_j(z,x) > 0, \pi_j(x,y) > 0 \text{ and } \pi_j(z,y) > 0, \forall j \in \mathrm{Supp}(\lambda) \setminus \{i\}$$
$$\pi_k(z,x) > 0, \pi_k(x,y) > 0 \text{ and } \pi_k(z,y) > 0, \forall k \notin \mathrm{Supp}(\lambda).$$

Since $\lambda$ is semidecisive for $x$ against $y$ and $\pi_j(x,y) > 0$ for all $j \in \mathrm{Supp}(\lambda)$, $\pi(x,y) > 0$. Since $|\mathrm{Supp}(\lambda)| = m$, it is not the case that $\pi(z,y) > 0$, or otherwise $\lambda'$ is semidecisive for $z$ against $y$, where $\mathrm{Supp}(\lambda') = \mathrm{Supp}(\lambda) \setminus \{i\}$. However, this contradicts the minimality of $m$ since $|\mathrm{Supp}(\lambda')| = m - 1$. Because $\pi$ is regular, $\pi(z,y) = 0$ implies $\rho(y,z) \geq \rho(z,y)$. Since $\rho(x,y) > \rho(y,z)$, $\rho(x,z) > \rho(z,x)$ by weak transitivity and Proposition 4.20. Hence $\pi(x,z) > 0$. By IIA-3, $\lambda^*$ is semidecisive for $x$ against $z$, where $\mathrm{Supp}(\lambda^*) = \{i\}$. However, this contradicts the fact the $m > 1$. $\qquad\square$

The added assumption of weak transitivity, rather than partial quasi-transitivity, in Theorem 4.22 allows Arrow's results to hold in the fuzzy framework with a general strict preference relation without putting added assumptions on individual preferences such as those in Fono and Andjiga (2005) and Mordeson and Clark (2009). However, we can relax the transitivity condition of the FPAR and still obtain similar results by specifying a strict preference relation. To do so, we make use of the following proposition.

**Proposition 4.23.** *Let $\rho$ be an FWPR on $X$. If $\rho$ is partially transitive, then $\rho$ is partially quasi-transitive with respect to $\pi = \pi_{(4)}$.*

*Proof.* Let $x, y, z \in X$. Suppose $\pi(x,y) > 0$ and $\pi(y,z) > 0$. Then $\rho(x,y) > 0$, $\rho(y,x) = 0$, $\rho(y,z) > 0$, and $\rho(z,y) = 0$. Hence, $\rho(x,z) > 0$. Suppose $\pi(x,z) = 0$. Then $\rho(z,x) > 0$. However, $\rho(y,z) > 0$ and $\rho(z,x) > 0$ implies $\rho(y,x) > 0$, a contradiction. Hence $\pi(x,z) > 0$.                                                             $\square$

Using this proposition, we can relax the transitivity condition on $\tilde{f}$ to partial transitivity when $\pi = \pi_{(4)}$.

Let $D_p$ denote the set of all partially transitive fuzzy weak orders.

**Theorem 4.24 (Fuzzy Arrow's Theorem 2).** *Let $\tilde{f} : D_p^n \to \mathscr{FR}$ be an FPAR. Suppose $\pi = \pi_{(4)}$. Let $\tilde{f}$ be weakly Paretian, partially transitive, and IIA-2. Then $\tilde{f}$ is dictatorial.*

*Proof.* Since $\tilde{f}$ is partially transitive and $\pi = \pi_{(4)}$, $\tilde{f}$ is partially quasi-transitive by Proposition 4.23. Further, because $\tilde{f}$ is weakly Paretian, there exists a decisive $\lambda$ for any pair of alternatives. Let $m(u,v)$ denote the size of the smallest $|\text{Supp}(\lambda)|$ for a $\lambda$ semidecisive for $u$ against $v$ in $X$. Let $m = \wedge\{m(u,v) \mid (u,v) \in X \times X\}$. Likewise, suppose $\lambda$ is semidecisive for $x$ against $y$ where $|\text{Supp}(\lambda)| = m$, and suppose $m > 1$. Now consider a $\bar{\rho} \in \mathscr{FR}^n$ such that $\bar{\rho}$ is identical to $\bar{\rho}$ in Theorem 4.24.

Then $\pi(x,y) > 0$ because $\lambda$ is semidecisive for $x$ against $y$ and $\pi_j(x,y) > 0$ for all $j \in \text{Supp}(\lambda)\backslash\{i\}$. In addition, $\pi(z,y) = 0$, else $\lambda'$ is semidecisive for $z$ against $y$, a contradiction of the minimality of $m$. Thus, $\rho(y,z) > 0$. Since $\tilde{f}$ is partially transitive, $\rho(x,y) > 0$ and $\rho(y,z) > 0$ imply $\rho(x,z) > 0$. Suppose $\pi(x,z) = 0$. Then $\rho(z,x) > 0$ by definition of $\pi_{(4)}$. However, $\rho(y,z) > 0$ and $\rho(z,x) > 0$ imply $\rho(y,x) > 0$ by the partial transitivity of $\tilde{f}$. This contradicts $\pi(x,y) > 0$. Hence $\pi(x,z) > 0$. By IIA-2, $\lambda^*$ is semidecisive for $x$ against $y$, where $\text{Supp}(\lambda^*) = \{i\}$. However, this contradicts $m > 1$.                                                             $\square$

Theorems 4.22 and 4.24 lay out the consequences of two specific combinations of assumptions on fuzzy aggregation rules. Given an FPAR that satisfies these definitions of transitivity, weak Paretianism and independence of irrelevant alternatives, the FPAR must be dictatorial under a variety of social strict preference relations. However, the implication of dictatorship cannot be generalized over all derivations of fuzzy Arrowian conditions. Thus, we now consider under what circumstances a nondictatorial FPAR can satisfy fuzzy Arrowian conditions. The key to these series of formal arguments lies in the concept of neutrality.

**Definition 4.25 (neutral).** Let $\tilde{f}$ be an FPAR. Then $\tilde{f}$ is said to be *neutral* if, for all $\bar{\rho}, \bar{\rho}' \in \mathscr{FR}^n$ and all $w, x, y, z \in X$, $\rho_i'(x,y) = \rho_i'(w,z)$, for all $i \in N$, implies $\tilde{f}(\bar{\rho})(x,y) = \tilde{f}(\bar{\rho}')(w,z)$.

In words, neutrality guarantees that an aggregation rule treats every pair of alternatives in a similar manner across preference profiles, i.e., the labeling of alternatives is arbitrary and does not affect the aggregation of preferences. In the exact case, neutrality has an important part in May's (1952) theorem characterizing the importance of majority rule as the only anonymous, neutral, and monotone choice function if there are two alternatives. In Arrowian context, Blau (1972) first noticed the logic of neutrality plays an important part in the formal arguments; however, he is unable to use neutrality to prove Arrow's theorem. Ubeda Ubeda (2003) first showed that IIA and weak Paretianism imply neutrality and that neutrality can be used in a more direct proof of Arrow's theorem. In the fuzzy case, this relationship no longer holds. This occurs because the concept of weak Paretianism is ordinal: for any two alternatives $x$ and $y$, $\rho_i(x,y) > \rho_i(y,x)$ for all $i \in N$ implies $\rho(x,y) > \rho(y,x)$ in the social preference relation when $\pi$ is regular. Yet neutrality, as defined in Definition 4.25, is cardinal in conception and weak Paretianism is insufficient to imply neutrality even when paired with IIA. Thus, we consider another characteristic of FPARs.

**Definition 4.26 (unanimous in acceptance).** Let $\tilde{f}$ be an FPAR. Then $\tilde{f}$ is said to be *unanimous in acceptance* if, for all $\bar{\rho} \in \mathscr{FR}^n$, $\rho_i(x,y) = 1$ for all $i \in N$ implies $\tilde{f}(\bar{\rho})(x,y) = 1$ Duddy et al. (2011).

Unanimity in acceptance is significantly less restrictive than unanimity (see Section 4.1.2) and requires the social preference to take a specific value only when all individuals definitely view one alternative as at least as good as another. Further, Definition 4.26 has no implications for a fuzzy aggregation rule when there exist some $x, y \in X$ and $\bar{\rho} \in \mathscr{FR}^n$ such that $\rho_i(x,y) = c$ for all $i \in N$ and $c \in [0,1)$. This seemingly insubstantial condition allows Duddy et al. (2011) to obtain the following relationship.

Let $\mathscr{FR}^*$ denote the set of all max-∗ transitive fuzzy weak orders.

**Proposition 4.27.** *Let* $\tilde{f} : \mathscr{FR}^{*n} \to \mathscr{FR}$ *be an FPAR. Suppose* $\tilde{f}$ *is* $max - *$ *transitive, IIA-1 and unanimous in acceptance. Then* $\tilde{f}$ *is neutral.*

*Proof.* The proof, which comes from Duddy et al. (2011), demonstrates that $\tilde{f}$ is neutral by considering all combinations of $(x,y)$, $(w,z) \in X \times X$.

Case 1: $(x,y) = (w,z)$. The proof follows immediately from the IIA-1 definition.
Case 2: $(x,y), (x,z) \in X \times X$. Let $\bar{\rho} \in \mathscr{FR}^{*n}$ be such that $\rho_i(y,z) = \rho_i(z,y) = 1$ for all $i \in N$. Then, by max-∗ transitivity of all individual weak orders, $\rho_i(x,y) \geq \rho_i(x,z) * \rho_i(z,y) = \rho_i(x,z)$ and $\rho_i(x,z) \geq \rho_i(x,y) * \rho_i(y,z) = \rho_i(x,y)$. Next, $\rho_i(x,y) \geq \rho_i(x,z)$ and $\rho_i(x,z) \geq \rho_i(x,y)$ imply $\rho_i(x,y) = \rho_i(x,z)$ for all $i \in N$. Similarly, by max-∗ transitivity, $\rho_i(y,x) \geq \rho_i(y,z) * \rho_i(z,x) = \rho_i(z,x)$ and $\rho_i(z,x) \geq \rho_i(z,y) * \rho_i(y,x) = \rho_i(y,x)$; and $\rho_i(y,x) = \rho_i(z,x)$, for all $i \in N$. Because $\rho_i(y,z) = \rho_i(z,y) = 1$ for all $i \in N$, $\rho(y,z) = \rho(z,y) = 1$. Hence, by the previous arguments, $\rho(x,y) \geq \rho(x,z)$, $\rho(x,z) \geq \rho(x,y)$, $\rho(y,x) \geq \rho(z,x)$, and $\rho(z,x) \geq \rho(y,x)$. Thus, $\rho(x,y) = \rho(x,z)$ and $\rho(y,x) = \rho(z,x)$ for the social preference as well.

The above arguments apply to all $\bar{\rho} \in \mathscr{FR}^{*n}$ such that $\rho_i(y,z) = \rho_i(z,y) = 1$ for all $i \in N$. Let $\mathscr{G}^n$ denote the set of all such profiles. Because the individual preferences between $y$ and $z$ are "irrelevant" so to speak, the proof now uses IIA-1 to prove the conclusion.

Case 3: Now for any profile $\bar{\rho} \in \mathscr{FR}^{*n}$ such that $\rho_i(x,y) = \rho_i(x,z)$ for all $i \in N$, there exists a $\bar{\rho}' \in \mathscr{G}^n$ such that $\rho_i(x,y) = \rho_i'(x,y) = \rho_i'(x,z) = \rho_i(z,x)$. IIA-1 implies $\rho(x,y) = \rho(y,x) = \rho'(y,x) = \rho'(x,y)$. For two distinct profiles $\bar{\rho}, \bar{\rho}' \in \mathscr{FR}^n$ such that $\rho_i(x,y) = \rho_i'(x,z)$ for all $i \in N$, there also exists a profie $\bar{\rho}^* \in \mathscr{G}^n$ such that $\rho_i(x,y) = \rho_i^*(x,y) = \rho_i^*(x,z) = \rho_i'(x,z)$. By IIA-1, $\rho(x,y) = \rho^*(x,y) = \rho^*(x,z) = \rho'(x,z)$.

Case 4: $(x,y),(w,y) \in X \times X$. The same conclusions can be proved using symmetric logic in Case 2. The first step is to assume $\rho_i(x,w) = \rho_i(w,x) = 1$ for all $i \in N$.

Case 5: $(x,y),(w,z) \in X \times X$. Let $\bar{\rho} \in \mathscr{FR}^{*n}$ such that $\rho_i(y,z) = \rho_i(z,y) = \rho_i(x,w) = \rho_i(w,x) = 1$ for all $i \in N$. Because $\rho_i$ is max-$*$ transitive, $\rho_i(x,y) \geq \rho_i(x,z)$ and $\rho_i(x,z) \geq \rho_i(x,y)$, and thus $\rho_i(x,y) = \rho_i(x,z)$, for all $i \in N$. Because $\tilde{f}$ satisfies max-$*$ transitivity and unanimity in acceptance, $\rho(x,y) = \rho(x,z)$. Further, $\rho_i(x,z) \geq \rho_i(x,w) * \rho_i(w,z)$ and $\rho_i(w,z) \geq \rho_i(w,x) * \rho_i(x,z)$ imply $\rho_i(x,z) = \rho_i(w,z)$, for all $i \in N$. To summarize, $\rho_i(x,y) = \rho_i(x,z) = \rho_i(w,z)$ for all $i \in N$. And because the conditions of unanimity in acceptance and max-$*$ transitivity have been met, an identical argument applies to the social preference relation and $\rho(x,y) = \rho(x,z) = \rho(w,z)$.

The above arguments apply to all $\rho_i$ that are max-$*$ transitive such that $\rho_i(y,z) = \rho_i(z,y) = \rho_i(x,w) = \rho_i(w,x) = 1$ for all $i \in N$. Let $\mathscr{G}^n$ denote the set of all such profiles. Because the individual preferences between $y$ and $z$ and between $x$ and $w$ are "irrelevant" so to speak, the proof now uses IIA-1 to prove the conclusion.

Case 6: Now for any profile $\bar{\rho} \in \mathscr{FR}^{*n}$ such that $\rho_i(x,y) = \rho_i(w,z)$ for all $i \in N$, there exists a $\bar{\rho}' \in \mathscr{G}^n$ such that $\rho_i(x,y) = \rho_i'(x,y) = \rho_i'(w,z) = \rho_i(w,z)$. IIA-1 implies $\rho(x,y) = \rho(w,z) = \rho'(w,z) = \rho'(x,y)$. For two distinct profiles $\bar{\rho}, \bar{\rho}' \in \mathscr{FR}^n$ such that $\rho_i(x,y) = \rho_i'(w,z)$ for all $i \in N$. Then there exists a profile $\bar{\rho}^* \in \mathscr{G}^n$ such that $\rho_i(x,y) = \rho_i^*(x,y) = \rho_i^*(w,z) = \rho_i'(w,z)$. By IIA-1, $\rho(x,y) = \rho^*(x,y) = \rho^*(w,z) = \rho'(w,z)$.

Case 7: $(x,y),(w,z) \in X \times X$ where $x = z$ or $y = w$. (This case is similar to Cases 2 and 3.) Let $a$ denote an arbitrary alternative that is distinct from $x$ and $w$. One exists because $|X| \geq 3$. Take any profile $\bar{\rho} \in \mathscr{FR}^{*n}$ where $\rho_i(a,y) = \rho_i(y,a) = \rho_i(x,w) = \rho_i(w,x) = \rho_i(z,a) = \rho_i(a,z) = 1$. Cases 2 and 3 imply $\rho_i(x,y) = \rho_i(x,a) = \rho_i(w,a) = \rho_i(w,z)$ and $\rho(x,y) = \rho(x,a) = \rho(w,a) = \rho(w,z)$ by unanimity in acceptance and max-$*$ transitivity.

Let $W^n$ denote the set of all such profiles. Let $(r_1,\ldots,r_n) \in \mathscr{FR}^{*n}$ be such that $r_j(x,y) = r_j(z,w)$ for all $j \in N$. Then there exists $(r_1',\ldots,r_n') \in W^n$ such that

$r_j(x,y) = r_j(z,w) = r'_j(x,y,) = r'_j(z,w)$ for all $j \in N$. IIA-1 implies that $\tilde{f}(\bar{\rho})(x,y) = \tilde{f}(\bar{\rho})(z,w) = \tilde{f}(\bar{\rho}')(x,y) = \tilde{f}(\bar{\rho}')(z,w)$ where $\bar{\rho} = (r_1,\ldots,r_n)$ and $\bar{\rho}' = (r'_1,\ldots,r'_n)$. Take any pair of distinct profiles $\bar{\rho}'' = (r''_1,\ldots,r''_n)$ and $\bar{\rho}^* = (r^*_1,\ldots,r^*_n)$ in $\mathscr{FR}^{*n}$ such that $r''_j(x,y) = r^*_j(z,w)$ for all $j \in N$. Then there exists $(r^{**}_1,\ldots,r^{**}_n) \in W^n$ such that $r''_j(x,y) = r''_j(z,w) = r^{**}_j(x,y,) = r^{**}_j(z,w)$ for all $j \in N$. IIA-1 implies $\tilde{f}(\bar{\rho}'')(x,y) = \tilde{f}(\bar{\rho}'')(z,w) = \tilde{f}(\bar{\rho}^{**}(x,y) = \tilde{f}(\bar{\rho}^{**})(z,w)$ □

While we do not use Proposition 4.25 to establish further results, it does illustrate that neutrality is not necessarily a strong restriction to place on an aggregation rule. As Proposition 4.27 demonstrates, neutrality arises naturally from the combination of max-∗ transitivity, IIA-1, and unanimity in acceptance. We have already discussed the importance of max-∗ transitivity and IIA-1; if one can justify Definition 4.26 and its application to fuzzy aggregation rules, neutrality is the natural conclusion. With a few more assumptions, we can use neutrality to derive a specific fuzzy aggregation rule.

In what follows, we show how neutrality can be used to classify a wide range of FPARs and determine whether these FPARs satisfy fuzzy Arrowian conditions. To do this, we need the following lemma.

**Lemma 4.28.** *Let $\tilde{f}$ be an FPAR. Then the following conditions are equivalent.*

*(1) $\tilde{f}$ is neutral;*
*(2) There exists a unique function $f_n : [0,1]^n \to [0,1]$ such that, for all $x,y \in X$ and all $\bar{\rho} \in \mathscr{FR}^n$, $f_n(\rho_1(x,y),\ldots,\rho_n(x,y)) = \tilde{f}(\bar{\rho})(x,y)$.*

*Proof.* (1) $\implies$ (2): Let $x,y \in X$. Let $(a_1,\ldots,a_n) \in [0,1]^n$. Then there exists $\bar{\rho} \in \mathscr{FR}^n$ such that $\rho_i(x,y) = a_i$ for all $i = 1,\ldots,n$. Define $f_n : [0,1]^n \to [0,1]$ as follows:

$$f_n((a_1,\ldots,a_n)) = \tilde{f}(\bar{\rho})(x,y).$$

It remains to be shown that $f_n$ is single-valued. Let $w,z \in X$. Then there exists a $\bar{\rho}' \in \mathscr{FR}^n$ such that $\rho'_i(w,z) = a_i$ for all $i = 1,\ldots,n$. Thus, $\rho_i(x,y) = \rho'_i(w,z)$ for all $i \in N$. Since $\tilde{f}$ is neutral, $\tilde{f}(\bar{\rho})(x,y) = \tilde{f}(\bar{\rho}')(w,z)$. Thus, $f_n$ is single-valued. In addition, uniqueness of $f_n$ is guaranteed by construction.

(2) $\implies$ (1): Let $\bar{\rho}, \bar{\rho}' \in \mathscr{FR}^n$ and $w,x,y,z \in X$. Suppose $\rho_i(x,y) = \rho'_i(w,z)$ for all $i \in N$. Then,

$$\tilde{f}(\bar{\rho})(x,y) = f_n(\rho_1(x,y),\ldots,\rho_n(x,y)) = f_n(\rho'_1(w,z),\ldots,\rho'_n(w,z))$$
$$= \tilde{f}(\bar{\rho})(w,z).$$

Thus, $\tilde{f}$ is neutral. □

In words, $f_n$ is the *auxillary function* associated with a specific FPAR $\tilde{f}$, and $a_i$ can be interpreted as the *weak preference intensity* of player $i$ for one alternative over another. By itself, Lemma 4.28 may seem unremarkable, but the lemma is an important step in examining the implications of neutrality on fuzzy aggregation rules. To derive a unique aggregation rule, we need one more definition.

**Definition 4.29.** Let $\tilde{f}$ be a neutral FPAR, and let $f_n$ be an auxillary function associated with $\tilde{f}$. Then $\tilde{f}$ is said to be

(1) *linearly decomposable* if, for all $(a_1,\ldots,a_n) \in [0,1]^n$, $f_n(a_1,\ldots,a_n) = a_1 f_n(1,0,\ldots,0) + \ldots + a_n f_n(0,,\ldots,0,1)$;
(2) *additive* if, for all $(a_1,\ldots,a_n),(b_1,\ldots,b_n) \in [0,1]^n$ such that $a_i + b_i \in [0,1]$, $i = 1,\ldots,n$, $f_n((a_1,\ldots,a_n)+(b_1,\ldots,b_n)) = f_n((a_1,\ldots,a_n)) + f_n((b_1,\ldots,b_n))$.

Linear decomposability implies two criteria. First, the condition requires that the collective preference between two alternatives is the sum of the $n$ collective preferences when only one individual preference is considered at a time by the FPAR. Second, the specific individual preference intensity ($a_i$) can be "removed" from the individual preference relation ($\rho_i$), and "reapplied" directly to the FPAR that only considers the preference of individual $i$. The stronger assumption of additivity requires that given a preference profile, the collective preference for one alternative over another can be created by first decomposing the preference intensities of the individuals, then applying the FPAR to those two profiles of preference intensities, and finally adding the two collective preferences.

The followinig lemma states the relationship between the conditions in Definition 4.29.

**Lemma 4.30.** *Let $\tilde{f}$ be a neutral FPAR. If $\tilde{f}$ is linearly decomposable, then $\tilde{f}$ is additive.*

*Proof.* Because $\tilde{f}$ is neutral, there exits an auxillary function $f_n$ associated with $\tilde{f}$ such that $f_n((a_1,\ldots,a_n)) = \tilde{f}(\bar{\rho})(x,y)$ for all $\bar{\rho} \in \mathscr{FR}^n$, $x,y \in X$, and $(a_1,\ldots,a_n) \in [0,1]^n$. Let $(a_1,\ldots,a_n),(b_1,\ldots,b_n) \in [0,1]^n$ be such that $a_i + b_i \in [0,1]$ for all $i = 1,\ldots,n$. Then,

$$
\begin{aligned}
f_n((a_1,\ldots,a_n)+(b_1,\ldots,b_n)) &= f_n((a_1+b_1,\ldots,a_n+b_n)) \\
&= (a_1+b_1)f_n((1,0,\ldots,0)) + \ldots \\
&\quad + (a_n+b_n)f_n(0,\ldots 0,1) \\
&= a_1 f_n((1,0,\ldots,0)) + \ldots \\
&\quad + a_n f_n((1,0,\ldots,0)) + b_1 f_n((1,0,\ldots,0)) \\
&\quad + \ldots + b_n f_n((1,0,\ldots,0)) \\
&= f_n((a_1,\ldots,a_2)) + f_n((b_1,\ldots,b_n))
\end{aligned}
$$

as desired.                                                                                    □

Finally, Theorem 4.31 and Corollary 4.32, which are simplified generalizations of García-Lapresesta and Llamazares (2000), illustrate the effects of a neutral and linear decomposable FPAR. To do so, it introduces the concept of restricting an auxillary function between the interval $[0,1]$, denoted $\hat{f}_n|_{[0,1]^n}$, because, under additivity, there is no guarantee that the sum of two n-tuples of preferences intensities will have components less than or equal to one. The restriction places no added assumptions on FPARs or individual preferences, but it allows us to obtain the following result.

**Theorem 4.31.** *Let $\tilde{f}$ be a neutral fuzzy aggregation rule. If $\tilde{f}$ is linearly decomposable, then there exists a unique linear transformation $\hat{f}_n$ of $\mathbb{R}^n$ into $\mathbb{R}$ such that $\hat{f}|_{[0,1]^n} = f_n$.*

*Proof.* Because $\tilde{f}$ is neutral, there exits an auxillary function $f_n$ associated with $\tilde{f}$ such that $f_n((a_1,\ldots,a_n)) = \tilde{f}(\bar{\rho})(x,y)$ for all $\bar{\rho} \in \mathscr{FR}^n$, $x,y \in X$, and $(a_1,\ldots,a_n) \in [0,1]^n$. For $i = 1,\ldots,n$, let $\bar{1}_i = (u_1,\ldots,u_n)$, where $u_i = 1$ and $u_j = 0$ for $j \neq i$. Then there exists a unique linear transformation $\hat{f}$ of $\mathbb{R}^n$ into $\mathbb{R}$ such that $\hat{f}(\bar{1}_i) = w_i$, where $w_i = f_n(\bar{1}_i)$ for all $i \in N$. Since $f_n$ is additive by the previous lemma,

$$\sum_{i=1}^n w_i = \sum_{i=1}^n f_n(\bar{1}_i)$$
$$= f_n((1,\ldots,1)) \leq 1 .$$

Now,

$$\hat{f}_n(\sum_{i=1}^n c_i \bar{1}_i) = \sum_{i=1}^n c_i \hat{f}_n(\bar{1}_i) .$$

Thus if $c_i \in [0,1]$, for $i \in N$, then

$$f_n(\sum_{i=1}^n c_i \bar{1}_i) = \sum_{i=1}^n c_i f_n(\bar{1}_i) \in [0,1]$$

because $\sum_{i=1}^n w_i \leq 1$. Let $((a_1,\ldots,a_n)) \in [0,1]^n$. Then

$$\hat{f}_n|_{[0,1]^n}((a_1,\ldots,a_n)) = \hat{f}_n((a_1,\ldots,a_n))$$
$$= = \sum_{i=1}^n a_i \hat{f}_n(\bar{1}_i)$$
$$= f_n((a_1,\ldots,a_n))$$

since $\tilde{f}$ is linearly decomposable. $\qquad\square$

**Corollary 4.32.** *Let $\tilde{f}$ be a neutral FPAR. If $\tilde{f}$ is linearly decomposable, then, for all $\bar{\rho} \in \mathscr{FR}^n$ and all $x,y \in X$,*

$$\tilde{f}(\bar{\rho})(x,y) = \sum_{i=1}^n w_i \rho_i(x,y) ,$$

$w_i = f_n(\bar{1}_i)$ *for all $i \in N$.*

According to Theorem 4.31 and Corollary 4.32, a neutral and linearly decomposable aggregation rule must be a weighted mean aggregation rule. A weighted mean FPAR is a generalization of Example 4.2(1). Such a generalization emphasizes two important distinctions between exact and fuzzy aggregation rules. First, there is a difference between the possible rules modeled under exact preferences and those modeled under the fuzzy framework. Corollary 4.32 allows scholars to consider committee or other voting bodies where individuals do not contribute equally to the social preference. In other words, some opinions are more relevant to the final collective preference than others. These situations can arise on any committee that produces a social fuzzy preference, which affected by seniority, professional rank, or any number of other social factors could influence the group's final decision. However, this rule is not necessarily anonymous, i.e. the labeling of the individuals does matter, because each individual has a preassigned weight to his or her preference. If the weighted mean is anonymous, then it is easily verified that $w_i = \frac{1}{n}$ for all $i \in N$.

Second and more importantly, neutrality does not imply a dictatorship. Unlike the findings in Ubeda (2003), fuzzy neutrality, when paired with linear decomposability, does not guarantee a non-dictatorial FPAR. This brings us one step closer to identifying conditions under which fuzzy social choice permits FPARs to satsify all Arrowian conditions.

**Definition 4.33 (weighted mean rule).** Let $\tilde{f}$ be an FPAR. Then $\tilde{f}$ is said to be the *weighted mean rule* if, for all $\bar{\rho} \in \mathscr{FR}^n$ and all $x, y \in X$,

$$\tilde{f}(\bar{\rho})(x,y) = \sum_{i=1}^{n} w_i \cdot \rho_i(x,y),$$

where $\sum_{i=1}^{n} w_i = 1$ and $w_i > 0$ for all $i \in N$.

Obviously, the weighted mean is non-dictatorial and independent of irrelevant alternatives under IIA-1. What remains to be shown is whether the FPAR satisfies weak Paretianism and max-∗ transitivity, which we now consider.

**Proposition 4.34.** *Let $\tilde{f}$ be an FPAR as defined in Definition 4.33. If $\pi$ is regular, then $\tilde{f}$ is weakly Paretian.*

*Proof.* Let $x, y \in X$. Suppose $\pi_i(x,y) > 0$ for all $i \in N$. Because $\pi$ is assumed to be regular, $\pi_i(x,y) > 0$ implies $\rho_i(x,y) > \rho_i(y,x)$. Further, $w_i \cdot \rho_i(x,y) > w_i \cdot \rho(y,x)$ for all $i \in N$ because $w_i \in (0,1]$. Hence,

$$\sum_{i=1}^{n} w_i \cdot \rho_i(x,y) > \sum_{i=1}^{n} w_i \cdot \rho_i(y,x).$$

Thus, $\tilde{f}(\bar{\rho})(x,y) > \tilde{f}(\bar{\rho})(y,x)$, and by regularity of the social strict preference, $\pi(x,y) > 0$. Hence, $\tilde{f}$ is weakly Paretian.                                           □

As a result of Proposition 4.34, the weighted mean is weakly Paretian. Further, it satisfies stronger Paretian conceptualizations as well.

**Definition 4.35 (positive responsiveness).** Let $\tilde{f}$ be an FPAR and let $\pi$ be the social strict preference with respect to $\tilde{f}(\bar{\rho})$, where $\bar{\rho} \in \mathscr{FR}^n$. Then $\tilde{f}$ satisfies *positive responsiveness* with respect to $\pi$ if, for all $\bar{\rho}, \bar{\rho}' \in \mathscr{F}\mathscr{R}^n$ and all $x, y \in X$, $\tilde{f}(\bar{\rho})(x,y) = \tilde{f}(\bar{\rho})(y,x)$ and there exists a $j \in N$ such that $\rho_i = \rho'_i$ for all $i \neq j$ and $(\pi_j(x,y) = 0$ and $\pi'_j(x,y) > 0$ or $\pi_j(y,x) > 0$ and $\pi'_j(y,x) = 0)$ imply $\pi'(x,y) > 0$.

In other words, positive responsiveness requires that given a preference profile in which there is no social strict preference between two alternatives $x$ and $y$, if one individual who has no strict preference for $x$ over $y$ acquires such a preference or who has a strict preference for $y$ over $x$ and loses such a preference, then the FPAR should "respond" and exhibit a social strict preference for $x$ over $y$. To show that the weighted mean satisfies positive responsiveness, we make use of the following proposition.

**Proposition 4.36.** *Let* $\rho \in \mathscr{F}\mathscr{R}$ *and let* $\pi$ *and* $\pi_{(*)}$ *be two different types of strict preference with respect to* $\rho$ *such that for all* $x, y \in X$, $\pi(x,y) > 0$ *if and only if* $\pi_{(*)}(x,y) > 0$. *Let* $\tilde{f}$ *be an FPAR and* $\bar{\rho} \in \mathscr{F}\mathscr{R}^n$. *Then* $\tilde{f}$ *satisfies positive responsiveness with respect to* $\pi$ *if and only if* $\tilde{f}$ *satisfies positive responsiveness with respect to* $\pi_{(*)}$.

*Proof.* Suppose $\tilde{f}$ satisfies positive responsiveness with respect to $\pi$. Suppose for all $\bar{\rho}, \bar{\rho}' \in \mathscr{F}\mathscr{R}^n$ and all $x, y \in X$, $\tilde{f}(\bar{\rho})(x,y) = \tilde{f}(\bar{\rho})(y,x)$ and there exists a $j \in N$ such that $\rho_i = \rho'_i$ for all $i \neq j$ and $(\pi_j(x,y) = 0$ and $\pi'_j(x,y) > 0$ or $\pi_j(y,x) > 0$ and $\pi'_j(y,x) = 0)$. Because $\pi(x,y) > 0$ if and only if $\pi_{(*)}(x,y) > 0$ and $\pi'(x,y) > 0$ if and only if $\pi'_{(*)}(x,y) > 0$ for all $x, y \in X$, for all $\bar{\rho}, \bar{\rho}' \in \mathscr{F}\mathscr{R}^n$ and all $x, y \in X$, $\tilde{f}(\bar{\rho})(x,y) = \tilde{f}(\bar{\rho})(y,x)$ and there exists a $j \in N$ such that $\rho_i = \rho'_i$ for all $i \neq j$ and $(\pi_{(*)j}(x,y) = 0$ and $\pi'_{(*)}(x,y) > 0$ or $\pi_{(*)j}(y,x) > 0$ and $\pi'_{(*)j}(y,x) = 0)$. Then $\pi(x,y) > 0$ since $\tilde{f}$ satisfies positive responsiveness with respect to $\pi$. Thus, $\pi_{(*)}(x,y) > 0$ by hypothesis. Hence, $\tilde{f}$ satisfies positive responsiveness with respect to $\pi_{(*)}$.

Using Proposition 4.36, we can characterize the weighted mean as satisfying positiveness responsiveness with respect to any regular $\pi$.

**Proposition 4.37.** *Let* $\tilde{f}$ *be an FPAR as defined in Definition 4.35. Then* $\tilde{f}$ *satisfies positive responsiveness with respect to any regular* $\pi$.

*Proof.* By Proposition 4.36, it suffices to show that the weighted mean rule satisfies positive responsiveness with respect to $\pi_{(3)}$, where $\pi_{(3)}(x,y) = \max\{0, (\rho(x,y) - \rho(y,x))\}$. Let $\bar{\rho}, \bar{\rho}' \in \mathscr{F}\mathscr{R}^n$ and $x, y \in X$. Suppose $\tilde{f}(\bar{\rho})(x,y) = \tilde{f}(\bar{\rho})(y,x)$ and $\rho_i = \rho'_i$ for all $i \in N\backslash\{j\}$. In addition, suppose either

*Proof.* $\pi_j(x,y) = 0$ and $\pi'_j(x,y) > 0$ or
$\pi_j(y,x) > 0$ and $\pi'_j(y,x) = 0$, $\qquad\qquad\qquad\qquad\qquad\qquad\qquad$ □

where strict reference is of type 3. Then $\pi'(x,y) = \max\{0, \rho'(x,y) - \rho'(y,x)\}$, and

$$\rho'(x,y) - \rho'(y,x) = \sum_{i=1}^{n}(w_i \cdot \rho_i'(x,y) - w_i \cdot \rho_i'(y,x))$$

$$= \sum_{i=1, i \neq j}^{n-1}(w_i \cdot \rho_i(x,y) - w_i \cdot \rho_i(y,x))$$
$$+ w_j \cdot \rho_j'(x,y) - w_j \cdot \rho_j'(y,x)$$

$$= \sum_{i=1}^{n}(w_i \cdot \rho_i(x,y) - w_i \cdot \rho_i(y,x)) - w_j \cdot (\rho_j(x,y) - \rho_j(y,x))$$
$$+ w_j \cdot (\rho_j'(x,y) - \rho_j'(y,x))$$

$$= -w_j \cdot (\rho_j(x,y) - \rho_j(y,x)) + w_j \cdot (\rho_j'(x,y) - \rho_j'(y,x))$$
$$> 0,$$

where the inequality holds if either (1) or (2) hold. Hence, the weighted mean satisfies positive responsiveness with respect to $\pi_{(3)}$. The desired result now follows from the definition of regularity and Proposition 4.36.                                         □

The weighted mean also satsifies the Pareto Condition under specific definitions of strict preference.

**Proposition 4.38.** *Let $\tilde{f}$ be an FPAR as defined in Definition 4.35. Then $\tilde{f}$ satisfies the Pareto Condition with respect to $\pi = \pi_{(1)}$ and $\pi = \pi_{(3)}$.*

*Proof.* Let $x,y \in X$. Let $m_{x,y} = \min_{i \in N}\{\pi_i(x,y)\}$. There is no loss in generality is assuming $m_{x,y} = \pi_1(x,y)$. If $m_{x,y} = 0$, the proof is complete. Suppose otherwise. Then,

$$1 \leq w_1 + \ldots + w_n + w_2(\frac{\pi_2(x,y)}{m_{x,y}} - 1) + \ldots + w_n(\frac{\pi_n(x,y)}{m_{x,y}} - 1)$$

$$= w_1 \cdot \frac{\pi_1(x,y)}{m_{x,y}} + \ldots + w_n \cdot \frac{\pi_n(x,y)}{m_{x,y}}.$$

Thus,

$$\pi_1(x,y) = m_{x,y} \leq w_1 \cdot \pi_1(x,y) + \ldots + w_n \cdot \pi_n(x,y). \tag{4.4}$$

Because $m_{x,y} > 0$, $\pi_i(x,y) > 0$ for all $i \in N$. Thus, $\rho_i(x,y) > \rho_i(y,x)$ for all $i \in N$. Hence,

$$\sum_{i=1}^{n}w_i \cdot \rho_i(x,y) > \sum_{i=1}^{n}w_i \cdot \rho_i(y,x) .$$

Suppose $\pi = \pi_{(1)}$. By (3.3), $\rho_1(x,y) \leq \sum_{i=1}^{n}w_i \cdot \rho_i(x,y)$, or $m_{x,y} \leq \pi(x,y) = \rho(x,y)$. Hence $\tilde{f}$ satisfies the Pareto Condition with respect to $\pi_{(1)}$.

Suppose $\pi = \pi_{(3)}$. Similarly by (3.3),

$$\rho_1(x,y) - \rho_1(y,x) \le \sum_{i=1}^{n} w_i \cdot [\rho_i(x,y) - \rho_i(y,x)] = \sum_{i=1}^{n} w_i \cdot \rho_i(x,y) - \sum_{i=1}^{n} w_i \cdot \rho_i(y,x),$$

or $m_{x,y} \le \rho(x,y) - \rho(y,x) = \pi(x,y)$. Hence $\tilde{f}$ satisfies the Pareto Condition with respect to $\pi_{(3)}$. □

Unlike Positive Responsiveness, the relationship between the weighted mean and the Pareto Condition in Proposition 4.38 cannot be generalized to the case of any regular strict preference relation. The reason for this is the lack of some type of behavioral assumptions on the relationship between the strict and weak preference relationship, such as monotonicity or $\rho = \iota \cup \pi$ for a specified t-conorm $\cup$. The following example presents a case where the weighted mean aggregation rule does not satisfy the Pareto Condition with respect to a regular strict preference rule.

*Example 4.39.* Let $X = \{x,y\}, N = \{1,2\}$ and $\tilde{f}$ be an FPAR as defined in Definition 4.35, where $w_i = \frac{1}{2}$ for all $i \in N$. Suppose the strict preference relation is defined as follows:

$$\pi(x,y) = \begin{cases} .3 & \text{if } \rho(x,y) = .6 \text{ and } \rho(y,x) = .4, \\ 0 & \text{if } \rho(y,x) > \rho(x,y), \\ 1 & \text{otherwise.} \end{cases}$$

It is obvious that $\pi$ is regular. Now consider a profile $\bar{\rho} \in \mathscr{FR}^2$ such that $\rho_1(x,y) = .5$, $\rho_2(x,y) = .7$, and $\rho_1(y,x) = \rho_2(y,x) = .4$. For this profile, the social preference relation is $\rho(x,y) = .6$ and $\rho(y,x) = .4$ because $\tilde{f}$ is the weighted mean. Then the individual and social strict preference relations are as follows:

$$\pi_1(x,y) = 1,$$
$$\pi_2(x,y) = 1,$$
$$\pi(x,y) = .3,$$

and the weighted mean does not satsify the Pareto Condition with respect to $\pi$ although $\pi$ is regular.

Currently, the only Arrowian condition unaccounted for is max-$*$ transitivity. Because the assumption max-$*$ transitivity is unusually general in the fuzzy framework, the weighted mean does not lend itself to developing one single formal argument detailing whether the FPAR satisfies the condition. Nonetheless, we can use the concept of a zero divsor to determine what type of transitivity conditions to consider.

*Example 4.40.* Let $\tilde{f}$ be an FPAR that is defined in 4.35. Let $N = \{1,2\}, X = \{x,y,z\}$ and $w_i = \frac{1}{2}$ for all $i \in N$. Suppose $\bar{\rho} \in \mathscr{FR}^2$ is defined as follows:

$$\rho_1(x,z) = \rho_2(x,z) = 0,$$
$$\rho_1(x,y) = \rho_2(y,z) = 1,$$
$$\rho_1(y,z) = \rho_2(x,y) = 0,$$
$$\rho_i(z,x) = \rho_i(y,x) = \rho_i(z,y) = 1,$$

for all $i \in N$. It is easily verified that $\rho_i$ is max-* transitive under all t-norms using the boundary conditions. Then the social preference relation, $\tilde{f}(\bar{\rho})$, is, by Definition 4.35, as follows:

$$\rho(x,z) = 0,$$
$$\rho(x,y) = \rho(y,z) = .5,$$
$$\rho(z,x) = \rho(y,x) = \rho(z,y) = 1.$$

If the social preference relation is to be max-*, then $\rho(x,z) \geq \rho(x,y) * \rho(y,z)$ for all $x,y,z \in X$. If * has no zero divisors, $\rho(x,y) * \rho(y,z) > 0$. However, $\rho(x,z) = 0 \ngeq \rho(x,y) * \rho(y,z)$, a contradiction. Thus, $\tilde{f}$ cannot be max-* transitive when * has no zero divisors.

Example 4.40 suggests that when considering max-* transitivity conditions for the weighted mean rule, we should consider definitions in which * has a zero divisor. If not, it is obvious then that the weighted mean will not satisfy the fuzzy Arrowian condition of transitivity. However, the converse of this relationship is not necessarily true as shown in the following example.

*Example 4.41.* Suppose $a * b = \begin{cases} \min\{a,b\} & \text{if } a+b > 1, \\ 0 & \text{otherwise.} \end{cases}$

In this case, * is the nilpotent minimum, and * has a zero divisor. Let $X = \{x,y,z\}$ and $N = \{1,2\}$. Suppose $\bar{\rho} = \{\rho_1,\rho_2\} \in \mathscr{F}\mathscr{R}^2$ and is defined as follows:

$$\rho_1(x,y) = .8$$
$$\rho_1(a,b) = .3, \forall (a,b) \in X \times X \setminus \{(x,y)\}, \text{ where } a \neq b$$
$$\rho_2(x,z) = .4; \ \rho_2(y,z) = .8$$
$$\rho_2(a,b) = .5, \forall (a,b) \in X \times X \setminus \{(x,z),(y,z)\}, \text{ where } a \neq b.$$

Suppose $\tilde{f}$ is an FPAR defined in Definition 4.35 and $w_i = \frac{1}{2}$ for all $i \in N$. Then $\tilde{f}(\bar{\rho})(x,z) = .35$, $\tilde{f}(\bar{\rho})(x,y) = .65$, and $\tilde{f}(\bar{\rho})(y,z) = .55$. However, $.35 \ngeq .65 * .55 = .55$. Hence, $\tilde{f}$ is not max-* trasitive when * is the nilpotent minimum.

Given this relationship, we illustrate two transitivity conditions that use a t-norm with zero divisors.

**Proposition 4.42.** *Let $\tilde{f}$ be a fuzzy aggregation ruled defined in Definition 4.35, and let $\bar{\rho} \in \mathscr{F}\mathscr{R}^n$ be max-* transitive. Then $\tilde{f}$ is max-* transitive if, for all $a,b \in [0,1]$, * is defined as follows for all $a,b \in [0,1]$:*

$$(1) \ a * b = \max\{a+b-1,0\}$$

*or*

$$(2) \ a * b = \begin{cases} a & \text{if } b = 1, \\ b & \text{if } a = 1, \\ 0 & \text{otherwise.} \end{cases}$$

*Proof.* (1) Let $x, y, z \in X$. Then by max-$*$ transitivity of $\rho_i$ and the definition of $*$, $\rho_i(x, z) \geq \rho_i(x, y) + \rho_i(y, z) - 1$ for all $i \in N$. By definition of $\tilde{f}$,

$$\sum_{i=1}^{n} w_i \cdot \rho_i(x, z) \geq \sum_{i=1}^{n} w_i \cdot \rho_i(x, y) + \sum_{i=1}^{n} w_i \cdot \rho_i(y, z) - \sum_{i=1}^{n} w_i$$
$$\rho(x, z) \geq \rho(x, y) + \rho(y, z) - 1.$$

(2) Let $x, y, z \in X$. Suppose $\rho(x, y) * \rho(y, z) = 0$. Then the proof is complete. Suppose $\rho(x, y) * \rho(y, z) > 0$. Then there are two cases to consider.

    a. First, suppose $\rho(x, y) = 1$ and $\rho(y, z) > 0$. Then, $\rho_i(x, y) = 1$ for all $i \in N$, by the definition of $\tilde{f}$. Further, by max-$*$ transitivity of $\rho_i$, $\rho_i(x, z) \geq \rho_i(x, y) * \rho_i(y, z)$ and $\rho_i(x, z) \geq \rho_i(y, z)$ for all $i \in N$. Hence, $\sum_{i=1}^{n} w_i \cdot \rho_i(x, z) \geq \sum_{i=1}^{n} w_i \cdot \rho_i(y, z)$. Thus, $\rho(x, y) * \rho(y, z) = \rho(y, z) \leq \rho(x, z)$.

    b. Second, a similar argument can be made for the case when $\rho(y, z) = 1$ and $\rho(x, y) > 0$. Hence, $\rho(x, z) \geq \rho(x, y) * \rho(y, z)$. $\qquad\square$

Proposition 4.42 provides two examples of t-norms under which the weighted mean is max-$*$ transitive. Proposition 4.42(1) uses the Łukasiewicz t-norm, and Proposition 4.42(2) uses the drastic t-norm. Let $H_L, H_D \subset \mathscr{FR}$ be such that $H_L$ and $H_D$ contain all the fuzzy preference relations that are max-$*$ transitive under the Łukasiewicz and drastic t-norm, respectively. We are now able to state two *possibility* results in the fuzzy Arrowian context.

**Theorem 4.43.** *Let strict preference be regular. Then there exists a nondictatorial* $\tilde{f} : H_L^n \to H_L$ *or* $\tilde{f} : H_D^n \to H_D$ *and satisfying IIA-1, Positive Responsiveness and weak Paretianism.*

*Proof.* Let $\tilde{f}$ be an FPAR as defined in 4.35. The result follows from Propositions 4.36, 4.37 and 4.42, and the immediacy of IIA-1 from the definition of the weighted mean.

    By specifying a strict preference relation we can obtain another possibility result that includes an FPAR satsifying the Pareto Condition. $\qquad\square$

**Theorem 4.44.** *Let strict preference be* $\pi_{(1)}$ *or* $\pi_{(3)}$. *Then there exists a nondictatorial* $\tilde{f} : H_L^n \to H_L$ *or* $\tilde{f} : H_D^n \to H_D$ *and satisfying IIA-1, Positive Responsiveness, weak Paretianism and the Pareto Condition.*

*Proof.* Let $\tilde{f}$ be an FPAR as defined in Definition 4.35. The result follows from Theorem 4.43 and Proposition 4.42. $\qquad\square$

The transitivity conditions in Theorems 4.43 and 4.44 are quite restrictive and can be relaxed given another FPAR.

**Definition 4.45.** Define the fuzzy aggregation rule $\tilde{f} : \mathscr{FR}^n \to \mathscr{FR}$ as follows. For all $\bar{\rho} \in \mathscr{FR}^n$, all $x, y \in X$ and all $\tau : \mathscr{FR}^n \to (0, 1)$,

$$\tilde{f}(\bar{\rho})(x,y) = \begin{cases} 1 & \text{if } x = y, \\ 1 & \text{if } \pi_i(x,y) > 0, \forall i \in N, \\ \tau(\bar{\rho}) & \text{otherwise.} \end{cases}$$

In words, Definition 4.45 is similar to a fuzzy Pareto rule, where the social strict preference for one alternative $x$ over another $y$ is positive if every individual strictly prefers $x$ to $y$ and the social strict preference is regular. The FPAR in Definition 4.45 is clearly reflexive, complete, weakly Paretian and IIA-1. It also satisfies IIA-3. To see that Definition 4.45 satisfies IIA-3 consider the following proposition.

**Proposition 4.46.** *Let strict preference be regular. Let $\tilde{f}$ be a fuzzy aggregation rule defined in Definition 4.45. Then $\tilde{f}$ is IIA-3.*

*Proof.* Let $\bar{\rho}, \bar{\rho}' \in \mathscr{FR}^n$ and $x,y \in X$. Suppose $\rho_i]_{\{x,y\}} \sim \rho_i']_{\{x,y\}}$ for all $i \in N$. Then by Proposition 3.13, $\rho_i(x,y) > \rho_i(y,x)$ if and only if $\rho_i'(x,y) > \rho_i'(x,y)$ for all $i \in N$. By the definition of $\tilde{f}$, $\tilde{f}(\bar{\rho})(x,y) = 1$ if and only if $\tilde{f}(\bar{\rho}')(x,y) = 1$, and $\tilde{f}(\bar{\rho})(y,x) = \tau(\bar{\rho})$ if and only if $\tilde{f}(\bar{\rho}')(y,x) = \tau(\bar{\rho})$. Hence, $\tilde{f}(\bar{\rho})]_{\{x,y\}} \sim \tilde{f}(\bar{\rho}')]_{\{x,y\}}$.                                          $\square$

To see that see when $\tilde{f}(\bar{\rho})$ in Definition 4.45 is max-$*$ transitive, we use a series of propositions that first consider max-min transitivity and then generalize to an arbitrary t-norm.

**Proposition 4.47.** *Let $\pi = \pi_{(1)}$ and $\rho \in \mathscr{FR}$. If $\rho$ is max-min transitive, then $\pi$ is max-min transitive.*

*Proof.* Let $\rho \in \mathscr{FR}$ be such that $\rho$ is max-min transitive, i.e.

$$\rho(x,z) \geq \min\{\rho(x,y), \rho(y,z)\}$$

for all $x,y,z \in X$. This proof will show that

$$\pi(x,z) \geq \min\{\pi(x,y), \pi(y,z)\}.$$

To do so, suppose contrary. Then there exists an $x,y,z \in Z$ such that

$$\rho(x,z) \geq \min\{\rho(x,y), \rho(y,z)\} \tag{4.5}$$

and

$$\pi(x,z) < \min\{\pi(x,y), \pi(y,z)\}. \tag{4.6}$$

Then

$$0 < \pi(x,y) = \rho(x,y) > \rho(y,x) \tag{4.7}$$

and

$$0 < \pi(y,z) = \rho(y,z) > \rho(z,y). \tag{4.8}$$

By Eqs. (4.7) and (4.8), $\rho(x,y) > 0$ and $\rho(y,z) > 0$, which implies $\rho(x,z) > 0$ by Eq. (4.5). Suppose $\pi(x,z) > 0$. Then by definition of $\pi_{(1)}$, $\pi(x,z) = \rho(x,z) \geq \min\{\rho(x,y), \rho(y,z)\} = \min\{\pi(x,y), \pi(y,z)\}$, where the latter equality holds by Eqs. (4.7) and (4.5). Since this contradicts Eq. (4.6), $\pi(x,z) = 0$. Hence,

$$\rho(z,x) \geq \rho(x,z). \tag{4.9}$$

There are now two cases to consider.

*Proof.* Suppose $\min\{\rho(x,y), \rho(y,z)\} = \rho(x,y)$. Then $\rho(y,z) \geq \rho(x,y)$. Hence, by transitivity,

$$\rho(x,z) \geq \rho(x,y). \tag{4.10}$$

By transitivity, $\rho(y,x) \geq \min\{\rho(y,z), \rho(z,x)\}$. Because $\rho(y,z) \geq \rho(x,y)$ and Eq. (3.8), $\min\{\rho(y,z), \rho(z,x)\} \geq \min\{\rho(x,y), \rho(x,z)\} = \rho(x,y)$. Then $\rho(y,x) \geq \rho(x,y)$; however, this contradicts Eq. (4.7).

Suppose $\min\{\rho(x,y), \rho(y,z)\} = \rho(y,z)$, which implies $\rho(x,y) \geq \rho(y,z)$. Then by transitivity,

$$\rho(x,z) \geq \rho(y,z) \tag{4.11}$$

By transitivity, $\rho(z,y) \geq \min\{\rho(z,x), \rho(x,y)\}$; and $\rho(z,y) \geq \rho(z,x)$, or $\rho(z,y) \geq \rho(x,y)$. If $\rho(z,y) \geq \rho(z,x)$, then $\rho(z,y) \geq \rho(z,x) \geq \rho(x,z) \geq \rho(y,z)$ by Eqs. (4.9) and (4.11). However, this contradicts Eq. (4.8). If $\rho(z,y) \geq \rho(x,y)$, then $\rho(z,y) \geq \rho(x,y) \geq \rho(y,z)$ by the assumption of $\rho(x,y) \geq \rho(y,z)$. However, this also contradicts Eq. (4.8). Thus, $\pi(x,z) \geq \min\{\pi(x,y), \pi(y,z)\}$, and $\pi$ is also max-min transitive. $\square$

$\square$

Proposition 4.47 demonstrates that when strict preference is of type one, max-min transitivity of an FWPR $\rho$ implies max-min transitivity of the strict preference relation derived from $\rho$. Like Proposition 4.38 and Example 4.39, this relationship between the max-min transitivity of $\rho$ and $\pi_{(1)}$ cannot be generalized to the case of all regular strict relations because the ordinal concept of strict preference is insufficient for the cardinal concept of max-min transitivity. Nonetheless, assuming that strict preference is of type one allows us to show the max-min transitivity of individual preference relations and obtain the following result.

**Proposition 4.48.** *Let $\pi = \pi_{(1)}$, $\bar{\rho} \in \mathscr{F}\mathscr{R}^n$ and $\tilde{f}$ be an FPAR defined in Definition (4.45). Suppose $\rho_i$ is max-min transitive for all $i \in N$. Then $\tilde{f}$ is max-min transitive.*

*Proof.* Let $x,y,z \in X$. Suppose $\bar{\rho} \in \mathscr{F}\mathscr{R}$ be such that $\rho_i$ is max-min transitive for all $i \in N$. If $\min\{\tilde{f}(\bar{\rho})(x,y), \tilde{f}(\bar{\rho})(y,z)\} = \tau(\bar{\rho})$, then the proof is complete. Suppose the contrary. Then $\min\{\tilde{f}(\bar{\rho})(x,y), \tilde{f}(\bar{\rho})(y,z)\} = 1$. By the definition of $\tilde{f}$, $\pi_i(x,y) > 0$ and $\pi_i(y,z) > 0$ for all $i \in N$. By Proposition 4.47, we have $\pi_i(x,z) \geq \min\{\pi_i(x,y), \pi_i(y,z)\}$ for all $i \in N$. Thus, $\pi_i(x,z) > 0$ for all $i \in N$ and $\tilde{f}(\bar{\rho})(x,z) = 1$. Hence, $\tilde{f}$ is max-min transitive. $\square$

To see when $\tilde{f}(\bar{\rho})$ in Definition 4.45 is max-$*$ transitive under any specified t-norm, consider the following proposition, which uses the boundary condition to prove the result.

**Proposition 4.49.** *Let* $\rho \in \mathscr{FR}$ *be such that* $\rho$ *is max-min transitive. Let* $*$ *be an arbitrary t-norm. Then* $\rho$ *is max-$*$ transitive.*

*Proof.* For any $a, b \in [0, 1]$, $a * b \leq a * 1 = a$ and $a * b \leq 1 * b = b$ by the boundary condition of $*$. Because $a * b \leq a$ and $a * b \leq b$, $a * b \leq \min\{a, b\}$. Let $x, y, z \in X$. By transitivity of $\rho$, $\rho(x, z) \geq \min\{\rho(x, y), \rho(y, z)\} \geq \rho(x, y) * \rho(y, z)$. Hence, $\rho$ is max-$*$ transitive. $\qquad\qquad\square$

We can now state another possibility result with less restrictive transitivity conditions.

**Theorem 4.50.** *Let* $\pi = \pi_{(1)}$. *Then there exists a nondictatorial* $\tilde{f} : \mathscr{FR}^{*n} \rightarrow \mathscr{FR}^*$ *satisfying IIA-1, IIA-3, weak Paretianism and the Pareto Condition.*

*Proof.* Let $\tilde{f}$ be defined by Definition 4.45. Clearly, $\tilde{f}$ is reflexive and complete, and it satisfies IIA-1, weak Paretianism and the Pareto Condition. By Proposition 4.46, $\tilde{f}$ is IIA-3, and Proposition 4.49 generalizes Propositions 4.47 and 4.48. Thus, $\tilde{f}$ is max-$*$ transitive. $\qquad\qquad\square$

Theorem 4.50 achieves a more general possibility result, but using Definition (4.45) has two important consequences. First, individual and social preferences must be max-$*$ transitive under the same t-norm definition. For example, given some $\bar{\rho} \in \mathscr{FR}^n$, it is impossible to guarantee the max-min transitivity of $\tilde{f}(\bar{\rho})$ when $\bar{\rho}$ is only max-$*$ transitive under the drastic t-norm. Second, as illustrated by Dutta (1987), adding the requirement of positive responsiveness to Theorem 4.50 will void the possibility results. This occurs because, when $\tilde{f}(\bar{\rho})(x, y) = \tilde{f}(\bar{\rho})(y, x)$ for some $x, y \in X$, an individual $i \in N$ switching from complete indifference between $x$ and $y$ ($\rho_i(x, y) = \rho_i(y, x)$) to some strict preference between the two ($\rho_i(x, y) \neq \rho_i(y, x)$) does not necessarily imply that the social preference will exhibit strict preference as well ($\tilde{f}(\bar{\rho})(x, y) \neq \tilde{f}(\bar{\rho})(y, x)$).

Even with these two considerations, the importance of Theorems 4.43, 4.44, and 4.50 remains: the fuzzy Arrowian framework allows for the nondictatorial aggregation of fuzzy preferences in a manner that satisfies normative democratic criteria. Further, as Theorem 4.31 demonstrates, the concept of a neutral FPAR can be used to derive an aggregation rule that is unique and not necessarily dictatorial when, in the exact case, neutrality implies dictatorship. Not only do the results in the fuzzy preference framework reveal substantive conclusions that are distinct from previous approaches using exact preferences, but also they suggest that the traditional, negative results of social choice theory are unsubstantiated when groups possess fuzzy preferences.

## 4.3 Empirical Application II: The Spatial Model and Fuzzy Aggregation

Section (4.2) discussed the difficulty that arises when using FWPRs in empirical analyses. Most often, researchers will not have the necessary data to create individual FWPRs for every member in a group of political actors. However, fuzzy numbers can be used to represent the degree to which an actor views an alternative as ideal i.e., the $\sigma$ function, and an FWPR can be estimated using such a function. This section further extends the analysis in Section (4.2) by illustrating how a fuzzy preference aggregation rule can be used to predict policy decisons of a group of actors.

In the spatial model, alternatives can be represented by $k$-dimensional Euclidean space or $\mathbb{R}^k$. When $k = 1$, $\sigma$ is identical to the fuzzy numbers presented in the previous empirical example, where, for some $x \in X$, $\sigma(x)$ denotes the degree to which $x$ is ideal. In this case, $\sigma_i : \mathbb{R}^1 \to [0, 1]$ for all $i \in N$. It is often assumed that $\sigma$ is *normal*, which requires there exists $x \in X$ such that $\sigma(x) = 1$. In words, normality ensures that every actor views at least one alternative as ideal. Let $\mathscr{FN}(X)$ denote all the fuzy subsets of $X$ such that the fuzzy subset is normal. When $N$ is the set of actors, it is assumed each actor prossesses a preference function, *preference function profile* can be written as $\bar{\sigma} = (\sigma_1, \ldots, \sigma_n)$.

**Table 4.1** Sigma Values of Four Alternatives

|           | $\sigma_1(\cdot)$ | $\sigma_2(\cdot)$ | $\sigma_3(\cdot)$ |
|-----------|-------------------|-------------------|-------------------|
| $w = .1$  | 0                 | 0                 | 0                 |
| $x = .5$  | 0.33              | 1.0               | 0                 |
| $y = .57$ | 0.1               | 1.0               | 0.1               |
| $z = .68$ | 0                 | 0.2               | 0.65              |

Let $N = \{1, 2, 3\}$ and let $X = [0, 1] \subset \mathbb{R}^1$. Figure 4.1 presents a fairly traditional profile of preference functions over the set of alternatives where no actor possesses more than three areas of discrete indifference. For example, player 1 is indifferent between all alternatives in the intervals $[0, .1]$ and $[.6, 1]$ ($\sigma_1(x) = 0$) and between all alternatives in the interval $[.35, .5]$ ($\sigma_1(x) = .33$). The fuzzy numbers presented in Figure 4.1 are sufficient to characterize the degree to which any alternative in $X$ is ideal for all three actors. Table 4.1 provides the sigma values of four alternatives in $X$. Here, $w \in X$ is outside the support of ideal alternatives for all three players, and $y \in X$ is in the support of ideal alternatives for all three players. In addition, $y$ is in the core of player two's set of ideal alternatives.

Given the fuzzy preference functions in Figure 4.1, we can create preference relations based on the degree to which each alternative is ideal. Section (4.2) gave two examples of such procedures. First, for all $x, y \in X$ and all $\sigma \in \mathscr{FN}(X)$,

$$\rho_{(G)}(x, y) = \vee\{t \in [0, 1] \mid \sigma(y) * t \le \sigma(x)\},$$

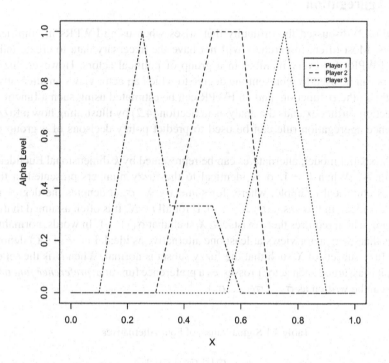

**Fig. 4.1** Example of a Three Player Fuzzy Spatial Model

which can be simplified to the following if $* = \min$:

$$\rho_{(G)}(x,y) = \begin{cases} 1 & \text{if } \sigma(x) \geq \sigma(y), \\ \sigma(x) & \text{otherwise.} \end{cases}$$

We have already shown that $\rho_{(G)}$ is reflexive, strongly connected and max-min transitive. Table 4.2 presents the preference profile $\bar{\rho}_{(G)}$ over the four alternatives selected in Table 4.1.

**Table 4.2** Inferred FWPRs Using $\rho_{(G)}$

| $i = 1$ | $w$ | $x$ | $y$ | $z$ | $i = 2$ | $w$ | $x$ | $y$ | $z$ | $i = 3$ | $w$ | $x$ | $y$ | $z$ |
|---|---|---|---|---|---|---|---|---|---|---|---|---|---|---|
| $w$ | 1 | 0 | 0 | 1 | $w$ | 1 | 0 | 0 | 0 | $w$ | 1 | 1 | 0 | 0 |
| $x$ | 1 | 1 | 1 | 1 | $x$ | 1 | 1 | 1 | 1 | $x$ | 1 | 1 | 0 | 0 |
| $y$ | 1 | .1 | 1 | 1 | $y$ | 1 | 1 | 1 | 1 | $y$ | 1 | 1 | 1 | .1 |
| $z$ | 1 | 0 | 0 | 1 | $z$ | 1 | .2 | .2 | 1 | $z$ | 1 | 1 | 1 | 1 |

A second procudure used for inferring an FWPR from a preference function is

$$
\rho_{(M)}(x,y) = \begin{cases} 1 & \text{if } x = y, \\ (\sigma(x) - \sigma(y) + c) \wedge 1 & \text{if } \sigma(x) \geq \sigma(y), \\ 1 - [\sigma(y) - \sigma(x) + 1 - c) \wedge 1] & \text{otherwise.} \end{cases}
$$

where $c \in [0,1]$ for all $x, y \in X$ and $\sigma \in \mathscr{FN}(X)$. It is obvious that $\rho_{(M)}$ is reflexive and complete when $c > 0$. We also know $\rho_{(M)}$ is weakly transitive (see Proposition (4.20)). If we set $c$ to a specific value, we can infer another preference profile as well. Let $c = .5$; Table 4.3 illustrates the preference profile $\bar{\rho}_{(M)}$ in this case. In contrast to $\bar{\rho}_{(G)}$, the image of $\bar{\rho}_{(M)}$ contains more elements for each actor than image of $\bar{\rho}_{(G)}$.

**Table 4.3** Inferred FWPRs Using $\rho_{(M)}$ when $c = .5$

| $i=1$ | $w$ | $x$ | $y$ | $z$ | $i=2$ | $w$ | $x$ | $y$ | $z$ | $i=3$ | $w$ | $x$ | $y$ | $z$ |
|---|---|---|---|---|---|---|---|---|---|---|---|---|---|---|
| $w$ | 1 | .17 | .4 | .5 | $w$ | 1 | 0 | 0 | .3 | $w$ | 1 | .5 | .4 | 0 |
| $x$ | .83 | 1 | .73 | .83 | $x$ | 1 | 1 | .5 | 1 | $x$ | .5 | 1 | .4 | 0 |
| $y$ | .6 | .27 | 1 | .6 | $y$ | 1 | .5 | 1 | 1 | $y$ | .6 | .6 | 1 | 0 |
| $z$ | .5 | .17 | .4 | 1 | $z$ | .7 | 0 | 0 | 1 | $z$ | 1 | 1 | 1 | 1 |

We can now apply an FPAR $\tilde{f}$ to the preference profiles $\bar{\rho}_{(G)}$ and $\bar{\rho}_{(M)}$. When $\tilde{f}$ is the weighted mean rule from Definition (4.33), assume $w_i = \frac{1}{3}$ for all $i \in N$. Then Table 4.4 illustrates $\tilde{f}(\bar{\rho}_{(G)})$ and $\tilde{f}(\bar{\rho}_{(M)})$ over the four alternatives $\{w, x, y, z\}$. When $\tilde{f}$ is the fuzzy Pareto rule from Definition (4.45), assume $\tau(\bar{\rho}_{(G)}) = (\bar{\rho}_{(M)}) = .5$, and Table 4.5 reports the results of the fuzzy Pareto rule over the same four alternatives.

**Table 4.4** The Weighted Mean Rule Using $\bar{\rho}_{(G)}$ and $\bar{\rho}_{(M)}$ when $w_i = \frac{1}{3}$

| $\tilde{f}(\bar{\rho}_{(G)})$ | $w$ | $x$ | $y$ | $z$ | $\tilde{f}(\bar{\rho}_{(M)})$ | $w$ | $x$ | $y$ | $z$ |
|---|---|---|---|---|---|---|---|---|---|
| $w$ | 1 | .33 | 0 | .33 | $w$ | 1 | .22 | .27 | .27 |
| $x$ | 1 | 1 | .67 | .67 | $x$ | .78 | 1 | .54 | .61 |
| $y$ | 1 | .7 | 1 | .7 | $y$ | .73 | .46 | 1 | .53 |
| $z$ | 1 | .4 | .4 | 1 | $z$ | .73 | .39 | .47 | 1 |

Tables 4.4 and 4.5 reveal an important distinction between the weighted mean and fuzzy Pareto rule. The weighted mean is more susceptible to the specific procedure chosen to infer fuzzy preference relations than the fuzzy Pareto rule. While the fuzzy Pareto rule returns two identical social preference relations regardless of how the inidividual preference relations were created, the weighted mean exhibits significant differences between the social preference relation from $\bar{\rho}_{(G)}$ and the one from $\bar{\rho}_{(M)}$.

We can also calculate the maximal sets from the four newly aggregated social preference relations. In Section (4.2), the fuzzy maximal set is defined as follows: for all $x \in X$,

**Table 4.5** The Fuzzy Pareto Rule Using $\bar{\rho}_{(G)}$ and $\bar{\rho}_{(M)}$ when $\tau(\bar{\rho}) = .5$

| $\tilde{f}(\bar{\rho}_{(G)})$ | w | x | y | z | $\tilde{f}(\bar{\rho}_{(M)})$ | w | x | y | z |
|---|---|---|---|---|---|---|---|---|---|
| w | 1 | .5 | .5 | .5 | w | 1 | .5 | .5 | .5 |
| x | .5 | 1 | .5 | .5 | x | .5 | 1 | .5 | .5 |
| y | 1 | .5 | 1 | .5 | y | 1 | .5 | 1 | .5 |
| z | .5 | .5 | .5 | 1 | z | .5 | .5 | .5 | 1 |

$$M(\rho,\mu)(x) = \mu(x) * (\circledast(\vee\{t \in [0,1] \mid \mu(w) * \rho(w,x) * t \leq \rho(x,w),\ \forall w \in \text{Supp}(\mu)\})),$$

where $\mu \in \mathscr{F}(X)$. $M(\rho,\mu)$ can be simplified by assuming that $\mu(x) = 1$ for all $x \in X$ and $* = \circledast = \min$. The first assumption acknowleges that all alternatives are fully possible. The second merely specifies a t-nrom. With these two assumptions, the maximal set can be written as

$$M(\rho,X)(x) = (\wedge(\vee\{t \in [0,1] \mid \rho(w,x) \wedge t \leq \rho(x,w),\ \forall w \in X\})).$$

As before, $M(\rho,X)(x)$ signifies the degree to which $x \in X$ is a maximal alternative given the FWPR $\rho$. Let $S = \{w,x,y,z\} \subseteq X$. Then Table 4.6 shows the final calculations for $M(\tilde{f}(\bar{\rho}_{(G)}),S)$ and $M(\tilde{f}(\bar{\rho}_{(M)}),S)$ where $\tilde{f}$ is either the weighted mean rule or the fuzzy Pareto rule. Furthermore, Figure 4.2 plots the four maximal sets over the entire set of alternatives. As before, the fuzzy Pareto rule returns identical results regardless of the specific profile, and the core of the fuzzy Pareto's maximal set is the support all three players' preference functions. In these cases (Figures 4.2(c) and 4.2(d)), the researcher could predict almost any alternative to be selected by the group of players. In contrast, the core of the weighted mean rule differs from $\bar{\rho}_{(G)}$ and $\bar{\rho}_{(M)}$, which lead to different predictions about what alternative would be selected. In Figure 4.2(a), the core of $\tilde{f}(\bar{\rho}_{(G)})$ is the alternative where all three players' fuzzy preference functions intersect at the maximum degree. In Figure 4.2(b), however, the core of $\tilde{f}(\bar{\rho}_{(M)})$ is the maximum intersection between players 2 and 3, which is the maximum intersection for any two players in the example. Hence, $\tilde{f}(\bar{\rho}_{(G)})$ appears to be more collegial and consensus-driven than $\tilde{f}(\bar{\rho}_{(M)})$ when $\tilde{f}$ is the wieghted mean rule.

**Table 4.6** Results for $M(\tilde{f}(\bar{\rho}_{(G)}),S)$ and $M(\tilde{f}(\bar{\rho}_{(M)}),S)$

|   | Weighted Mean | | Fuzzy Pareto | |
|---|---|---|---|---|
|   | $M(\tilde{f}(\bar{\rho}_{(G)}),S)$ | $M(\tilde{f}(\bar{\rho}_{(M)}),S)$ | $M(\tilde{f}(\bar{\rho}_{(G)}),S)$ | $M(\tilde{f}(\bar{\rho}_{(M)}),S)$ |
| w | 0 | .22 | .5 | .5 |
| x | 0.67 | 1.0 | 1 | 1 |
| y | 1.0 | .46 | 1 | 1 |
| z | 0.4 | .39 | 1 | 1 |

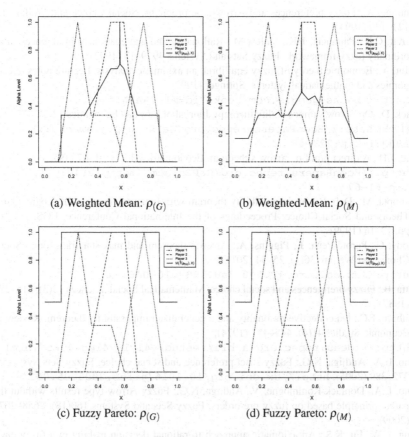

(a) Weighted Mean: $\rho_{(G)}$      (b) Weighted-Mean: $\rho_{(M)}$

(c) Fuzzy Pareto: $\rho_{(G)}$      (d) Fuzzy Pareto: $\rho_{(M)}$

**Fig. 4.2** Maximal Set

Using either FPAR, the procedures described in the definitions of $\rho_{(G)}$ and $\rho_{(M)}$ allow for easy estimation of individual FWPRs without requiring researchers to gather data concerning the degree to which an individual prefers every alternative over every other alternative. When aggregating the individual preference relations, the researcher can choose any number of FPARs, and the maximal set can clearly relate the social preference relation back to individual preference functions. In the example presented in this section, the weighted mean rule generates a maximal set with one alternative in its core while the fuzzy Pareto rule results in a maximal set whose core spans the support of the individual preference function.

## References

Arrow, K.: Social Choice and Individual Values. Wiley, New York (1951)
Austen-Smith, D., Banks, J.S.: Positive Political Theory I: Collective Preference. University of Michigan Press, Ann Arbor (1999)

Banerjee, A.: Fuzzy preferences and Arrow-type problems. Social Choice and Welfare 11, 121–130 (1994)

Barrett, C.R., Pattanaik, P.K., Salles, M.: Rationality and aggregation of preferences in an ordinally fuzzy framework. Fuzzy Sets and Systems 49, 9–13 (1992)

Billot, A.: Economic theory of fuzzy equilibria: an axiomatic analysis. Lecture notes in economics and mathematical systems. Springer (1992),
http://books.google.com/books?id=ml-7AAAAIAAJ

Black, D.: On Arrow's impossibility theorem. Journal of Law and Economics 12(2), 227–248 (1969), http://EconPapers.repec.org/RePEc:ucp:jlawec:v:12:y: 1969:i:2:p:227-48

Blau, J.H.: A direct proof of Arrow's theorem. Econometrica 40(1), 61–67 (1972),
http://EconPapers.repec.org/RePEc:ecm:emetrp:v:40:y:1972:i: 1:p:61-67

Dasgupta, M., Deb, R.: An impossibility theorem with fuzzy preferences. In: Logic, Game Theory and Social Choice: Proceedings of the International Conference, LGS, vol. 99, pp. 13–16 (1999)

Duddy, C., Perote-Peña, J., Piggins, A.: Arrow's theorem and max-star transitivity. Social Choice and Welfare 36(1), 25–34 (2011),
http://dx.doi.org/10.1007/s00355-010-0461-x

Dutta, B.: Fuzzy preferences and social choice. Mathematical Social Sciences 13(3), 215–229 (1987)

Fishburn, P.C.: On collective rationality and a generalized impossibility theorem. Review of Economic Studies 41(4), 445–457 (1974),
http://ideas.repec.org/a/bla/restud/v41y1974i4p445-57.html

Fono, L.A., Andjiga, N.G.: Fuzzy strict preference and social choice. Fuzzy Sets Syst. 155, 372–389 (2005), http://dx.doi.org/10.1016/j.fss.2005.05.001

Fono, L.A., Donfack-Kommogne, V., Andjiga, N.G.: Fuzzy Arrow-type results without the pareto principle based on fuzzy pre-orders. Fuzzy Sets and Systems 160(18), 2658–2672 (2009)

Fung, L.W., Fu, K.S.: An axiomatic approach to rational decision making in a fuzzy environment. In: Zadah, L.A., Fu, K.S., Tanaka, K., Shimura, M. (eds.) Fuzzy Sets and Their Applications to Cognitive and Decision Processes, ch. 10, pp. 227–256. Academic Publishers, New York (1975)

García-Lapresta, J.L., Llamazares, B.: Aggregation of fuzzy preferences: Some rules of the mean. Social Choice and Welfare 17(4), 673–690 (2000),
http://dx.doi.org/10.1007/s003550000048

Inada, K.I.: Alternative incompatible conditions for a social welfare function. Econometrica 23(4), 396–399 (1955)

Little, I.M.D.: Social choice and individual values. Journal of Political Economy 60, 422 (1952)

May, K.O.: A set of independent necessary and sufficient conditions for simple majority decision. Econometrica 20(4), 680–684 (1952),
http://dx.doi.org/10.2307/1907651

Mordeson, J.N., Clark, T.D.: Fuzzy Arrow's theorem. New Mathematics and Natural Computation 05(02), 371–383 (2009),
http://www.worldscientific.com/doi/abs/10.1142/ S1793005709001362

Ovchinnikov, S.: Social choice and łukasiewicz logic. Fuzzy Sets and Systems 43(3), 275–289 (1991), Aggregation and Best Choices of Imprecise Opinions

Perote-Peña, J., Piggins, A.: Strategy-proof fuzzy aggregation rules. Journal of Mathematical Economics 43(5), 564–580 (2007)

Richardson, G.: The structure of fuzzy preferences: Social choice implications. Social Choice and Welfare 15, 359–369 (1998)

Salles, M.: Fuzzy utility. In: Handbook of Utility Theory: vol. 1: Principles, p. 321. Springer (1998)

Sen, A.K.: Liberty, unanimity and rights. Economica 43(171), 217–245 (1976), http://dx.doi.org/10.2307/2553122

Skala, H.: Arrow's impossibility theorem: Some new aspects. In: Gottinger, H., Leinfellner, W. (eds.) Decision Theory and Social Ethics, Theory and Decision Library, vol. 17, pp. 215–225. Springer, Netherlands (1978), http://dx.doi.org/10.1007/978-94-009-9838-4_11

Ubeda, L.: Neutrality in arrow and other impossibility theorems. Economic Theory 23(1), 195–204 (2004), http://dx.doi.org/10.1007/s00199-002-0353-0

# Chapter 5
# Characteristics of Strategy-Proof Fuzzy Social Choice

**Abstract.** The Gibbard-Satterthwaite theorem, which states that a social choice function over three or more alternatives that does not incentivize individuals to misrepresent their sincere preferences must be dictatorial, under a fuzzy framework requires the specification of a fuzzy social choice function that selects some type of outcome. This chapter considers the strategic manipulation of fuzzy social choice functions where both individuals and groups can choose alternatives to various degrees and shows that with minimal assumptions on individual preferences, strategy-proof fuzzy social choice functions satisfy fuzzy versions of peak-only, weak Paretianism, and monotonicity. Furthermore, strategy-proofness is necessary and sufficient for the augmented median voter rule which is considered in chapter 6.

## Introduction

The Gibbard-Satterthwaite Theorem (G-S in what follows) states that a social choice function over three or more alternatives that does not incentivize individuals to misrepresent their sincere preferences must be dictatorial (Gibbard, 1973; Satterthwaite, 1975). It follows that voters in collective choice institutions will manipulate the voting procedure to obtain a more preferred social outcome by reporting insincere preferences. Hence, designers of democratic institutions must accept that the system's rules will encourage dishonesty in the voting population. Social choice scholars have tried to avoid this conclusion by relaxing several of the original assumptions of the G-S Theorem. One approach restricts the domain of individual preferences to single-peaked profiles. Under single-peaked profiles there exists a strict ordering of all possible alternatives, individuals possess a single ideal alternative, and strict preference decreases monotonically in both directions from the ideal point. Under this assumption, the augmented median voter rule emerges as a non-manipulable and non-dictatorial choice function Barberá (2001); Ching (1997); Moulin (1980); Sprumont (1991).

While some scholars Dryzek and List (2003); Mackie (2003) hold that this restriction voids the normatively negative results of the G-S Theorem, Penn, Patty, and Gailmard and Patty (2011) extend the G-S results to a general case by demonstrating that even though individuals possess single-peaked preferences, there exist opportunities to manipulate the social choice when individuals report insincere preferences that violate the natural ordering of the alternatives. The crux of their argument rests on the empirical observation that no real-world voting rule actually has a ballot restriction that forces individuals to submit single-peaked preferences to the social choice function; hence, individuals will submit insincere, non-single-peaked preferences when it manipulates the social choice. Under these assumptions, a strategy-proof rule must be dictatorial.

What has remained absent from the discussion is the effect of fuzzy preferences and fuzzy social choice on the conclusion of the G-S Theorem. In the fuzzy framework, individuals can prefer one alternative over another to a certain degree instead of only possessing strict preference or indifference between the two Banerjee (1995); Orlovsky (1978). The addition of fuzzy preferences then requires the specification of a fuzzy social choice function that selects some type of outcome. Past efforts, that have explored situations where actors have fuzzy preferences but, as a group, must select one alternative unequivocally, have only confirmed the G-S conclusion Abdelaziz et al. (2008); CÃŽrte-Real (2007); Tang (1994). The strategic manipulation of truly fuzzy social choice functions, where society chooses alternatives to varying degrees, has yet to be considered.

This chapter addresses this lacuna in the manipulation literature. We consider the strategic manipulation of fuzzy social choice functions where both individuals and groups can choose alternatives to various degrees. We demonstrate that with very minimal assumptions on individual preferences, strategy-proof fuzzy social choice functions satisfy fuzzy versions of peak-only, weak Paretianism, and monotonicity. Moreover, strategy-proofness is necessary and sufficient for the augmented median voter rule. We also illustrate the implications of this framework in the spatial model.

The results of this chapter are relevant to the strategic manipulation literature, which remains divided as to whether choice functions can be both non-manipulable and non-dictatorial when restricting individual preferences to a single-peaked domain Mackie (2003); Penn et al. (2011). They suggest that social choice can be both strategy-proof and non-dictatorial if alternatives are chosen to various degrees. Nonetheless, the findings require that an individual's choice intensity for one alternative is independent of her choice intensity for all other alternatives. The chapter proceeds as follows. Section one reviews the literature discussing fuzzy manipulation. Section two presents the main concepts and definitions of fuzzy social choice. Section three details the main findings of the chapter. Section four offers a discussion and a critique of the social choice model in the context of the spatial model, and section five summarizes the chapter's conclusions.

## 5.1 Fuzzy Choice and Manipulation

Most efforts incorporating fuzzy mathematics into social choice functions start with a fuzzy preference relation, which is a function $\rho : X \times X \to [0, 1]$, where $X$ is the set of alternatives. In words, $\rho(x,y)$ refers to the degree to which $x$ is at least as good as $y$. If $\rho(x,y) = 1$, then alternative $x$ is said to be definitely at least as good as alternative $y$; if $\rho(x,y) = 0$, then $x$ is said to be definitely not at least as good as $y$. When $\rho(x,y) \in (0,1)$, the weak preference for $x$ over $y$ is said to be vague or ambiguous.[1] For a set of $n$ actors, $N$, previous definitions of fuzzy choice functions (Definition 5.20) maps an $n$-tuple of fuzzy preference relations to one alternative in $X$ Abdelaziz et al. (2008); CÃŽrte-Real (2007); Orlovsky (1978); Tang (1994).

Because a fuzzy preference relation is not directly comparable to a subset of alternatives, scholars have considered various mechanisms to aggregate individual preference relations into a social choice (see Chapter 3 for a more thorough discussion of these mechanisms). Initial studies assumed that individuals possess fuzzy preferences but must make "crisp" individual choices over the set of alternatives, and the choice function associates a set of alternatives to these crisp choices CÃŽrte-Real (2007); Orlovsky (1978). Such situations arise when actors, who possess fuzzy preference relations, must vote "yes" or "no" for an amendment or select only one candidate among many. Later research examined choice functions that aggregate a collection of individual fuzzy preference relations into a social preference relation and then associate an alternative with the fuzzy social preference Abdelaziz et al. (2008). These models depict situations where sets of actors, such as political parties or groups of states, do not need to produce a transitive ranking of alternatives and, instead, produce a set of vague opinions about policy, such as a platform or a treaty. The following example illustrates the difference between the two approaches.

*Example 5.1.* Let $X = \{a,b\}$ and $N = \{1,2,3\}$. Suppose $\rho_1(a,b) = .4$, $\rho_2(a,b) = .4$, and $\rho_3(a,b) = .9$. In words, $\rho_i$ is the fuzzy individual preference relation associated with $i \in N$. Furthermore, suppose there is reciprocity in preferences, and accordingly, $\rho_i(b,a) = 1 - \rho_i(a,b)$. Under this set up, three players have preferences over two alternatives and must decide as a group what alternative to select.

*Orlovsky Rule.* The Orlovsky (1978) rule demonstrates the first approach to fuzzy choice, where each actor must make a crisp decision. The Orlovsky rule—or a variation of it—is an individual choice function (*IC*) that, given a specified $n$-tuple of preference relations, maps a set of two alternatives into the set $\{0,1\}$, formally, $IC_i : X^2 \rightarrow \{0,1\}$ for $i = 1,\ldots,n$. More specifically, let $x,y \in X$ and the Orlovsky rule be defined as follows:

$$IC_i(\rho_i)(x,y) = \begin{cases} 1 & \text{if } \rho_i(x,y) > \rho_i(y,x), \\ 0 & \text{else.} \end{cases}$$

In words, individual $i$ votes for or chooses $x$ over $y$ if and only if $IC_i(\rho_i)(x,y) = 1$ and $IC_i(\rho_i)(y,x) = 0$. Considering the above example,

---

[1] For a more thorough review of fuzzy preferences and how they relate to tradition preferences see Orlovsky (1978), Dutta (1987), Richardson (1998) and Llamazares (2005).

$$IC(a,b) = = (IC_1(\rho_1)(a,b), IC_2(\rho_2)(a,b), IC_3(\rho_3)(a,b))$$
$$= (0,0,1)$$
$$IC(b,a) = IC(b,a) = (IC_1(\rho_1)(b,a), IC_2(\rho_2)(b,a), IC_3(\rho_3)(b,a))$$
$$= (1,1,0),$$

because actors 1 and 2 choose $b$ and actor 3 chooses $a$. We can apply various voting rules, but under majority rule $b$ is the outcome.

*Example 5.2. Mean Aggregation Rule.* The mean aggregation rule is a fuzzy aggregation rule and demonstrates the second type of fuzzy choice function, where actors need not make crisp decisions because fuzzy individual preference relations are aggregated into a social one, denoted $\rho_S$. The mean aggregation rule is defined as follows:

$$\rho_S(x,y) = \frac{1}{n} \sum_{i=1}^{n} \rho_i(x,y).$$

Using the mean aggregation rule, we can specify the social fuzzy preference relation, which is $\rho_S(a,b) = .567$ and $\rho_S(b,a) = .433$. When we use the Orlovsky rule on $\rho_S$, the social choice becomes $a$ because $\rho_S(a,b) > \rho_S(b,a)$.

In both conceptualizations, the group only selects one alternative even though the choice functions are said to be "fuzzy." Further, they both return identical results to the G-S theorem where a choice function is non-manipulable if and only if it is dictatorial Abdelaziz et al. (2008); Tang (1994). Côrte-Real (2007) demonstrates the Orlovsky rule is strategy-proof but considers the case of only two alternatives. Nonetheless, new results may be obtained when considering the fuzzy choice functions proposed by Dasgupta and Deb (1991) and Banerjee (1995). Under their framework, choice is represented as a fuzzy subset of the set of alternatives, i.e. $\beta : X \to [0,1]$. For any alternative $x \in X$, $\beta(x)$ denotes the degree to which $x$ is chosen. A fuzzy social choice function, as used in this book, associates a fuzzy subset of the set of alternatives with a collection of individuals and their preferences. While the conceptualization of fuzzy choice has received a great deal of attention in revealed preference theory Georgescu (2005, 2007), the possibility of manipulating these types of choice functions has yet to be considered.

Informally, a social choice function is manipulable by an actor if the actor can unilaterally change the social choice in her favor by submitting an insincere or false preference. To address this formally, it follows that there exists some mechanism to compare the social choice with an individual's preferences. In the exact case, each individual possesses a transitive ranking of the alternatives, and a choice is manipulable if there exists an individual who can unilaterally move the social choice further up her ranking. In the fuzzy case, this mechanism relating individual preferences to the social choice is more complicated. When comparing individual fuzzy preference relations and an exact social choice, Abdelaziz, Figueira and Meddeb (2008) utilize four different procedures that determine whether an individual prefers one alternative over another, hence four definitions of manipulability. In contrast, Côrte-Real

(2007) first assumes that individuals make exact choices between pairs of alternatives, which are easily compared to an exact social choice, thus simplifying the analysis. Perote-Peña and Piggins (2007) offer another solution to the problem by considering the manipulation of fuzzy aggregation rules, where an $n$-tuple of fuzzy preference relations are aggregated into a single social preference relation. While this setup has only confirmed the G-S theorem in the fuzzy framework (Duddy et al., 2011; Perote-Peña and Piggins, 2007), modeling individual preferences as fuzzy subsets of the set of alternatives rather than fuzzy preferences relations simplifies the analysis.

Further, representing individuals in this manner is not completely divorced from fuzzy preference relations. Dasgupta and Deb (1991) and Georgescu (2007) illustrate how fuzzy subsets can be related to fuzzy preference relations using concepts similar to the $R$-maximality and $R$-greatness concepts in revealed preference theory (Sen, 1971; Suzumura, 1976). In addition, Clark, Larson, Mordeson, Potter and Wierman (2008) discuss several substantive interpretations of fuzzy subsets of the set of alternatives as representations of individual preference. For example, let $\beta$ be a fuzzy subset of $X$ and $x \in X$. When $\beta(x)$ refers to the degree to which $x$ is ideal, actors are uncertain how ideal each alternative is; however, they are quite certain whether $x$ is better, or preferred to, another alternative $y \in X$, $\beta(x) > \beta(y)$ in this case.

## 5.2 Fuzzy Social Choice: Definitions and Concepts

This section details the fuzzy social choice framework and introduces the concepts of strategy-proofness used in this chapter. Let $X$ be a finite set of alternatives such that $|X| \geq 3$, and $N = \{1, \ldots, n\}$ be a set of $n$ actors, where $n \geq 2$. A fuzzy preference relation, $\rho$, is a function $\rho : X \times X \to [0, 1]$. It is assumed throughout that each $i \in N$ possesses a fuzzy preference relation, $\rho_i$, that is reflexive, i.e. $\rho_i(x, x) = 1$, and complete, i.e. $\rho_i(x, y) = 0$ implies $\rho_i(y, x) > 0$. Recall that we call $\rho_i$ a *fuzzy weak order* and $\mathscr{FR}$ denotes the set of all fuzzy weak orders. Then a *preference profile* is an $n$-tuple of fuzzy weak orders, $\bar{\rho} = (\rho_1, \ldots, \rho_n) \in \mathscr{FR}^n$ and describes the fuzzy preferences of all individuals. For any non-empty $S \subseteq X$, let $\bar{\rho}\rceil_S = (\rho_1|_{S \times S}, \ldots, \rho_n|_{S \times S})$. In words, $\bar{\rho}\rceil_S$ denotes the restriction of the preference profile to the subset $S$ and, accordingly, $\bar{\rho}\rceil_S$ describes only $\rho(x, y)$ and $\rho(y, x)$ for $x, y \in S$ and every $i \in N$. In addition, for any FWPR $\rho$ and all $\alpha \in [0, 1]$, $\rho^\alpha = \{(x, y) \in X \times X \mid \rho(x, y) \geq \alpha\}$. Often, $\rho^\alpha$ is called the $\alpha$-cut of $\rho$.

The reader will recall Definition 3.13, which we restate here for clarity.

**Definition 5.3 (regular).** Let $\rho$ be an FWPR. Then $\pi$, the asymmetric relation with respect to $\rho$, is said to be *regular* if for all $x, y \in X$,

$$\pi(x, y) > 0 \iff \rho(x, y) > \rho(y, x).$$

We assume $\pi$ is regular throughout the remainder of the chapter. Given any $\bar{\rho} \in \mathscr{FR}^n$, define $\tilde{P}, \tilde{R} : X \times X \to \mathscr{F}(N)$ for all $x, y \in X$ and $i \in N$, by the formulas

$$\tilde{P}(x, y; \bar{\rho})(i) = \begin{cases} \pi_i(x, y) & \text{if } \pi_i(x, y) > 0 \\ 0 & \text{otherwise,} \end{cases}$$

$$\tilde{R}(x, y; \bar{\rho})(i) = \begin{cases} \rho_i(x, y) & \text{if } \pi_i(y, x) = 0 \\ 0 & \text{otherwise.} \end{cases}$$

**Definition 5.4.** An FWPR $\rho$ on $X$ is

(1) **(consistent)** (also called weakly transitive) if, for all $x, y, z \in X$,

$$\rho(x, y) \geq \rho(y, x) \text{ and } \rho(y, z) \geq \rho(z, y) \text{ imply } \rho(x, z) \geq \rho(z, x),$$

(2) **(partially quasi-transitive)** if, for all $x, y, z \in X$,

$$\pi(x, y) \wedge \pi(y, z) > 0 \text{ implies } \pi(x, z) > 0.$$

We use the notation $\mathscr{FCR} = \{\rho \in \mathscr{FR} \mid \rho \text{ is consistent}\}$ for the set of all consistent preference relation (also termed consistent weak orders).

**Definition 5.5.** A function $\mathscr{C} : \mathscr{FCR}^n \to X$ is called a *fuzzy social choice function* (FSCF).

Throughout, we assume that any fuzzy social choice function $\mathscr{C}$ satisfies full range (i.e., for any $x \in X$ there exists a $\bar{\rho} \in \mathscr{FCR}^n$ such that $\mathscr{C}(\bar{\rho}) = x$). Such an FSCF must also be nontrivial (i.e., it is not the case that $\mathscr{C}$ is a constant function). For any $\bar{\rho} \in \mathscr{FR}^n$ and any $i \in N$, $\bar{\rho}_{-i} = \bar{\rho}_{N \setminus \{i\}} = (\rho_1, \ldots, \rho_{i-1}, \rho_{i+1}, \ldots, \rho_n)$. In words, $\bar{\rho}_{-i}$ denotes the removal of $\rho_i$ from $\bar{\rho}$.

**Definition 5.6 (dictatorial).** Let $\mathscr{C}$ be a fuzzy choice function. Then $\mathscr{C}$ is *dictatorial* if there exists an $i \in N$ such that for all $\bar{\rho} \in \mathscr{FCR}^n$ and all $x \in X$,

$$\pi_i(x, \mathscr{C}(\bar{\rho})) = 0.$$

**Definition 5.7 (manipulable).** Let $\bar{\rho} \in \mathscr{FR}^n$. A fuzzy social choice function C is *manipulable* at $\bar{\rho}$ if there exists $i \in N$, $\rho_i \in \mathscr{FR}$, $\rho_i' \in \mathscr{FR}$ such that

$$\pi_i(C(\bar{\rho}_{-i}, \rho_i'), C(\rho)) > 0.$$

**Definition 5.8 (strategy-proof).** A fuzzy social choice function C is *strategy-proof* if it is not manipulable.

**Theorem 5.9.** *Let $\mathscr{C}$ be a fuzzy social choice function. Then $\mathscr{C}$ is strategy-proof if and only if it is dictatorial.*

We conclude this section with two proofs of the G-S Theorem in the fuzzy preference framework.

The first proof fuzzifies the logic found in Austen-Smith and Banks (2005), and the second proof uses the logic presented in Abdelaziz, Figueira and Meddeb (2008). Both are semi-novel efforts. The proof in Austen-Smith and Banks has not been extended to the fuzzy case, and our framework would relax the completeness in Abdelaziz et al. (2008) while also covering the case of weakly transitive fuzzy pre-orders. Abdelaziz et al. (2008) assume max-min transitivity, and our results demonstrate that there is no relationship between the max-min and weakly transitive conditions.

### 5.2.1 Fuzzifying ASB II

**Definition 5.10 (monotonic).** Let $\mathscr{C}$ be an FSCF. Then $\mathscr{C}$ is *monotonic* if, for all $x, y \in X$ and all $\bar{\rho}, \bar{\rho}' \in \mathscr{FCR}^n$,

$$[\mathrm{Supp}(\tilde{P}(x,y;\bar{\rho})) \subseteq \mathrm{Supp}(\tilde{P}(x,y;\bar{\rho}')) \text{ and } \mathscr{C}(\bar{\rho}) = x] \implies \mathscr{C}(\bar{\rho}') \neq y.$$

A function $\tilde{f} : \mathscr{FCR}^n \to \mathscr{FCR}$ is called a *fuzzy preference aggregation rule* (FPAR).

In what follows, let $\mathscr{FCR}_\pi$ denote the set of consistent fuzzy weak orders where, for all $x, y \in X$ such that $x \neq y$, $\pi(x,y) > 0$ or $\pi(y,x) > 0$. Hence, individuals who possess preference relations in $\mathscr{FWR}_\pi$ are never indifferent between two alternatives, i.e., there does not exist $x, y \in X$ such that $x \neq y$ and $\rho(x,y) = \rho(y,x)$ when $\rho \in \mathscr{FWR}_\pi$. The goal of the following argument is show that a strategy-proof FSCF with a range of $\mathscr{FWR}_\pi$ is dictatorial, and that this result implies that a strategy-proof FSCF with a range of $\mathscr{FWR}$ is dictatorial as well.

**Lemma 5.11.** *Let $\mathscr{C}$ be an FSCF such that $\mathscr{C} : \mathscr{FCR}_\pi^n \to X$. Let $\bar{\rho} \in \mathscr{FWR}^n$ and non-empty $S \subset X$ be such that for all $(x,y) \in S \times (X \setminus S)$, $\pi_i(x,y) > 0$ for all $i \in N$. If $\mathscr{C}$ is strategy-proof, then $\mathscr{C}(\bar{\rho}) \in S$.*

*Proof.* Suppose $\mathscr{C}$ is strategy-proof and, contrary to the hypothesis, $\mathscr{C}(\bar{\rho}) \notin S$. Because $\mathscr{C}$ satisfies full range, there exists a $\bar{\rho}' \in \mathscr{FCR}_\pi^n$ such that $\mathscr{C}(\bar{\rho}') \in S$. Now construct the sequence of profiles $\{\bar{z}_0, \bar{z}_1, \ldots, \bar{z}_n\}$ such that

$$\bar{z}_0 = (\rho_1, \ldots, \rho_i, \ldots, \rho_n)$$
$$\bar{z}_1 = (\rho_1', \ldots, \rho_i, \ldots, \rho_n)$$
$$\vdots$$
$$\bar{z}_i = (\rho_1', \ldots, \rho_i', \ldots, \rho_n)$$
$$\vdots$$
$$\bar{z}_n = (\rho_1', \ldots, \rho_i', \ldots, \rho_n').$$

In words $\bar{z}_i$ signifies the profile where the $i$th individual has switched from $\rho_i$ to $\rho_i'$ from $\bar{z}_{i-1}$. In the argument that follows, let $\bar{z}_{j,k}$ signify $\rho_k$ in profile $\bar{z}_j$, and $\bar{z}_{j\backslash k}$ denotes the removal of $\rho_k$ from $\bar{z}_j$. By construction, $\mathscr{C}(\bar{z}_0) \notin S$ and $\mathscr{C}(\bar{z}_n) \in S$. Thus, there exists $\bar{z}_{i-1}$ and $\bar{z}_i$ such that $\mathscr{C}(\bar{z}_{i-1}) \notin S$ and $\mathscr{C}(\bar{z}_i) \in S$. However, in the profile $\bar{z}_{i-1}$, $i$ has a sincere preference of $\pi_i(\mathscr{C}(\bar{z}_i), \mathscr{C}(\bar{z}_{i-1})) > 0$ since $C(\bar{z}_i \in S$. Thus, $i \in N$ can manipulate $\mathscr{C}$ at $\bar{z}_{i-1}$ by submitting $\bar{z}_{i,i}$ instead of $\bar{z}_{i-1,i}$ because $\bar{z}_i = (\bar{z}_{i-1\backslash i}, \bar{z}_{i,i})$. However, this contradicts the strategy-proofness of $\mathscr{C}$. Hence, $\mathscr{C}(\bar{\rho}) \in S$.     $\square$

**Lemma 5.12.** *Let $\mathscr{C}$ be an FSCF such that $\mathscr{C} : \mathscr{FCR}_\pi^n \to X$. If $\mathscr{C}$ is strategy-proof, then there exists a weakly Paretian fuzzy preference aggregation rule $\tilde{f} : \mathscr{FWR}_\pi^n \to \mathscr{FWR}_\pi$ such that $\tilde{f}$ is independent of irrelevant of alternatives IIA-3.*

*Proof.* Throughout the proof, let $\alpha, \beta \in [0,1]$ be such that $\alpha > \beta$. [Note: $\alpha$ and $\beta$ can vary across $\bar{\rho}$, not within $\bar{\rho}$.] Suppose $\mathscr{C}$ is strategy proof. For any $\bar{\rho} \in \mathscr{FCR}_\pi^n$ and $x, y \in X$, define the new profile $\bar{\rho}^{(x,y)} = (\rho_1^{(x,y)}, \ldots, \rho_n^{(x,y)})$ as follows, for all $i \in N$:

$$\rho_i^{(x,y)}(x,w) = \rho_i^{(x,y)}(y,w) = \alpha, \forall w \in X \backslash \{x,y\};$$

$$\rho_i^{(x,y)}(w,x) = \rho_i^{(x,y)}(w,y) = \beta, \forall w \in X \backslash \{x,y\};$$

$$\rho_i^{(x,y)}(x,y) = \rho_i(x,y) \text{ and } \rho_i^{(x,y)}(y,x) = \rho_i(y,x);$$

$$\rho_i^{(x,y)}(w,z) = \rho_i(w,z), \forall w,z \in X \backslash \{x,y\}; \text{ and}$$

$$\rho_i^{(x,y)}(w,w) = 1, \forall w \in X.$$

By the regularity of $\pi$ and the definition of $\bar{\rho}^{(x,y)}$, it follows that $\pi_i^{(x,y)}(x,w) > 0$ and $\pi_i^{(x,y)}(y,w) > 0$ for all $w \in X \backslash \{x,y\}$ and all $i \in N$. Further, $\pi_i^{(x,y)}(x,y) > 0$ if and only if $\rho_i(x,y) > \rho_i(y,x)$ for all $i \in N$, and $\pi_i^{(x,y)}(w,z) > 0$ if and only if $\rho_i(w,z) > \rho_i(z,w)$ for all $w,z \in X \backslash \{x,y\}$ and all $i \in N$. Now define the fuzzy preference aggregation rule $\tilde{f}_\mathscr{C}$ as, for all $x,y \in X$ and all $\bar{\rho} \in \mathscr{FCR}^n$,

$$\tilde{f}_\mathscr{C}(\bar{\rho})(x,y) = \begin{cases} 1 & \text{if } x = y, \\ \alpha & \text{if } \mathscr{C}(\bar{\rho}^{(x,y)}) = x \text{ and } x \neq y, \\ \beta & \text{if } \mathscr{C}(\bar{\rho}^{(x,y)}) = y \text{ and } x \neq y. \end{cases}$$

By Lemma 5.11, $\tilde{f}_\mathscr{C}$ is weakly Paretian.

To see that $\tilde{f}_\mathscr{C}$ is IIA-3, suppose that there exist $\bar{\rho}, \bar{\rho}' \in \mathscr{FCR}_\pi^n$ and $x,y \in X$ such that $\rho_i]_{\{x,y\}} \sim \rho_i']_{\{x,y\}}$ for all $i \in N$ and

$$\tilde{f}_\mathscr{C}(\bar{\rho})]_{\{x,y\}} \not\sim \tilde{f}_\mathscr{C}(\bar{\rho}')]_{\{x,y\}}.$$

By construction, $Im(\tilde{f}_\mathscr{C}(\bar{\rho})) = \{1, \alpha, \beta\} = Im(\tilde{f}_\mathscr{C}(\bar{\rho}'))$. It follows that there exists $x,y \in X$ such that $\pi(x,y) > 0$ and $\pi'(y,x) > 0$. Then by the definition of $\tilde{f}_\mathscr{C}$, $\mathscr{C}(\bar{\rho}^{(x,y)}) = x$ and $\mathscr{C}(\bar{\rho}'^{(x,y)}) = y$. Now construct a sequence of profiles $\{\bar{z}_0, \bar{z}_1, \ldots, \bar{z}_n\}$ such that

$$\bar{z}_0 = (\rho_1^{(x,y)}, \ldots, \rho_i^{(x,y)}, \ldots, \rho_n^{(x,y)})$$
$$\bar{z}_1 = (\rho_1'^{(x,y)}, \ldots, \rho_i^{(x,y)}, \ldots, \rho_n^{(x,y)})$$
$$\vdots$$
$$\bar{z}_i = (\rho_1'^{(x,y)}, \ldots, \rho_i'^{(x,y)}, \ldots, \rho_n^{(x,y)})$$
$$\vdots$$
$$\bar{z}_n = (\rho_1'^{(x,y)}, \ldots, \rho_i'^{(x,y)}, \ldots, \rho_n'^{(x,y)}),$$

where $\bar{z}_{j,k}$ signifies $\rho_k$ in profile $\bar{z}_j$, and $\bar{z}_{j-k}$ denotes the removal of $\rho_k$ from $\bar{z}_j$. By construction, $\mathscr{C}(\bar{z}_0) = x$ and $\mathscr{C}(\bar{z}_n) = y$. Thus, there must exist a $\bar{z}_{i-1}$ and $\bar{z}_i$ such that $\mathscr{C}(\bar{z}_{i-1}) = x$ and $\mathscr{C}(\bar{z}_i) \neq x$. There are two cases to consider.

Case 1: Suppose $\pi_i^{(x,y)}(\mathscr{C}(\bar{z}_i), x) > 0$. Thus, $i \in N$ can manipulate $\mathscr{C}$ at $\bar{z}_{i-1}$ by submitting $\bar{z}_{i,i}$ rather than $\bar{z}_{i-1,i}$. (Note $\bar{z}_i = (\bar{z}_{i-1\setminus i}, \bar{z}_{i,i})$.)

Case 2: Suppose $\pi_i^{(x,y)}(x, \mathscr{C}(\bar{z}_i)) > 0$. By the assumption,

$$\rho_i]_{\{x,y\}} \sim \rho_i']_{\{x,y\}}$$

for all $i \in N$, this implies $\pi_i'^{(x,y)}(x, \mathscr{C}(\bar{z}_i)) > 0$. Thus, $i \in N$ can manipulate $\mathscr{C}$ at $\bar{z}_i$ by submitting $\bar{z}_{i-1,i}$ rather than $\bar{z}_{i,i}$. (Note $\bar{z}_{i-1} = (\bar{z}_{i\setminus i}, \bar{z}_{i-1,i})$.) In both cases, a contradiction emerges. Thus, $\tilde{f}_{\mathscr{C}}$ is IIA-3.

What remains to be shown is that $\tilde{f}_{\mathscr{C}}(\bar{\rho}) \in \mathscr{FCR}_\pi$ for all $\bar{\rho} \in \mathscr{FCR}_\pi^n$. To see this, first note that by definition $\tilde{f}(\bar{\rho})$ must be complete and reflexive for all $\bar{\rho} \in \mathscr{FWR}_\pi^n$. Let $x, y \in X$ be such that $x \neq y$ and $\bar{\rho} \in \mathscr{FCR}_\pi^n$. Because $\mathscr{C}$ satisfies the relationship in Lemma 5.11, $\mathscr{C}(\bar{\rho}^{(x,y)}) = x$ or $\mathscr{C}(\bar{\rho}^{(x,y)}) = y$. Thus,

$$\tilde{f}_{\mathscr{C}}(\bar{\rho})(x,y) > \tilde{f}_{\mathscr{C}}(\bar{\rho})(y,x)$$

or

$$\tilde{f}_{\mathscr{C}}(\bar{\rho})(y,x) > \tilde{f}_{\mathscr{C}}(\bar{\rho})(x,y),$$

and, by the regularity of $\pi$, $\pi(x,y) > 0$ or $\pi(y,x) > 0$.

Finally, we must show $\tilde{f}_{\mathscr{C}}(\bar{\rho})$ is consistent for all $\bar{\rho} \in \mathscr{FCR}_\pi^n$.

To do so, suppose that there exist $x, y, z \in X$ and $\bar{\rho} \in \mathscr{FCR}_\pi^n$ such that

$$\tilde{f}_{\mathscr{C}}(\bar{\rho})(x,y) > \tilde{f}_{\mathscr{C}}(\bar{\rho}(y,x),$$

$$\tilde{f}_{\mathscr{C}}(\bar{\rho})(y,z) > \tilde{f}_{\mathscr{C}}(\bar{\rho}(z,y),$$

and

$$\tilde{f}_{\mathscr{C}}(\bar{\rho})(z,x) > \tilde{f}_{\mathscr{C}}(\bar{\rho})(x,z).$$

(Note $\tilde{f}_{\mathscr{C}}(\bar{\rho})(z,x) \neq \tilde{f}_{\mathscr{C}}(\bar{\rho})(x,z)$ unless $x = z$.) The proof now demonstrates this leads to a contradiction. Let $\bar{\rho}^{(a,b,c)} \in \mathscr{FCR}_\pi^n$ be defined in the same manner as above. Specifically, for all $i \in N$,

$$\forall (a,b) \in \{x,y,z\} \times X \backslash \{x,y,z\}, \ \rho_i^{(x,y,z)}(a,b) = \alpha;$$

$$\forall (b,a) \in X \backslash \{x,y,z\} \times \{x,y,z\}, \ \rho_i^{(x,y,z)}(b,a) = \beta;$$

$$\forall (a,b) \in \{x,y,z\} \times \{x,y,z\}, \ \rho_i^{(x,y,z)}(a,b) = \rho_i(a,b);$$

$$\forall (a,b) \in X \backslash \{x,y,z\} \times X \backslash \{x,y,z\}, \ \rho_i^{(x,y,z)}(a,b) = \rho_i(a,b); \text{ and}$$

$$\forall a \in X, \ \rho_i^{(x,y,z)}(a,a) = 1.$$

By Lemma 5.11, $\mathscr{C}(\bar{\rho}^{(x,y,z)}) \in \{x,y,z\}$. Thus, consider the three possible cases.

Case 1:  Suppose $\mathscr{C}(\bar{\rho}^{(x,y,z)}) = x$. Now define the profile $\bar{\rho}^{(x,y,z)(x,z)}$ in the above manner from the profile $\bar{\rho}^{(x,y,z)}$. Therefore,

$$\rho_i^{(x,y,z)(x,z)}\rceil_{\{x,z\}} \sim \rho_i^{(x,z)}\rceil_{\{x,z\}}$$

for all $i \in N$. Because $\tilde{f}_{\mathscr{C}}$ is IIA-3 and because we have assumed $\tilde{f}_{\mathscr{C}}(\bar{\rho})(z,x) > \tilde{f}_{\mathscr{C}}(\bar{\rho}(x,z)$, then

$$\tilde{f}_{\mathscr{C}}(\bar{\rho}^{(x,y,z)(x,z)})(z,x) > \tilde{f}_{\mathscr{C}}(\bar{\rho}^{(x,y,z)(x,z)})(x,z) ,$$

which implies $\mathscr{C}(\bar{\rho}^{(x,y,z)(x,z)}) = z$. Now construct a sequence of profiles $\{\bar{z}_0, \bar{z}_1, \ldots, \bar{z}_n\}$ such that

$$\bar{z}_0 = (\rho_1^{(x,y,z)}, \ldots, \rho_i^{(x,y,z)}, \ldots, \rho_n^{(x,y,z)})$$
$$\bar{z}_1 = (\rho_1^{(x,y,z)(x,z)}, \ldots, \rho_i^{(x,y,z)}, \ldots, \rho_n^{(x,y,z)})$$
$$\vdots$$
$$\bar{z}_i = (\rho_1^{(x,y,z)(x,z)}, \ldots, \rho_i^{(x,y,z)(x,z)}, \ldots, \rho_n^{(x,y,z)})$$
$$\vdots$$
$$\bar{z}_n = (\rho_1^{(x,y,z)(x,z)}, \ldots, \rho_i^{(x,y,z)(x,z)}, \ldots, \rho_n^{(x,y,z)(x,z)}),$$

where $\bar{z}_{j,k}$ signifies $\rho_k$ in profile $\bar{z}_j$, and $\bar{z}_{j-k}$ denotes the removal of $\rho_k$ from $\bar{z}_j$. By construction, $\mathscr{C}(\bar{z}_0) = x$ and $\mathscr{C}(\bar{z}_n) = z$. Thus, there exist some $\bar{z}_{i-1}$ and $\bar{z}_i$ such that $\mathscr{C}(\bar{z}_{i-1}) = x$ and $\mathscr{C}(\bar{z}_i) \neq x$. Then by Lemma 5.11, $\mathscr{C}(\bar{z}_i) \in \{y,z\}$. Suppose $\mathscr{C}(\bar{z}_i) = y$. Because $\pi_i^{(x,y,z)(x,z)}(x,y) > 0$, $i \in N$ can manipulate $\mathscr{C}(\bar{z}_i)$ by submitting $\bar{z}_{i-1,i}$ rather than $\bar{z}_{i,i}$. (Remember, $\pi_i^{(x,y,z)(x,z)}(x,y) > 0$ is $i$'s sincere preference under $\bar{z}_i$ by the construction of $\bar{\rho}^{(x,y,z)(x,z)}$.) Suppose $\mathscr{C}(\bar{z}_i) = z$. Suppose $\pi_i^{(x,y,z)}(z,x) > 0$. Then $i \in N$ can manipulate $\mathscr{C}$ at $\bar{z}_{i-1}$ by submitting $\bar{z}_{i,i}$ rather than $\bar{z}_{i-1,i}$. Suppose $\pi_i^{(x,y,z)}(x,z) > 0$. Because

$$\rho_i^{(x,y,z)(x,z)}\rceil_{\{x,z\}} \sim \rho_i^{(x,y,z)}\rceil_{\{x,z\}}$$

for all $i \in N$, $\pi_i^{(x,y,z)(x,z)}(x,z) > 0$. Then $i \in N$ can manipulate $\mathscr{C}$ at $\bar{z}_i$ by submitting $\bar{z}_{i-1,i}$ rather than $\bar{z}_{i,i}$.

Case 2: Suppose $\mathscr{C}(\bar{\rho}^{(x,y,z)}) = y$. Now define the profile $\bar{\rho}^{(x,y,z)(x,y)}$ in the above manner from the profile $\bar{\rho}^{(x,y,z)}$. Therefore,

$$\rho_i^{(x,y,z)(x,y)}\rceil_{\{x,y\}} \sim \rho_i^{(x,y)}\rceil_{\{x,y\}}$$

for all $i \in N$. Because $\tilde{f}_{\mathscr{C}}$ is IIA-3 and because we have assumed $\tilde{f}_{\mathscr{C}}(\bar{\rho})(x,y) > \tilde{f}_{\mathscr{C}}(\bar{\rho}(y,x))$,

$$\tilde{f}_{\mathscr{C}}(\bar{\rho}^{(x,y,z)(x,y)})(x,y) > \tilde{f}_{\mathscr{C}}(\bar{\rho}^{(x,y,z)(x,y)})(y,x) \,,$$

which implies $\mathscr{C}(\bar{\rho}^{(x,y,z)(x,y)}) = x$. The fact that this leads to a contradiction is evident from the previous case.

Case 3: Suppose $\mathscr{C}(\bar{\rho}^{(x,y,z)}) = z$. Now define the profile $\bar{\rho}^{(x,y,z)(y,z)}$ in the above manner from the profile $\bar{\rho}^{(x,y,z)}$. Therefore,

$$\rho_i^{(x,y,z)(y,z)}\rceil_{\{y,z\}} \sim \rho_i^{(y,z)}\rceil_{\{y,z\}}$$

for all $i \in N$. Because $\tilde{f}_{\mathscr{C}}$ is IIA-3 and because we have assumed $\tilde{f}_{\mathscr{C}}(\bar{\rho})(y,z) > \tilde{f}_{\mathscr{C}}(\bar{\rho}(z,y))$,

$$\tilde{f}_{\mathscr{C}}(\bar{\rho}^{(x,y,z)(x,y)})(y,z) > \tilde{f}_{\mathscr{C}}(\bar{\rho}^{(x,y,z)(x,y)})(z,y),$$

which implies $\mathscr{C}(\bar{\rho}^{(x,y,z)(x,y)}) = y$. The fact that this leads to a contradiction is evident from the previous cases. In all three cases, there exist some $i \in N$ who can manipulate $\mathscr{C}$, which contradicts the assumption that $\mathscr{C}$ is strategy-proof. Thus, not $\tilde{f}_{\mathscr{C}}(\bar{\rho})(z,x) > \tilde{f}_{\mathscr{C}}(\bar{\rho})(x,z)$. By previous argument, $\tilde{f}_{\mathscr{C}}(\bar{\rho})(x,z) > \tilde{f}_{\mathscr{C}}(\bar{\rho})(z,x)$, and the consistency condition is established. $\quad\square$

**Lemma 5.13.** *Let $\mathscr{C}$ be an FSCF. If $\mathscr{C}$ is dictatorial when $\mathscr{C} : \mathscr{FCR}_\pi^n \to X$, then $\mathscr{C}$ is dictatorial when $\mathscr{C} : \mathscr{FCR}^n \to X$.*

*Proof.* Let $j \in N$ be the dictator when $\mathscr{C} : \mathscr{FCR}_\pi^n \to X$. Now suppose $j \in N$ is not a dictator when $\mathscr{C} : \mathscr{FCR}^n \to X$. Then there must exist a profile $\bar{\rho} \in \mathscr{FCR}^n$ and $x,y \in X$ such that $\mathscr{C}(\bar{\rho}) = y$ and $\pi_j(x,y) > 0$. Let $J \subset X$ be such that $J = \{x \in X \mid \pi_j(y,x) = 0, \forall y \in X\}$. Consider some $\bar{\rho}' \in \mathscr{FCR}_\pi^n$ such that, for all $x \in J$ and $y \in X \backslash J$, $\pi_i'(x,y) > 0$ and $\pi_i'(y,x) > 0$, where $i \neq j$. Now construct the sequence of profiles $\{\bar{z}_0, \bar{z}_1, \ldots, \bar{z}_n\}$ such that

$$\bar{z}_0 = (\rho_1, \ldots, \rho_i, \ldots, \rho_n)$$
$$\bar{z}_1 = (\rho_1', \ldots, \rho_i, \ldots, \rho_n)$$
$$\vdots$$
$$\bar{z}_i = (\rho_1', \ldots, \rho_i', \ldots, \rho_n)$$
$$\vdots$$
$$\bar{z}_n = (\rho_1', \ldots, \rho_i', \ldots, \rho_n'),$$

where $\bar{z}_{j,k}$ signifies $\rho_k$ in profile $\bar{z}_j$, and $\bar{z}_{j-k}$ denotes the removal of $\rho_k$ from $\bar{z}_j$. Because $\mathscr{C}$ is dictatorial when $\mathscr{C} : \mathscr{FCR}_\pi^n \to X$, $\mathscr{C}(\bar{z}_n) \in J$. Because $\mathscr{C}(\bar{z}_0) = y \notin J$, there exist some $l \in N$ such that $\mathscr{C}(\bar{z}_{l-1}) \notin J$ and $\mathscr{C}(\bar{z}_l) \in J$.

If $l = j$, first note that $\pi_j(\mathscr{C}(\bar{z}_{j-1}), \mathscr{C}(\bar{z}_j)) = 0$. Also, because $\mathscr{C}(\bar{z}_{j-1}) \notin J$, there exists $w \in J$ such that $\pi_j(w, \mathscr{C}(\bar{z}_{j-1})) > 0$, i.e. $\rho_j(w, \mathscr{C}(\bar{z}_{j-1})) > \rho_j(\mathscr{C}(\bar{z}_{j-1}), w)$. Because $\rho_i(\mathscr{C}(\bar{z}_j), w) = \rho_i(w, \mathscr{C}(\bar{z}_j))$, consistency then implies $\pi_j(\mathscr{C}(\bar{z}_j), \mathscr{C}(\bar{z}_{j-1})) > 0$. However, because $\pi_j(\mathscr{C}(\bar{z}_l), \mathscr{C}(\bar{z}_{l-1})) > 0$ is $j$'s sincere preference at $\bar{z}_{j-1}$, $j$ can can manipulate $\mathscr{C}$ at $\bar{z}_{j-1}$ by submitting $\bar{z}_{j,j}$ rather than $\bar{z}_{j-1,j}$.

If $l \neq j$, then $l$ can manipulate $\mathscr{C}$ at $\bar{z}_l$ by submitting $\bar{z}_{l-1,l}$ rather than $\bar{z}_{l,l}$ because $\pi_l'(\mathscr{C}(\bar{z}_{l-1}), \mathscr{C}(\bar{z}_l)) > 0$ by construction. Either case contradicts the strategy-proofness of $\mathscr{C}$. □

**Proof of Fuzzy Gibbard-Satterthwaite 5.9.** First, note that any FSCF $\mathscr{C}$ that is dictatorial must be strategy-proof. Second, suppose $\mathscr{C}$ is an FSCF. Note that Lemma 5.11, Lemma 5.12, and Fuzzy Arrow's Theorem imply that $\mathscr{C}$ is dictatorial on the domain $\mathscr{FCR}_\pi^n$. Lemma 5.13 then implies $\mathscr{C}$ must be dictatorial on the full domain $\mathscr{FWR}^n$.

### 5.2.2   Relaxing the Conditions of Abdelaziz et. al.

Here we assume that G-S holds for the crisp case and show that it holds for the fuzzy case.

**Definition 5.14.** Let $\mathscr{CR}$ be the set of all $\rho \in \mathscr{FR}$ such that

(1) **(crisp)** for all $x, y \in X$, $\rho(x, y) \in \{0, 1\}$;
(2) **(reflexive)** for all $x \in X$, $\rho(x, x) = 1$;
(3) **(crisp complete)** for all $x, y \in X$, $\rho(x, y) = 1$ or $\rho(y, x) = 1$;
(4) **(transitive)** for all $x, y, z \in X$, $\rho(x, y) = 1$ and $\rho(y, z) = 1$ implies $\rho(x, z) = 1$.

**Lemma 5.15.** $\mathscr{CR} \subset \mathscr{FCR}$.

*Proof.* Suppose $\rho \in \mathscr{CR}$. We must show that $\rho$ is reflexive, complete and weakly transitive. First, reflexivity follows immediately from Definition 5.14(2). Second, completeness follows from Definition 5.14(3). Third, let $x, y, z \in X$ be such that

$\rho(x,y) \geq \rho(y,x)$ and $\rho(y,z) \geq \rho(z,y)$. Then, by Definition 5.14(1) and Definition 5.14(2), $\rho(x,y) = 1$ and $\rho(y,z) = 1$. (If not, then $\rho(x,y) = 0$, and Definition 5.14(3) implies $\rho(y,x) = 1$, a contradiction.) By Definition 5.14(4), $\rho(x,z) = 1$, which implies $\rho(x,z) \geq \rho(z,x)$.                                                              □

**Theorem 5.16.** *[Gibbard (1973); Satterthwaite (1975); Austen-Smith and Banks II (2005)] Suppose $\mathscr{C} : \mathscr{CR} \to X$. Then $\mathscr{C}$ is strategy-proof if and only if it is dictatorial. (See Section 2.3).*

**Definition 5.17 (non-dominated set).** For any $x \in X$, define $nd : X \to [0,1]$ as $nd(x) = 1 - \max_{y \in X}\{\pi(y,x)\}$. Then define $ND(\pi,X) = \{x \in X \mid nd(x) = 1\}$.

**Lemma 5.18.** *Let $A, D \subseteq X$ such that $A \neq \emptyset$, and let $\rho : D \times D \to \{0,1\}$. If $\rho$ satisfies conditions*

*(1) $\rho \in \mathscr{CR}$ when $D = A$,*
*(2) $\rho \in \mathscr{CR}$ when $D = X \backslash A$,*
*(3) $\pi(x,y) > 0$ for any $(x,y) \in A \times (X \backslash A)$,*

*then $\rho \in \mathscr{CR}$ when $D = X$ and $ND(\pi,X) \subseteq A$, where $\pi$ is the strict preference relation with respect to $\rho$ when $D = X$.*

*Proof.* First, the proof shows that $\rho \in \mathscr{CR}$ when $D = X$. To do so, we consider the four criteria in Definition 5.14.

Crisp  This condition holds by definition of $\rho : D \times D \to \{0,1\}$.
Reflexive  For any $x \in A$, $\rho(x,x) = 1$ by condition (1). For any $x \in X \backslash A$, $\rho(x,x) = 1$ by condition (1).
Crisp Complete  For any $x,y \in A$, $\rho(x,y) = 1$ or $\rho(y,x) = 1$ . For any $x \in A$ and $y \in X \backslash A$, $\rho(x,y) = 1$. For any $x,y \in X \backslash A$, $\rho(x,y) = 1$ or $\rho(y,x) = 1$.
Transitive  Suppose $\rho(x,y) = 1$ and $\rho(y,z) = 1$ and consider the following memberships of $x,y,z \in X$.

  (i)  Suppose $x,y,z \in A$ or $x,y,z \in X \backslash A$. Then $\rho(x,z) = 1$ because $\rho$ is transitive over $A$ and $X \backslash A$, respectively.
  (ii)  Suppose $x \in A$ and $z \in X \backslash A$ . Then by condition (3), $\pi(x,z) > 0$. The regularity of $\pi$ then implies $\rho(x,z) \geq \rho(z,x)$, which guarantees $\rho(x,z) = 1$ by the completeness and definition of $\rho$, regardless of whether $y \in A$ or $y \in X \backslash A$.
  (iii)  Suppose $y \in X \backslash A$ and $z \in A$. Then $\pi(z,y) > 0$, which contradicts the assumption $\rho(y,z) = 1$ by the regularity of $\pi$, regardless $x \in A$ or $x \in X \backslash A$.
  (iv)  Suppose $x \in X \backslash A$ and $y \in A$. Then $\pi(y,x) > 0$, which contradicts the assumption $\rho(x,y) = 1$ by the regularity of $\pi$, regardless of $z \in A$ or $z \in X \backslash A$.

Thus, $\rho \in \mathscr{CR}$. Now take an $x \in ND(\pi,X)$, and suppose, contrary to the hypothesis, that $x \notin A$. Because $A$ is nonempty, there exist $y \in A$. By condition (3), $\pi(y,x) > 0$. Thus, $x \notin ND(\pi,X)$, a contradiction. Hence, the conclusion is now established.    □

**Theorem 5.19.** *Suppose $\mathscr{C} : \mathscr{FCR}^n \to X$. Then $\mathscr{C}$ is strategy-proof if and only if it is dictatorial.*

*Proof.* First, note that any FSCF $\mathscr{C}$ that is dictatorial must be strategy-proof. Suppose $\mathscr{C}$ is strategy-proof. Then let $\mathscr{C}' : \mathscr{CR}^n \to X$ be such that for all $\bar{\rho} \in \mathscr{CR}^n$, $\mathscr{C}'(\bar{\rho}) = \mathscr{C}(\bar{\rho})$. Thus, $\mathscr{C}'$ is strategy-proof because $\mathscr{C}$ is strategy-proof on a larger domain. According to Theorem 5.16, $\mathscr{C}'$ is dictatorial. Let $h \in N$ be the dictator for $\mathscr{C}'$. The goal of the proof is to show that $h \in N$ is a dictator for $\mathscr{C}$. Now suppose $h$ is not a dictator when $\mathscr{C} : \mathscr{FCR}^n \to X$. Then there must exist a profile $\bar{\rho} \in \mathscr{FCR}^n$ such that, for some $x, y \in X$, $\mathscr{C}(\bar{\rho}) = y$ and $\pi_h(x,y) > 0$. Let $H \subset X$ be such that $H = \{x \in X \mid \pi_h(y,x) = 0, \forall y \in X\}$. Consider some $\bar{\rho}' \in \mathscr{CR}^n$ that, for all $x \in H$ and $y \in X \backslash H$, $\pi_h'(x,y) > 0$ and $\pi_j'(y,x) > 0$ where $h \neq j$. (Lemma 5.18 verifies that such a profile exists.) Now construct the sequence of profiles $\{\bar{z}_0, \bar{z}_1, \dots, \bar{z}_n\}$ such that

$$\bar{z}_0 = (\rho_1, \dots, \rho_i, \dots, \rho_n)$$
$$\bar{z}_1 = (\rho_1', \dots, \rho_i, \dots, \rho_n)$$
$$\vdots$$
$$\bar{z}_i = (\rho_1', \dots, \rho_i', \dots, \rho_n)$$
$$\vdots$$
$$\bar{z}_n = (\rho_1', \dots, \rho_i', \dots, \rho_n'),$$

$\bar{z}_{j,k}$ signifies $\rho_k$ in profile $\bar{z}_j$, and $\bar{z}_{j-k}$ denotes the removal of $\rho_k$ from $\bar{z}_j$. Because $\mathscr{C}$ is dictatorial when $\mathscr{C} : \mathscr{CR}^n \to X$, $\mathscr{C}(\bar{z}_n) \in H$. Because $\mathscr{C}(\bar{z}_0) = y \notin J$, there exists some $i \in N$ such that $\mathscr{C}(\bar{z}_{i-1}) \notin J$ and $\mathscr{C}(\bar{z}_i) \in J$.

If $i = h$, first note that $\pi_h(\mathscr{C}(\bar{z}_{h-1}), \mathscr{C}(\bar{z}_h)) = 0$. Also, because $\mathscr{C}(\bar{z}_{h-1}) \notin J$, there exists $w \in J$ such that $\pi_h(w, \mathscr{C}(\bar{z}_{h-1})) > 0$, i.e. $\rho_h(w, \mathscr{C}(\bar{z}_{h-1})) > \rho_h(\mathscr{C}(\bar{z}_{h-1}), w)$. Because $\rho_h(\mathscr{C}(\bar{z}_h), w) = \rho_h(w, \mathscr{C}(\bar{z}_h))$, consistency then implies $\pi_h(\mathscr{C}(\bar{z}_h), \mathscr{C}(\bar{z}_{h-1})) > 0$. However, because $\pi_h(\mathscr{C}(\bar{z}_h), \mathscr{C}(\bar{z}_{h-1})) > 0$ is $h$'s sincere preference at $\bar{z}_{h-1}$, $h$ can can manipulate $\mathscr{C}$ at $\bar{z}_{h-1}$ by submitting $\bar{z}_{h,h}$ rather than $\bar{z}_{h-1,h}$.

If $i \neq h$, then $i$ can manipulate $\mathscr{C}$ at $\bar{z}_i$ by submitting $\bar{z}_{i-1,i}$ rather than $\bar{z}_{i,i}$ because $\pi_i'(\mathscr{C}(\bar{z}_{i-1}), \mathscr{C}(\bar{z}_i)) > 0$ by construction. Either case contradicts the strategy-proofness of $\mathscr{C}$.                                                                                $\square$

## 5.3   Findings

To this point, we have used a definition of a fuzzy social choice function that permits a mapping of n-tuples of fuzzy preferences into a crisp set of alternatives. In this section, we change the definition of a fuzzy social choice function to permit a mapping of fuzzy subsets into a fuzzy subset of alternatives. We then consider a new definition of manipulation.

As we are using it, $(\sigma_{N \backslash i}, \sigma_i')$ represents the profile of individual preference functions where $(\sigma_{N \backslash i}, \sigma_i') = (\sigma_1, \sigma_2, \dots, \sigma_i', \dots, \sigma_n)$ and $(\sigma_{N \backslash i}(x), \sigma_i(x))$ is the profile's restriction to some $x \in X$. Essentially, $(\sigma_{N \backslash i}, \sigma_i')$ formally represents the case where

$i \in N$ submits an insincere choice intensity $\sigma_i'$ to the fuzzy choice function rather than the sincere preference $\sigma_i$.

**Definition 5.20 (fuzzy choice function).** A function $\mathscr{C} : \mathscr{F}(X)^n \to \mathscr{F}(X)$ is called a *fuzzy choice function*.

**Definition 5.21 (manipulable).** A fuzzy choice function $\mathscr{C}$ is *manipulable* if there exists $x \in X$, $\sigma \in \mathscr{F}(X)^n$, $i \in N$ and $\sigma_i' \in \mathscr{F}(X)$ such that

(1) $\mathscr{C}(\sigma)(x) < \sigma_i(x)$ and $\mathscr{C}(\sigma)(x) < \mathscr{C}(\sigma_{N \setminus i}, \sigma_i')(x)$,

*or*

(2) $\mathscr{C}(\sigma)(x) > \sigma_i(x)$ and $C(\sigma)(x) > \mathscr{C}(\sigma_{N \setminus i}, \sigma_i')(x)$.

In words, we sometimes say $i \in N$ can manipulate $\mathscr{C}$ at profile $\bar{\rho}$ by submitting the *insincere* preference $\rho_i'$ instead of the *sincere* preference $\rho_i$.

According to Definition 5.21, a fuzzy choice function is manipulable if an individual $i \in N$ is able to move the degree of social choice for an alternative in the direction of her sincere choice intensity by submitting an insincere choice intensity for the same alternative. In other words, any change in an individual's fuzzy individual choice intensities will be to her detriment or will not have any effect on the final social choice intensity, and, identically, there is no benefit to submitting an insincere choice intensity. It is important to note that Definition 5.21 assumes that $i \in N$ manipulates the social choice intensity for one alternative even though it may adversely affect the social choice intensity of another alternative. Hence, the model assumes that an individual's choice intensities are *separable*. Because it is quite possible that an individual's choice intensity for $x$ depends on his social choice intensity for $y$ (for example politicians may choose some economic policies when the country engages in war and other polices in times of peace), the separability assumption may not be ideal. Nonetheless, Le Breton and Sen (1999) suggest that some degree of separability is necessary for any model of social choice to be strategy-proof and non-dictatorial.

**Definition 5.22 (strategy-proof).** (*Strategy-Proof (SP)*). A fuzzy social choice function $\mathscr{C}$ is said to be strategy-proof if it is not manipulable.

*Example 5.23.* Let $\sigma \in \mathscr{F}(X)^n$ and let $\mathscr{C}$ be a fuzzy choice function. Suppose for some $x \in X$ there exists an $i \in N$ such that $\sigma_i(x) = .4$. Suppose $\mathscr{C}(\sigma)(x) = .3$ and for some $\sigma_i' \in \mathscr{F}(X)$, $\mathscr{C}(\sigma_{N \setminus i}, \sigma_i')(x) = .9$. Hence, according to Definition 5.21, $i$ can manipulate $\mathscr{C}$ at $\sigma$ by submitting $\sigma_i'$ even though $|\sigma_i(x) - \mathscr{C}(\sigma)(x)| < |\sigma_i(x) - \mathscr{C}(\sigma_{N \setminus i}, \sigma_i')(x)|$.

**Definition 5.24 (weakly Paretian).** A fuzzy choice function $\mathscr{C}$ is said to be weakly Paretian if for all $\sigma \in \mathscr{F}(X)^n$ and all $x \in X$,

$$\max_{i \in N}\{\sigma_i(x)\} \geq \mathscr{C}(\sigma)(x) \geq \min_{i \in N}\{\sigma_i(x)\}.$$

In words, weak Paretianism guarantees that the degree to which a fuzzy choice function selects an alternative is (1) not greater than the choice intensity of the individual who chooses the alternative to the most intense degree and (2) not less than the choice intensity of the individual who chooses the alternative to the least intense degree.

**Definition 5.25 ($\sigma$-only).** A fuzzy choice function $\mathscr{C}$ is said to satisfy the $\sigma$-only condition if, for all $\sigma, \sigma' \in \mathscr{F}(X)^n$ and all $x \in X$,

$$\sigma_i(x) = \sigma_i'(x), \forall i \in N \implies \mathscr{C}(\sigma)(x) = \mathscr{C}(\sigma')(x).$$

In words, the $\sigma$-only condition guarantees for all $x \in X$ that the degree to which a fuzzy choice function chooses $x$ is independent of the choice intensities assigned to the other alternatives in $X$.

**Definition 5.26 (monotonic).** A fuzzy choice function $\mathscr{C}$ is said to be monotonic if, for all $x \in X$, and all $\sigma, \sigma' \in \mathscr{F}(X)^n$,

$$\sigma_i(x) \leq \sigma_i'(x), \forall i \in N \implies \mathscr{C}(\sigma)(x) \leq \mathscr{C}(\sigma')(x)$$

Monotonicity requires that increasing the degree to which individuals choose a specific alternative will not decrease the degree of choice for that alternative.

To characterize the properties of fuzzy choice functions, the following formal arguments utilize several reinterpretations of the pivotal voter theorem presented in Reny (2001).

**Proposition 5.27.** *If a fuzzy choice function $\mathscr{C}$ is strategy-proof, then it satisfies the $\sigma$-only condition.*

*Proof.* Assume $\mathscr{C}$ is SP. Now suppose $\mathscr{C}$ is not $\sigma$-only. This proof will show that this leads to a contradiction. Because $\mathscr{C}$ is not $\sigma$-only, there must exists $\sigma, \sigma' \in \mathscr{F}(X)^n$ and an $x \in X$ such that $\sigma_i(x) = \sigma_i'(x) \forall i \in N$ and $\mathscr{C}(\sigma)(x) \neq \mathscr{C}(\sigma')(x)$.

Now construct the following sequence of profiles $Z = \{z_0, z_1, \ldots, z_i, \ldots, z_n\}$ such that the following hold:

$$
\begin{aligned}
z_0 &= (\sigma_1, \ldots, \sigma_i, \ldots, \sigma_n) \\
z_1 &= (\sigma_1', \ldots, \sigma_i, \ldots, \sigma_n) \\
&\vdots \\
z_i &= (\sigma_1', \ldots, \sigma_i', \ldots, \sigma_n) \\
&\vdots \\
z_n &= (\sigma_1', \ldots, \sigma_i', \ldots, \sigma_n').
\end{aligned}
$$

In words $z_i$ signifies the profile where the $i$th individual has switch from $\sigma_i$ to $\sigma_i'$ from $z_{i-1}$. In the argument that follows let $z_{j,k}$ signify $\sigma_k$ in profile $z_j$ and $z_{j\setminus k}$ denotes the removal of $\sigma_k$ from $z_j$.

By construction $z_0 = \sigma$ and $z_n = \sigma'$. By assumption, $\mathscr{C}(\sigma)(x) \neq \mathscr{C}(\sigma')(x)$ so $\mathscr{C}(z_0)(x) \neq \mathscr{C}(z_n)(x)$. Thus, it is apparent that there exist some $z_{i-1}, z_i \in Z$ such that $\mathscr{C}(z_{i-1})(x) = \mathscr{C}(\sigma)(x)$ and $\mathscr{C}(z_i)(x) \neq \mathscr{C}(\sigma)(x)$. This proof will show that $i$ can manipulate $\mathscr{C}$ at $z_{i-1}$ and $z_i$. Now there are two cases to consider.

Case 1: $\mathscr{C}(z_{i-1})(x) < \mathscr{C}(z_i)(x)$. First, suppose $\sigma_i(x) < \mathscr{C}(z_i)(x)$. Here,

$$\mathscr{C}(z_i)(x) > \mathscr{C}(z_{i\setminus i}, z_{i-1,i}) = \mathscr{C}(z_i)(x) .$$

Thus, $i$ can manipulate $\mathscr{C}$ at $z_i$ by submitting $z_{i-1,i}$ rather than $z_{i,i}$. Second, suppose $\sigma_i(x) \geq \mathscr{C}(z_i)(x)$. Because $\mathscr{C}(z_{i-1})(x) < \mathscr{C}(z_i)(x)$, then $\sigma_i(x) > \mathscr{C}(z_{i-1})(x)$. However,

$$\mathscr{C}(z_{i-1})(x) < \mathscr{C}(z_{i-1\setminus i}, z_{i,i}) = \mathscr{C}(z_i)(x) .$$

Again $i$ can manipulate $\mathscr{C}$ at $z_{i-1}$ by submitting $z_{i,i}$ rather than $z_{i-1,i}$. This is a contradiction. Hence, $\mathscr{C}(z_{i-1})(x) \geq \mathscr{C}(z_i)(x)$.

Case 2: $\mathscr{C}(z_{i-1})(x) > \mathscr{C}(z_i)(x)$. First, suppose $\sigma_i(x) < \mathscr{C}(z_{i-1})(x)$. Here,

$$\mathscr{C}(z_{i-1})(x) > \mathscr{C}(z_{i-1\setminus i}.z_{i,i}) = \mathscr{C}(z_i)(x) .$$

Thus, $i$ can manipulate $\mathscr{C}$ at $z_{i-1}$ by submitting $z_{i,i}$ rather than $z_{i-1,i}$. Second, suppose $\sigma_i(x) \geq \mathscr{C}(z_{i-1})(x)$. Because $\mathscr{C}(z_{i-1})(x) > \mathscr{C}(z_i)(x)$, then $\sigma_i(x) > \mathscr{C}(z_i)(x)$. However,

$$\mathscr{C}(z_i)(x) < \mathscr{C}(z_{i\setminus i}, z_{i-1,i})(x) = \mathscr{C}(z_i)(x) ,$$

which is another contradiction. Hence, $\mathscr{C}(z_{i-1})(x) = \mathscr{C}(z_i)(x)$.

Thus, for any $z_i, z_j \in Z$, $\mathscr{C}(z_i)(x) = \mathscr{C}(z_j)(x)$. Accordingly, $\mathscr{C}(\sigma)(x) = \mathscr{C}(z_0)(x) = \mathscr{C}(z_n)(x) = \mathscr{C}(\sigma')(x)$. Hence, $\mathscr{C}$ satisfies the $\sigma$-only condition. □

**Proposition 5.28.** *If a fuzzy choice function $\mathscr{C}$ is strategy-proof, then it is weakly Paretian.*

*Proof.* Assume $\mathscr{C}$ is SP. Now, suppose $\mathscr{C}$ is not weakly Paretian. This proof will show that this leads to a contradiction. There are two cases for consideration.

Case 1: Suppose there exists $x \in X$ and $\sigma \in \mathscr{F}(X)^n$ such that $\mathscr{C}(\sigma)(x) < \min_{i \in N}\{\sigma_i(x)\}$. By full range, we know there also exists a $\sigma' \in \mathscr{F}(X)^n$ such that $\mathscr{C}(\sigma')(x) \geq \min_{i \in N}\{\sigma_i'(x)\}$. Now construct a sequence of profiles $Z = \{z_0, z_1, \ldots, z_i, \ldots, z_n\}$ such that

$$z_0 = (\sigma_1(x), \ldots, \sigma_i(x), \ldots, \sigma_n(x))$$
$$z_1 = (\sigma_1'(x), \ldots, \sigma_i(x), \ldots, \sigma_n(x))$$
$$\vdots$$
$$z_i = (\sigma_1'(x), \ldots, \sigma_i'(x), \ldots, \sigma_n(x))$$
$$\vdots$$
$$z_n = (\sigma_1'(x), \ldots, \sigma_i'(x), \ldots, \sigma_n'(x)),$$

where $z_{j,k}$ signifies $\sigma_k$ in profile $z_j$ and $z_{j\setminus k}$ denotes the removal of $\sigma_k$ from $z_j$. By construction $z_0 = \sigma$ and $z_n = \sigma'$. By assumption, $\mathscr{C}(\sigma)(x) < \mathscr{C}(\sigma')(x)$ so $\mathscr{C}(z_0)(x) < \mathscr{C}(z_n)(x)$. Then it is obvious there exists $z_{i-1}, z_i \in Z$ such that $\mathscr{C}(z_{i-1})(x) < \mathscr{C}(z_i)(x)$. Suppose $\sigma_i(x) > \mathscr{C}(z_{i-1})(x)$. Then

$$\mathscr{C}(z_{i-1})(x) < \mathscr{C}(z_{i-1\setminus i}, z_{i,i})(x) = \mathscr{C}(z_i)(x) .$$

Thus, $i$ can manipulation $\mathscr{C}$ at $z_{i-1}$ by submitting $z_{i,i}$ rather than $z_{i-1,i}$. Now suppose $\sigma_i(x) \leq \mathscr{C}(z_{i-1})(x)$, then $\sigma_i(x) < \mathscr{C}(z_i)(x)$. However,

$$\mathscr{C}(z_i)(x) > \mathscr{C}(z_{i\setminus i}, z_{i-1,i})(x) = \mathscr{C}(z_{i-1})(x) .$$

Thus, $i$ can manipulate $\mathscr{C}$ at $z_i$ by submitting $z_{i-1,i}$ rather than $z_{i,i}$. This is a contradiction.

Case 2: Suppose there exists $x \in X$ and $\sigma \in \mathscr{F}(X)^n$ such that $\mathscr{C}(\sigma)(x) > \max_{i \in N}\{\sigma_i(x)\}$. By full range, we know there also exists a $\sigma' \in \mathscr{F}(X)^n$ such that $\mathscr{C}(\sigma')(x) \leq \max_{i \in N}\{\sigma_i'(x)\}$. Now construct a sequence of profiles $Z = \{z_0, z_1, \ldots, z_i, \ldots, z_n\}$ in a manner detailed above. By construction $z_0 = \sigma$ and $z_n = \sigma'$. By assumption, $\mathscr{C}(\sigma)(x) > \mathscr{C}(\sigma')(x)$ so $\mathscr{C}(z_0)(x) > \mathscr{C}(z_n)(x)$. Likewise, there must exist $z_{i-1}, z_i \in Z$ such that $\mathscr{C}(z_{i-1})(x) > \mathscr{C}(z_i)(x)$. Suppose $\sigma_i(x) < \mathscr{C}(z_{i-1})(x)$. Then

$$\mathscr{C}(z_{i-1})(x) > \mathscr{C}(z_{i-1\setminus i}, z_{i,i}) = \mathscr{C}(z_i)(x) .$$

Thus, $i$ can manipulate $\mathscr{C}$ at $z_{i-1}$ by submitting $z_{i,i}$ rather than $z_{i-1,i}$. Now suppose $\sigma_i(x) \geq \mathscr{C}(z_{i-1})(x)$, then $\sigma_i(x) > \mathscr{C}(z_i)(x)$. However, $\mathscr{C}(z_i)(x) < \mathscr{C}(z_{i\setminus i}, z_{i-1,i})(x) = \mathscr{C}(z_{i-1})(x)$, another contradiction.

Hence, $\max_{i \in N}\{\sigma_i(x)\} \geq \mathscr{C}(\sigma)(x) \geq \min_{i \in N}\{\sigma_i(x)\}$, and $\mathscr{C}$ is weakly Paretian. $\qquad \square$

*Example 5.29.* Let $\mathscr{C}$ be a fuzzy choice function such that $\mathscr{C}(\sigma)(x) = c$ for all $x \in X$ and all $\sigma \in \mathscr{F}(X)^n$, where $c \in [0,1]$. In this case, $\mathscr{C}$ is SP, i.e. $\mathscr{C}(\sigma_{N \setminus i}, \sigma_i')(x) = c$ for all $i \in N$ and all $\sigma_i' \in \mathscr{F}(X)$, but is not weakly Paretian if $c < \sigma_i(x)$ for all $i \in N$ or $c > \sigma_i(x)$ for all $i \in N$.

Hence, if a fuzzy choice function $\mathscr{C}$ is strategy-proof but does not satisfy full range, then it does not satisfy weak Paretianism. Together, Proposition 5.28 and

Example 5.29 demonstrate the equivalence of weak Paretianism and full range under strategy-proof social choice functions. The next proposition and the subsequent example highlight the relationship between strategy-proofness and monotonicity.

**Proposition 5.30.** *If a fuzzy choice function $\mathscr{C}$ is strategy-proof, then it is monotonic.*

*Proof.* Assume $\mathscr{C}$ is SP. Now suppose $\mathscr{C}$ is not monotonic. This proof will illustrate that this leads to a contradiction. Because $\mathscr{C}$ is not monotonic, there exists an $x \in X$ and $\sigma, \sigma' \in \mathscr{F}(X)^n$ such that $\sigma_i(x) \leq \sigma_i'(x)$, for all $i \in N$ and $\mathscr{C}(\sigma)(x) > \mathscr{C}(\sigma')(x)$. Because $\mathscr{C}$ is $\sigma$-only, $\sigma \neq \sigma'$, so there exists at least one $i \in N$ such that $\sigma_i(x) < \sigma_i'(x)$. Now construct the following sequence of profiles $Z = \{z_0, z_1, \ldots, z_i, \ldots, z_n\}$ such that

$$z_0 = (\sigma_1(x), \ldots, \sigma_i(x), \ldots, \sigma_n(x))$$
$$z_1 = (\sigma_1'(x), \ldots, \sigma_i(x), \ldots, \sigma_n(x))$$
$$\vdots$$
$$z_i = (\sigma_1'(x), \ldots, \sigma_i'(x), \ldots, \sigma_n(x))$$
$$\vdots$$
$$z_n = (\sigma_1'(x), \ldots, \sigma_i'(x), \ldots, \sigma_n'(x)),$$

where $z_{j,k}$ signifies $\sigma_k$ in profile $z_j$ and $z_{j \setminus k}$ denotes the removal of $\sigma_k$ from $z_j$. Because $\mathscr{C}$ is $\sigma$-only by Proposition 5.27, $\mathscr{C}(z_{i-1})(x) = \mathscr{C}(z_i)(x)$ if $z_{i-1}(x) = z_i(x)$. By the construction of $Z$, there exists $i \in N$ such that $z_{i,i}(x) > z_{i-1,i}(x)$ and $\mathscr{C}(z_i)(x) < \mathscr{C}(z_{i-1})(x)$.

Suppose $\sigma_i(x) > \mathscr{C}(z_i)(x)$. Then, $\mathscr{C}(z_i)(x) < \mathscr{C}(z_{i \setminus i}, z_{i-1,i})$, then $i$ can manipulate $\mathscr{C}$ at $z_i$ by submitting $z_{i-1,i}$ rather than $z_{i,i}$. Second suppose $\sigma_i(x) \leq \mathscr{C}(z_i)(x)$. Then by assumption, $\sigma_i(x) < \mathscr{C}(z_{i-1})(x)$, and $\mathscr{C}(z_{i-1})(x) > \mathscr{C}(z_{i-1 \setminus i}, z_{i,i})$. Thus, $i$ can manipulate $\mathscr{C}$ at $z_{i-1}$ by submitting $z_{i,i}$ rather than $z_{i-1,i}$. Hence, $\mathscr{C}(\sigma)(x) \leq \mathscr{C}(\sigma')(x)$.

Proposition 5.30 demonstrates that strategy-proofness is sufficient for a monotonic fuzzy choice function. However, previous research in crisp preference relations has shown that strategy-proofness is necessary and sufficient for a monotonic condition Muller and Satterthwaite (1977). This does not hold in the fuzzy framework as the following example demonstrates.

*Example 5.31.* For all $x \in X$ and all $\sigma \in \mathscr{F}(X)^n$, define $\mathscr{C}$ as $\mathscr{C}(\sigma)(x) = (\frac{1}{n}) \sum_{\forall i \in N} \sigma_i(x)$. It is easy to verify that $\mathscr{C}$ is a monotonic choice function. Now let $x \in X$. Suppose $N = \{1, 2, 3\}$ and $\sigma \in \mathscr{F}(X)^n$ is such that $\sigma(x) = (.4, .1, .6)$. In this case, $\mathscr{C}(\sigma)(x) = \frac{1}{3}(.4 + .1 + .6) = .367$. Obviously, some $i \in \{1, 3\}$ could manipulate $\mathscr{C}$ with some $\sigma_i'(x) > \sigma_i(x)$. Also, $i = 2$ can manipulate $\mathscr{C}$ by submitting some $\sigma_3'(x) > \sigma_3(x) = .6$.

The following definition is necessary to characterize the domain of strategy-proof choice functions.

**Definition 5.32 (augmented median voter rule).** Let $M : \mathscr{F}(X)^n \to \mathscr{F}(X)$ be a fuzzy choice function defined as follows:

$$M(\sigma)(x) = \text{med}\{p_1, \ldots, p_{n-1}, \sigma_1(x), \ldots, \sigma_n(x)\},$$

where $p_i \in [0,1]$, $i = 1, \ldots, n-1$.[2]

Here $\{p_1, \ldots, p_{n-1}\}$ is a set of predefined "phantom" alternatives that serve two purposes. First, it allows $M(\sigma)$ to be generalized to any type of rank-selecting function such as minimum or maximum. Second, the set also ensures an odd number of alternatives, so a median can always be selected. If not, the median becomes an average of two choice intensities, and this procedure is manipulable as demonstrated in Example 5.31.

**Lemma 5.33.** *M is a strategy-proof fuzzy choice function.*

*Proof.* Assume $M$ is defined as in Definition 5.32. Suppose $M$ is not strategy-proof. This leads to a contradiction.

Let $x \in X$, $i \in N$ and $\sigma \in \mathscr{F}(X)^n$. The are two cases to consider. First, suppose $\sigma_i(x) < M(\sigma)(x)$. Then there exists a $\sigma_i' \in \mathscr{F}(X)$ such that $M(\sigma)(x) > M(\sigma_{N\setminus i}, \sigma_i')(x)$ by assumption. For clarity, let $M(\sigma)(x) = a$ and $M(\sigma_{N\setminus i}, \sigma_i')(x) = b$. Obviously, $a > b$.

Note that $a \in \{p_1, \ldots, p_{n-1}, \sigma_1(x), \ldots, \sigma_i(x), \ldots, \sigma_n(x)\}$. Because $\sigma_i(x) < a$, $\sigma_i'(x) \not\leq \sigma_i(x)$, else

$$\text{med}\{p_1, \ldots, p_{n-1}, \sigma_1(x), \ldots, \sigma_i(x), \ldots, \sigma_n(x)\}$$
$$= \text{med}\{p_1, \ldots, p_{n-1}, \sigma_1(x), \ldots, \sigma_i(x), \ldots, \sigma_n(x)\},$$

Thus, $\sigma_i'(x) > \sigma(x)$. This implies then

$$b = \text{med}\{p_1, \ldots, p_{n-1}, \sigma_1(x), \ldots, \sigma_i'(x), \ldots, \sigma_n(x)\}$$
$$\geq a,$$

However, $M(\sigma_{N\setminus i}, \sigma_i')(x) = b$ and $a > b$. This is a contradiction.

Second, suppose $\sigma_i(x) > M(\sigma)(x)$. Then there exists a $\sigma_i' \in \mathscr{F}(X)$ such that $M(\sigma)(x) < M(\sigma_{N\setminus i}, \sigma_i')$. Again, let $M(\sigma)(x) = a$ and $M(\sigma_{N\setminus i}, \sigma_i') = b$. Obviously, $a \neq b$ and $a < b$. Because $\sigma_i(x) > a$, $\sigma_i'(x) \not\geq \sigma_i(x)$, else

$$\text{med}\{p_1, \ldots, p_{n-1}, \sigma_1(x), \ldots, \sigma_i(x), \ldots, \sigma_n(x)\}$$
$$= \text{med}\{p_1, \ldots, p_{n-1}, \sigma_1(x), \ldots, \sigma_i'(x), \ldots, \sigma_n(x)\}.$$

Thus, $\sigma_i'(x) < \sigma_i(x)$. However, as before,

---

[2] Several studies characterize the augmented median voter rule using $\{p_1, \ldots, p_{n+1}\}$ Austen-Smith and Banks (2005); Barberá (2001); Moulin (1980). In this case, however, by setting $p_1 = 0$ and $p_{n+1} = 1$, the rule can be more succinctly written using $n-1$ alternatives.

$$b = \mathrm{med}\{p_1, \ldots, p_{n-1}, \sigma_1(x), \ldots, \sigma_i'(x), \ldots, \sigma_n(x)\}$$
$$\leq a,$$

a contradiction. The desired result now follows.

In words, Lemma 5.33 demonstrates that if $i \in N$ attempts to manipulate the value of $M(\sigma)(x)$ with $\sigma_i'$, one of two events will happen. Either the new manipulated social choice will be identical to the original social choice intensity or the new manipulated social choice intensity will move further away from $i$'s sincere intensity for $x \in X$. Hence, $i \in N$ will not be better off by reporting any $\sigma_i' \neq \sigma_i$.

The chapter's main theorem that $\mathscr{C}$ is strategy-proof if and only if $\mathscr{C} = M$ (Definition 5.32) follows the logic in Ching and Serizawa (1998) and makes use of the following lemma.

**Lemma 5.34.** *Let $\bar{\sigma}_i(x) = 1$ and $\underline{\sigma}_i(x) = 0$ for all $x \in X$ and some $i \in N$. A fuzzy choice function $\mathscr{C}$ is strategy-proof if and only if, for all $\sigma \in \mathscr{F}(X)^n$, all $x \in X$ and all $i \in N$, the following holds:*

$$\mathscr{C}(\sigma)(x) = med\{\sigma_i(x), \mathscr{C}(\sigma_{N\setminus i}, \underline{\sigma}_i)(x), \mathscr{C}(\sigma_{N\setminus i}, \bar{\sigma}_i)(x)\}.$$

*Proof.* Suppose $\mathscr{C}$ is a strategy-proof fuzzy choice function. Let $\sigma \in \mathscr{F}(X)^n$ and $x \in X$. By monotonicty and Proposition 5.30, $\mathscr{C}(\sigma_{N\setminus i}, \bar{\sigma}_i)(x) \geq \mathscr{C}(\sigma_{N\setminus i}, \underline{\sigma}_i)(x)$. There are three cases to consider to prove the relationship in the lemma.

Case 1: Suppose

$$\sigma_i(x) \in (\mathscr{C}(\sigma_{N\setminus i}, \underline{\sigma}_i)(x), \mathscr{C}(\sigma_{N\setminus i}, \bar{\sigma}_i)(x)).$$

Further, suppose $\mathscr{C}(\sigma)(x) < \sigma_i(x)$, then $i$ can submit $\bar{\sigma}_i(x)$ where

$$\mathscr{C}(\sigma_{N\setminus i}, \bar{\sigma}_i)(x) > \sigma_i(x) > \mathscr{C}(\sigma)(x).$$

Thus, $\mathscr{C}(\sigma)(x) < \mathscr{C}(\sigma_{N\setminus i}, \bar{\sigma}_i)(x)$, and $\mathscr{C}$ is manipulable, a contradiction. Now suppose $\mathscr{C}(\sigma)(x) > \sigma_i(x)$. Similarly,

$$\mathscr{C}(\sigma_{N\setminus i}, \underline{\sigma}_i)(x) < \sigma_i(x) < \mathscr{C}(\sigma)(x).$$

Because $\mathscr{C}(\sigma_{N\setminus i}, \underline{\sigma}_i)(x) < \mathscr{C}(\sigma)(x)$, $\mathscr{C}$ is manipulable, a contradiction. Hence, $\mathscr{C}(\sigma)(x) = \sigma_i(x)$ when

$$\sigma_i(x) \in (\mathscr{C}(\sigma_{N\setminus i}, \underline{\sigma}_i)(x), \mathscr{C}(\sigma_{N\setminus i}, \bar{\sigma}_i)(x)).$$

Case 2: Suppose $\sigma_i(x) \leq \mathscr{C}(\sigma_{N\setminus i}, \underline{\sigma}_i)(x)$. To see that $\mathscr{C}(\sigma)(x) = \mathscr{C}(\sigma_{N\setminus i}, \underline{\sigma}_i)(x)$, suppose $\mathscr{C}(\sigma)(x) < \mathscr{C}(\sigma_{N\setminus i}, \underline{\sigma}_i)(x)$. Then

$$\underline{\sigma}_i(x) \leq \mathscr{C}(\sigma)(x) < \mathscr{C}(\sigma_{N\setminus i}, \underline{\sigma}_i)(x).$$

In this case, $i$ can manipulate $\mathscr{C}$ at $(\sigma_{N\setminus i}, \underline{\sigma}_i)$ by submitting $\sigma_i$ rather than $\underline{\sigma}_i$, a contradiction. Likewise, suppose

$$\mathscr{C}(\sigma)(x) > \mathscr{C}(\sigma_{N\setminus i}, \underline{\sigma}_i)(x) \geq \sigma_i(x) .$$

Now, $i$ can manipulate $\mathscr{C}$ at $\sigma$ by submitting $\underline{\sigma}_i$ rather than $\sigma_i$. Hence, $\mathscr{C}(\sigma)(x) = \mathscr{C}(\sigma_{N\setminus i}, \underline{\sigma}_i)(x)$ when $\sigma_i(x) \leq \mathscr{C}(\sigma_{N\setminus i}, \underline{\sigma}_i)(x)$.

Case 3: Suppose $\bar{\sigma}_i(x) \geq \mathscr{C}(\sigma_{N\setminus i}, \bar{\sigma}_i)(x)$ and $\mathscr{C}(\sigma)(x) > \mathscr{C}(\sigma_{N\setminus i}, \bar{\sigma}_i)(x)$. Then

$$\bar{\sigma}_i(x) \geq \mathscr{C}(\sigma)(x) > \mathscr{C}(\sigma_{N\setminus i}, \bar{\sigma}_i)(x) ,$$

where $i$ can manipulate $\mathscr{C}$ at $(\sigma_{N\setminus i}, \bar{\sigma}_i)$ by submitting $\sigma_i$ rather than $\bar{\sigma}_i$. Now suppose $\mathscr{C}(\sigma)(x) < \mathscr{C}(\sigma_{N\setminus i}, \bar{\sigma}_i)(x)$. Then

$$\mathscr{C}(\sigma)(x) < \mathscr{C}(\sigma_{N\setminus i}, \bar{\sigma}_i)(x) \leq \sigma_i(x) ,$$

and $i$ can manipulate $\mathscr{C}$ at $\sigma$ by submitting $\bar{\sigma}_i$ rather than $\sigma_i$. Hence, $\mathscr{C}(\sigma)(x) = \mathscr{C}(\sigma_{N\setminus i}, \bar{\sigma}_i)(x)$ when $\sigma_i(x) \geq \mathscr{C}(\sigma_{N\setminus i}, \bar{\sigma}_i)(x)$.

The preceding arguments prove that if a fuzzy choice function $\mathscr{C}$ is strategy-proof, then

$$\mathscr{C}(\sigma)(x) = \text{med}\left\{\sigma_i(x), \mathscr{C}(\sigma_{N\setminus i}, \underline{\sigma}_i)(x), \mathscr{C}(\sigma_{N\setminus i}, \bar{\sigma}_i)(x)\right\} .$$

The fact that

$$\mathscr{C}(\sigma)(x) = \text{med}\left\{\sigma_i(x), \mathscr{C}(\sigma_{N\setminus i}, \underline{\sigma}_i)(x), \mathscr{C}(\sigma_{N\setminus i}, \bar{\sigma}_i)(x)\right\}$$

is strategy-proof for all $\sigma \in \mathscr{F}(X)$, all $x \in X$ and all $i \in N$ is easily obtained from Lemma 5.33.    □

In words, the arguments in Lemma 5.34 demonstrate that if $\mathscr{C}$ is a strategy-proof fuzzy social choice function, an individual $i$ cannot move the social choice intensity of an alternative in the direction of her sincere choice intensity by submitting either of the extreme functions $\bar{\sigma}_i$ or $\underline{\sigma}_i$. Similar to the logic in Lemma 5.34, submitting one of these profiles will either have no effect on the social choice intensity or will move the social intensity further from the individual's sincere intensity. Because strategy-proofness is monotonic and $\sigma$-only, individuals who are not able to manipulate the fuzzy social choice with extreme profiles will not be able to manipulate the social choice with less extreme profiles. Thus, the following result now emerges.

**Theorem 5.35.** *Any fuzzy choice function $\mathscr{C}$ is strategy-proof if and only if it is a fuzzy augmented median voter rule.*

*Proof.* Once we have established that strategy-proofness implies $\sigma$-only (Proposition 5.27) and the relationship in Lemma 5.34, Ching (1997) shows that $\mathscr{C}$ must be the augmented median voter rule defined in Definition 5.32. Lemma 5.33 demonstrates the strategy-proofness of the fuzzy augmented median voter rule.    □

## 5.4 Implications for the Spatial Model

**Definition 5.36 (coalition).** A nonempty subset $C$ of $N$ is called a *coalition* if and only if $\exists x \in X$ such that $\sigma_c(x) > 0, \forall c \in C$. Let $\hat{C}$ denote the set of all possible majority-supported coalitions. Then a subset $K$ of $N$ is called a *collegium* if every member of $K$ is in $\hat{C}$.

*Example 5.37.* This example illustrates the problem of group decision making when the selection process is based upon the preference functions described in Chapter 3. This spatial model illustrates how, under the definition presented in Chapter 1, the fuzzy maximal set can be manipulated by certain actors.

Consider Figure 5.1, where

$$PS_N(R) = \{(.25, 1, 0), (0, .75, .25), (0, .5, .5), (1, 0, 0), (0, 0, 1)\},$$

and the shaded grey areas represent possible coalition formations. Currently, we aggregate the above fuzzy preferences as follows:

1). Individual fuzzy preferences are aggregated into a fuzzy choice function, $MG_i(\sigma_i, \rho_i)(x)$, which signifies the degree to which $i \in N$ chooses alternative $x$. In the case of our previously defined binary preference relation, $\rho_i$, $MG_i(\sigma_i, \rho_i)(x) = \sigma_i(x)$.
2). For every $C \subseteq N$, such that $C$ controls a majority of legislative seats, we then take the minimum of $\sigma_i(x)$ for $\forall i \in C$ and $\forall x \in X$.
3). Finally, for every $C \subseteq N$ we take the maximum across the values calculated in 2. This final value refers to the degree to which each coalition is chosen. The coalition with the highest choice score is predicted to form.

In Figure 5.1, we would compute the following coalitions to their respective degrees: $\{1, 2\}$ at .25, $\{1, 3\}$ at 0, $\{2, 3\}$ at .5, $\{1, 2, 3\}$ at 0. Since $\vee\{.25, 0, .5, 0\} = .5$, the maximal set forms at $\{2, 3\}$.

But, since actor 2 is a member of every decisive and possible coalition (i.e., a collegium where $K = 2$), he can manipulate the fuzzy maximal set by advertising an artificially narrow profile around his ideal region. For instance, it would be in actor 2's interests to advertise an insincere profile ending at $x_2$, thus eliminating his least-desirable possible outcome $\{0, .5, .5\}$. In doing this, actor 2 forces actor 3 to coalesce around a region closer to actor 2's ideal range (from $\sigma_i = .5$ to $\sigma_i = .75$).

Figure 5.1 illustrates the role of $M(\sigma)$ in avoiding this manipulation.

Lemma 5.34 and Theorem 5.35 demonstrate that a fuzzy social choice function is strategy-proof if and only if it is a form of the fuzzy augmented median voter rule from Definition 5.32. Further, in contrast to previous results using traditional preference relations, this relationship holds without restricting the domain of individual preferences, $\mathscr{F}(X)^n$. While the representation of an individual's preferences with the $\sigma_i$ function can be used to produce a transitive weak preference relations $R_i$, where $xR_iy \iff \sigma_i(x) \geq \sigma_i(y)$, the use of the $\sigma_i$ creates substantive differences between the structure of traditional and fuzzy strategy-proof choice functions. The reason these differences emerge is that the group of individuals no longer decides what alternative to select but rather decides the degree to which the group chooses each alternative.

To illustrate this difference, the model presented in this chapter is applied to the spatial model, where the set of alternatives $X$ becomes some subset of $k$-dimensional

Euclidean space or $\mathbb{R}^k$. When $k = 1$, $\sigma_i$ can be presented by a traditional fuzzy number, i.e. $\sigma_i : \mathbb{R}^1 \to [0,1]$ for $i = 1,\dots,n$, which is a similar definition to that of a fuzzy subset. Further, it is often assumed that $\sigma_i$ is *normal*, which requires that there exists $x \in X$ such that $\sigma_i(x) = 1$, for $i = 1,\dots,n$. In words, normality ensures that every actor views at least one alternative as ideal. While the condition seems innocuous and strongly related to the standard assumptions of spatial models, it is not necessary to the framework presented here.

Figure 5.1 illustrates a three-player fuzzy preference profile where each $\sigma_i$ is represented by a normal fuzzy number in one-dimensional space. It is obvious that the fuzzy number representation allows for greater variation in individual choice than a traditional single-peaked profile. In this example, not only are the fuzzy choice intensities able to capture the single-plateau characteristics of concern to some scholars Berga and Moreno (2009); Ching and Serizawa (1998); Massó and Neme (2001), but they also allow for non-single-peaked preferences (player 2), which is one substantive difference between exact and fuzzy choice. Further, the shaded areas show the social choice intensities induced by the fuzzy median rule.

To see that the social choice is indeed strategy-proof, even with non-single-peaked preferences, consider $x_1 \in X$. Here, $\sigma_1(x_1) > 0$, $\sigma_2(x_1) = 0$, and $\sigma_3(x_1) = 0$. Regardless of any $\sigma_1' \in \mathscr{F}(X)$ and any possible values of $\sigma_1'(x_1)$, $M(\sigma)(x_1) = 0$ because $M(\sigma)(x_1) = \text{med}\{\sigma_1'(x_1), \sigma_2(x_1), \sigma_3(x_1)\} = \text{med}\{\sigma_1'(x_1), 0, 0\}$. In addition, consider $x_2 \in X$, where $\sigma_1(x_2) < M(\sigma)(x_2)$, $\sigma_2(x_2) > M(\sigma)(x_2)$, and $\sigma_3(x_2) = M(\sigma)(x_2)$. Similarly, player 1 cannot manipulate the fuzzy choice for $x_2$. For any $\sigma_1' \in \mathscr{F}(X)$ and any specific value of $\sigma_1'(x_2)$, $M(\sigma_{N\setminus 1}, \sigma_1')(x_2) \geq M(\sigma)(x_2)$. Thus, player 1 can only move the degree of social choice further away from her sincere choice intensity for $x_2$.

When working in multidimensional space, the framework of the $\sigma_i$ function remains largely the same, where $\sigma_i : \mathbb{R}^k \to [0,1]$. When $k = 2$, we are interested in fuzzy subsets where every point in the set $\sigma_i^{-1}((0,1])$ is in the interior or on the

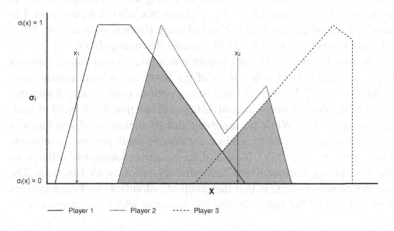

— Player 1      ---- Player 2      ····· Player 3

**Fig. 5.1** Three-player Example of the Fuzzy Median Rule in One-Dimensional Space

boundary of a simple closed curve. A simple closed curve is a curve for which there is a one-to-one continuous function from the unit circle onto it. In addition, a simple closed curve has an interior that is bounded and an exterior, but there is no need for the curve to be convex. Finally, we can restrict $\sigma_i$ in a particular way such that, for all $t \in \text{Im}(\sigma_i)\backslash\{0\}$, $\{x \in X \mid \sigma_i(x) = t\}$ forms a compact set.

Figure 5.2 presents a three-player fuzzy preference profile in two-dimensional space, and the $\sigma_i$ function becomes a third dimension perpendicular to both the X and Y policy dimensions. In this case, $\text{Im}(\sigma_i) = \{0, .25, .5, .75, 1.0\}$, where $\sigma_i(x) = 1$ can be represented by individual $i$'s inner-most indifference circle and $\sigma_i(x) = 0$ signifies the area outside $i$'s outer-most indifference circle. When the individual fuzzy choice intensities are constructed in this manner, they are similar to a Likert scale. As in the previous example, the shaded gray areas show the social choice induced by the fuzzy median rule, and darker areas represent a more intense social choice. Unlike the exact case, the fuzzy median rule remains strategy-proof in two-dimensional space without using the dimension-by-dimension median rule. It is easy to extrapolate that this result holds even if there exists a $t \in \text{Im}(\sigma_i)\backslash\{0\}$ such that $\{x \in X \mid \sigma_i(x) = t\}$ forms a concave set.

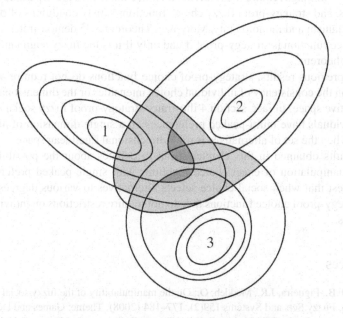

**Fig. 5.2** Fuzzy Median Rule in Two-Dimensional Space

Finally, another substantive difference occurs when there does not exist an $x \in X$ such that $\sigma_i(x) > 0$ for all $i \in S$, where $S \subseteq N$ such that $|S| > 1$. In this case, players have non-intersecting $\text{Im}(\sigma_i)$. Under this scenario, $M(\sigma)(x) = 0$ for all $x \in X$ unless the phantom alternatives are arranged such that $M(\sigma)(x) = \max_{i \in N}\{\sigma_i(x)\}$, and the group of players rejects all possible alternatives. In this case, it is unclear as to what the social choice is. In the traditional approach, a choice function associates an alternative to all possible combination of individual preferences that are transitive and complete relations. However, in the fuzzy case, when the social choice function is designating a choice intensity to each alternative, it is possible that a strategy-proof choice function assigns a zero intensity to all alternatives. This is not necessarily a misrepresentation of the original intention of strategy-proof choice functions if rejecting all alternatives is some type of social choice.

## 5.5 Conclusions

This chapter proposed a framework for characterizing strategy-proof fuzzy social choice functions in which individual preferences and the social choice are represented by fuzzy subsets of the set of alternatives. Essentially, actors are deciding to what degree the group chooses each alternative given an $n$-tuple of fuzzy individual choice intensities rather than choosing a specific alternative, which is the approach taken in previous studies of both exact and fuzzy social choice. Similar to previous results, these findings require that individual choice intensities are separable across alternatives, and strategy-proof fuzzy choice functions satisfy conditions of $\sigma$-only, weak Paretianism and monotonicity. Moreover, Theorem 5.35 demonstrates a fuzzy social choice function is strategy-proof if and only if it is the fuzzy augmented median voter theorem.

Unlike previous results, strategy-proof choice functions do not require any restrictions on the consistency of individual choice intensities or the dimensionality of the alternative space. In fact, section 4 illustrates strategy-proof fuzzy social choice when individuals have multi-peaked preferences on a single dimension of alternatives and when the set of alternatives is multidimensional Euclidean space.

The results obtained in this chapter speak to debates about the possibility of strategic manipulation of exact choice functions with single-peaked preferences. They suggest that when social choice selects alternatives to various degrees there exists strategy-proof choice functions that do not require restrictions on individuals' preferences.

## References

Abdelaziz, F.B., Figueira, J.R., Meddeb, O.: On the manipulability of the fuzzy social choice functions. Fuzzy Sets and Systems 159(2), 177–184 (2008), Theme: Games and Decision

Austen-Smith, D., Banks, J.S.: Positive Political Theory II: Strategy and Structure. University of Michigan Press, Ann Arbor (2005)

Banerjee, A.: Fuzzy choice functions, revealed preference and rationality. Fuzzy Sets and Systems 70(1), 31–43 (1995)

Barberá, S.: An introduction to strategy-proof social choice functions. Social Choice and Welfare 18(4), 619–653 (2001), http://dx.doi.org/10.1007/s003550100151

Berga, D., Moreno, B.: Strategic requirements with indifference: single-peaked versus single-plateaued preferences. Social Choice and Welfare 32(2), 275–298 (2009), http://dx.doi.org/10.1007/s00355-008-0323-y

Breton, M.L., Sen, A.: Separable preferences, strategyproofness, and decomposability. Econometrica 67(3), 605–628 (1999), http://ideas.repec.org/a/ecm/emetrp/v67y1999i3p605-628.html

Ching, S.: Strategy-proofness and "median voters". International Journal of Game Theory 26(4), 473–490 (1997), http://dx.doi.org/10.1007/BF01813886

Ching, S., Serizawa, S.: A maximal domain for the existence of strategy-proof rules. Journal of Economic Theory 78(1), 157–166 (1998), http://www.sciencedirect.com/science/article/pii/S0022053197923371

Clark, T.D., Larson, J.M., Mordeson, J.N., Potter, J.D., Wierman, M.J. (eds.): Applying Fuzzy Mathematics to Formal Models in Comparative Politics. STUDFUZZ, vol. 225. Springer, Heidelberg (2008)

Côrte-Real, P.P.: Fuzzy voters, crisp votes. International Game Theory Review 9(1), 67–86 (2007)

Dasgupta, M., Deb, R.: Fuzzy choice functions. Social Choice and Welfare 8(2), 171–182 (1991), http://dx.doi.org/10.1007/BF00187373

Dryzek, J.S., List, C.: Social choice theory and deliberative democracy: A reconciliation. British Journal of Political Science 33, 1–28 (2003), http://journals.cambridge.org/article_S0007123403000012

Duddy, C., Perote-Peña, J., Piggins, A.: Arrow's theorem and max-star transitivity. Social Choice and Welfare 36(1), 25–34 (2011), http://dx.doi.org/10.1007/s00355-010-0461-x

Dutta, B.: Fuzzy preferences and social choice. Mathematical Social Sciences 13(3), 215–229 (1987)

Georgescu, I.: Rational Choice and Revealed Preference: A Fuzzy Approach. Ph.D. thesis, Abo Akademi University Turku Centre for Computer Science, Lemminkainengatan 14B Fin-20520 Abo, Finland (2005)

Georgescu, I.: Similarity of fuzzy choice functions. Fuzzy Sets and Systems 158(12), 1314–1326 (2007), http://www.sciencedirect.com/science/article/pii/S0165011407000346

Gibbard, A.: Manipulation of voting schemes: A general result. Econometrica 41(4), 587–601 (1973)

Llamazares, B.: Factorization of fuzzy preferences. Social Choice and Welfare 24(3), 475–496 (2005), http://dx.doi.org/10.1007/s00355-003-0311-1

Mackie, G.: Democracy Defended. Cambridge University Press, New York (2003)

Massó, J., Neme, A.: Maximal domain of preferences in the division problem. Games and Economic Behavior 37(2), 367–387 (2001), http://www.sciencedirect.com/science/article/pii/S0899825601908504

Moulin, H.: On strategy-proofness and single peakedness. Public Choice 35(4), 437–455 (1980), http://dx.doi.org/10.1007/BF00128122

Muller, E., Satterthwaite, M.A.: The equivalence of strong positive association and strategy-proofness. Journal of Economic Theory 14(2), 412–418 (1977), http://ideas.repec.org/a/eee/jetheo/v14y1977i2p412-418.html

Orlovsky, S.: Decision-making with a fuzzy preference relation. Fuzzy Sets and Systems 1, 155–167 (1978)

Penn, E.M., Patty, J.W., Gailmard, S.: Manipulation and singlepeakedness: a general result. American Journal of Political Science 55, 436–439 (2011)

Perote-Peña, J., Piggins, A.: Strategy-proof fuzzy aggregation rules. Journal of Mathematical Economics 43(5), 564–580 (2007)

Reny, P.J.: Arrow's theorem and the gibbard-satterthwaite theorem: a unified approach. Economics Letters 70(1), 99–105 (2001),
http://ideas.repec.org/a/eee/ecolet/v70y2001i1p99-105.html

Richardson, G.: The structure of fuzzy preferences: Social choice implications. Social Choice and Welfare 15, 359–369 (1998)

Satterthwaite, M.A.: Strategy-proofness and Arrow's conditions: Existence and correspondence theorems for voting procedures and social welfare functions. Journal of Economic Theory 10(2), 187–217 (1975)

Sen, A.K.: Choice functions and revealed preference. Review of Economic Studies 38(115), 307–317 (1971)

Sprumont, Y.: The division problem with single-peaked preferences: A characterization of the uniform allocation rule. Econometrica 59(2), 509–519 (1991)

Suzumura, K.: Rational choice and revealed preference. Review of Economic Studies 43(1), 149–158 (1976)

Tang, F.F.: Fuzzy preferences and social choice*. Bulletin of Economic Research 46(3), 263–269 (1994),
http://dx.doi.org/10.1111/j.1467-8586.1994.tb00591.x

# Chapter 6
# Fuzzy Black's Median Voter Theorem

**Abstract.** This chapter focuses on Black's Median Voter theorem which states that the median voter's ideal alternative will be the socially preferred to other alternatives under majority rule when the following strict conditions hold: 1) all alternatives can be strictly ordered; 2) each voter strictly prefers one alternative to all other alternatives; and 3) each voter's strict preferences decrease monotonically from that alternative. This chapter shows that when fuzzy strict, rather than purely strict, preferences are applied Black's Median Voter theorem holds; but, it does not hold when fuzzy weak preferences are applied. However, a potential problem arises when using fuzzy strict preferences in cases where the maximal set, while not empty, may contain more alternatives than the median voter's ideal alternative.

## 6.1 The Structure of Fuzzy Rules and Strict Preference

The social choice literature has given considerable attention to Black's Median Voter Theorem (1969). The Theorem states that the median voter's ideal alternative will be socially preferred to other alternatives under majority rule when a set of rather strict conditions holds. These conditions collectively define what is referred to in the literature as single-peaked preference profiles. They are

(1) all alternatives can be strictly ordered,
(2) each voter strictly prefers one alternative to all other alternatives, and
(3) each voter's strict preferences decrease monotonically from that alternative.

The theorem's focus on voting rules that produce nonempty maximal sets and thereby avoids the negative conclusions associated with both Arrow's Theorem (1951) and Gibbard-Satterthwaite's Theorem (1973; 1975) has drawn substantial attention from social choice scholars. Moreover, while it is difficult to find situations in which political actors might have single-peaked preference profiles, the theorem nonetheless results in testable hypotheses (Kiewiet and McCubbins, 1988; Romer and Rosenthal, 1979).

M.B. Gibilisco et al., *Fuzzy Social Choice Theory*,
Studies in Fuzziness and Soft Computing 315,
DOI: 10.1007/978-3-319-05176-5_6, © Springer International Publishing Switzerland 2014

Despite the attention devoted by social choice scholars to Black's Median Voter Theorem, the fuzzy social choice literature thus far has devoted limited consideration to the result. Among the exceptions are Gibilisco et al (2012) and Mordeson et al (2010). Gibilisco et al. (2012) follows the lead of traditional preference relations as presented in Austen-Smith and Banks (1999) and demonstrates the type of single-peaked preferences that guarantee a non-empty fuzzy maximal set for fuzzy simple rules and fuzzy voting rules under partial (Mordeson and Clark, (2009); Mordeson et al, (2010)) and regular strict preferences (Fono and Andjiga, 2005). The paper's main results are that Black's Median Voter Theorem holds for both partial and regular fuzzy strict preferences. However, the result no longer holds for fuzzy weak preferences relations. Nonetheless, the problem emerges that while the maximal set is no longer empty, it may contain more alternatives than the median voter's ideal alternative.

This chapter revises and extends the arguments in Gibilisco et al. (2012). The results confirm those in Gibilisco et al. (2012): Black's Median Voter Theorem does not hold under all conceptualizations of the fuzzy maximal set. In what follows, we repeat much of the argument in Gibilisco et al. (2012) for the sake of clarity. The next section presents the basic notation and concepts behind fuzzy preference relations, fuzzy aggregation rules, and decisive sets. Section three then introduces a class of fuzzy simple rules and fuzzy voting rules that allow for regular strict preference relations. Section four considers the types of single-peaked preferences fuzzy voting and simple rules that produce a non-empty maximal set. Section five then illustrates the conditions under which the Median Voter Theorem holds in the fuzzy framework. Section six presents an empirical application, and Section seven adds further considerations that we will visit in the next chapter.

## 6.2   Basic Definitions and Concepts

Social choice theorists frequently concern themselves with a strict preference relation $\pi : X \times X \to [0,1]$ defined with respect to a fuzzy weak preference relation $\rho$ (a FWPR, i.e., a function $\rho : X \times X \to [0,1]$). The relation $\pi$ is irreflexive ($\pi(x,x) = 0$ for all $x \in X$), antisymmetric ($\pi(x,y) > 0$ implies $\pi(y,x) = 0$ for all distinct $x,y \in X$), and not necessarily complete.

The results in this chapter depend on the regularity of $\pi$. However, below we present several generalized definitions from the literature in which this assumption is not necessary Mordeson et al. (2010). Several of these definitions have been briefly discussed in earlier chapters as well as in Gibilisco et al. (2012), but are repeated here for the sake of clarity.

**Definition 6.1 (partial strict).** Let $\rho$ be an FWPR. Then $\pi$, the strict preference relation with respect to $\rho$, is said to be *partial* if, for all $x,y \in X$,

$$\pi(x,y) > 0 \iff \rho(y,x) = 0.$$

Given any $\bar{\rho} \in \mathscr{F}\mathscr{R}^n$, define $\tilde{P}, \tilde{R} : X \times X \to \mathscr{F}(N)$ by for all $x,y \in X$ and $i \in N$,

$$\tilde{P}(x,y;\bar{\rho})(i) = \begin{cases} \pi_i(x,y) & \text{if } \pi_i(x,y) > 0 \\ 0 & \text{otherwise,} \end{cases}$$

$$\tilde{R}(x,y;\bar{\rho})(i) = \begin{cases} \rho_i(x,y) & \text{if } \pi_i(y,x) = 0 \\ 0 & \text{otherwise.} \end{cases}$$

The reader is reminded that $\bar{\rho}$ is the preference profiles for all individuals. Thus $\tilde{P}$ and $\tilde{R}$ create two different kinds of fuzzy values; $\tilde{P}$ relates to to the set of strict preferences and $\tilde{R}$ relates to the set of weak preference relations. In these definitions of $\tilde{P}$ and $\tilde{R}$ the membership grade is focused on the actor $i$ rather than on the fixed alternatives $x$ and $y$.

**Definition 6.2 (regular).** An FWPR $\rho$ on $X$ is *regularly acyclic* if, for all $\{x_1, x_2, \ldots, x_n\} \in X$, $\pi(x_1, x_2) \wedge \pi(x_2, x_3) \wedge \ldots \wedge \pi(x_{n-1}, x_n) > 0$ implies $\pi(x_n, x_1) = 0$.

A definition for quasi-transitivity of a FWPR is given in Def 3.34 and partially quasi-transitivity of a FWPR is given in 4.3.

**Proposition 6.3.** *Let $\rho$ be an FWPR. If $\pi$ is regular, then the following properties hold:*

*(1) $\rho$ is max-min transitive implies $\rho$ is partially quasi-transitive.*
*(2) $\rho$ is weakly transitive implies $\rho$ is partially quasi-transitive.*
*(3) $\rho$ is partially quasi-transitive implies $\rho$ is regularly acyclic.*

*Proof.* (1) Let $x, y, z \in X$ and suppose $\rho$ is max-min transitive. Suppose $\pi(x,y) > 0$, $\pi(y,z) > 0$, and, contrary to the hypothesis, $\pi(x,z) = 0$. Since $\pi$ is regular,

$$\rho(x,y) > \rho(y,x) \tag{6.1}$$
$$\rho(y,z) > \rho(z,y) \tag{6.2}$$
$$\rho(z,x) \geq \rho(x,z). \tag{6.3}$$

Since $\rho$ is max-min transitive,

$$\rho(x,z) \geq \rho(x,y) \wedge \rho(y,z). \tag{6.4}$$

There are two cases to consider.

    a. First, suppose $\rho(x,y) > \rho(y,z)$. Then $\rho(x,z) \geq \rho(y,z)$ by (6.4), and thus $\rho(z,x) \geq \rho(y,z)$ by (6.3). By max-min transitivity, $\rho(z,y) \geq \rho(z,x) \wedge \rho(x,y)$. By (6.2), it follows that $\rho(y,z) > \rho(z,x) \wedge \rho(x,y)$. However, we already have shown $\rho(y,z) \leq \rho(z,x)$ and $\rho(y,z) < \rho(x,y)$, a contradiction.

    b. Second, suppose $\rho(y,z) \geq \rho(x,y)$. Then $\rho(x,z) \geq \rho(x,y)$ by (6.4), and $\rho(z,x) \geq \rho(x,y)$ by (6.3). By max-min transitivity, we have $\rho(y,x) \geq \rho(y,z) \wedge \rho(z,x)$. However, $\rho(y,z) \geq \rho(x,y)$ and $\rho(z,x) \geq \rho(x,y)$. Thus, $\rho(y,x) \geq \rho(x,y)$, which contradicts (6.1).

(2) Let $x,y,z \in X$ and suppose $\rho$ is weakly transitive. Suppose $\pi(x,y) > 0$, $\pi(y,z) > 0$, and, contrary to the hypothesis, $\pi(x,z) = 0$. Since $\pi$ is regular, $\rho(z,x) \geq \rho(x,z)$. However, $\rho(z,x) \geq \rho(x,z)$ and $\rho(x,y) > \rho(y,x)$ implies $\rho(z,y) \geq \rho(y,z)$ which contradicts $\pi(y,z) > 0$.

(3) The result is immediate from Definition 6.2. $\qquad\qquad\qquad\qquad\square$

Since we are interested in how individual's preferences ultimately result in a social choice, we need to consider ways to transform a set o FWPRs for a group into an aggregate preferene relation.

**Definition 6.4 (FPAR).** In this chapter, we call a function $\tilde{f} : FR^n \to FR$ a fuzzy preference aggregation rule (FPAR).

*Note 6.5.* We will often suppress the notation $\tilde{f}(\bar{\rho})$ and let $\rho$ denote the social preference relation. In this manner, we can derive $\rho$ 's components $\iota$ and $\pi$, which correspond to the social fuzzy indifference and social fuzzy strict preference relations, respectively.

**Definition 6.6.** Let $\tilde{f}$ be an FPAR. Then $\tilde{f}$ is

(1) **(max-min transitive)**, if for all $\bar{\rho} \in \mathscr{F}\mathscr{R}^n$, $\tilde{f}(\bar{\rho})$ is max-min transitive;
(2) **(regularly acyclic)**, if for all $\bar{\rho} \in \mathscr{F}\mathscr{R}^n$, $\tilde{f}(\bar{\rho})$ is regularly acyclic.

It is also useful to identify some properties of coalitions. Fuzzy coalitions are denoted here by $\mathscr{L}$ and $\mathscr{L}$ is a subsets of the set of all fuzzy subsets of $N$, which is denoted by $\mathscr{F}(N)$.

**Definition 6.7.** Let $\mathscr{L} \in \mathscr{F}(N)$. Then

(1) **(monotonic)** $\mathscr{L}$ is said to be *monotonic* if, for all $\lambda \in \mathscr{L}$ and all $\lambda' \in \mathscr{F}(N)$, Supp$(\lambda) \subseteq$ Supp$(\lambda')$ implies $\lambda' \in \mathscr{L}$.
(2) **(proper)** $\mathscr{L}$ is said to be *proper* if, for all $\lambda \in \mathscr{L}$ and all $\lambda' \in \mathscr{F}(N)$, Supp$(\lambda) \cap$ Supp$(\lambda') = \emptyset$ implies $\lambda' \notin \mathscr{L}$.

The property of monotonicity indicates that if a set of coalitions, $\mathscr{L}$, contains a coalition $\lambda$ that has positive membership for a group of actors then it contains all coalitions that have positive membership for that group of actors. Proper means that if a set of coalitions, $\mathscr{L}$, contains a coalition $\lambda$ that has positive membership for a group of actors then it does not contain any coalitions that contain none of those actors.

**Definition 6.8 (decisive).** Let $\tilde{f}$ be a FPAR and $\lambda \in \mathscr{F}(N)$. Then $\lambda$ is decisive for $\tilde{f}$ if $\forall \bar{\rho} \in \mathscr{F}\mathscr{R}^n$, $\pi_i(x,y) > 0 \ \forall i \in$ Supp$(\lambda)$ implies $\pi(x,y) > 0$.

A decisive coalition decides the social choice. We are of course interested in all decisive coalitions.

**Definition 6.9 (decisive set).** Let

$$\mathscr{L}(\tilde{f}) = \{\lambda \in \mathscr{F}(N) \mid \lambda \text{ is decisive for } \tilde{f}\}.$$

Then $\mathscr{L}(\tilde{f})$ is the set of decisive coalitions for the fuzzy preference aggregation rule $\tilde{f}$.

We will now show that the set of decisive coalitions is monotonic and proper.

**Proposition 6.10.** *Let $\tilde{f}$ be an FPAR. Then*

*(1) $\mathscr{L}(\tilde{f})$ is monotonic;*

*(2) $\mathscr{L}(\tilde{f})$ is proper.*

*Proof.* (1) Suppose $\lambda, \lambda' \in \mathscr{F}(N)$ are such that $\mathrm{Supp}(\lambda) \subset \mathrm{Supp}(\lambda')$. Suppose $\lambda'$ is not decisive for $\tilde{f}$. Then there exists $\bar{\rho} \in \mathscr{F}\mathscr{R}^n$ and $x, y \in X$ such that $\pi_i(x,y) > 0$ for all $i \in \mathrm{Supp}(\lambda')$, but not $\pi(x,y) > 0$. Thus $\pi_i(x,y) > 0$ for all $i \in \mathrm{Supp}(\lambda)$, but not $\pi(x,y) > 0$. Hence $\lambda$ is not decisive for $\tilde{f}$.

(2) Let $\lambda \in \mathscr{L}(\tilde{f})$. Let $x, y \in X$ and $\bar{\rho} \in \mathscr{F}\mathscr{R}^n$ be such that $\pi_i(x,y) > 0$ for all $i \in \mathrm{Supp}(\lambda)$ and $\pi_i(y,x) > 0$ for all $i \in N \setminus \mathrm{Supp}(\lambda)$. Then $\pi(x,y) > 0$. Let $\lambda' \in \mathscr{F}\mathscr{P}(N)$. Suppose that $\mathrm{Supp}(\lambda) \cap \mathrm{Supp}(\lambda') = \emptyset$. If $\lambda' \in \mathscr{L}(\tilde{f})$, then $\pi(y,x) > 0$, a contradiction. Hence $\lambda'$ is not decisive for $\tilde{f}$ for all $\lambda' \in \mathscr{F}(N)$ such that $\mathrm{Supp}(\lambda') \cap \mathrm{Supp}(\lambda) = \emptyset$. Thus $\mathscr{L}(\tilde{f})$ is proper. $\square$

## 6.3 New and Old Fuzzy Voting Rules

As we briefly discussed earlier, there are cases in which the social preference relations for a given FPAR can be entirely characterized by its set of decisive coalitions. For instance, suppose we wish to determine whether an alternative $x \in X$ is strictly preferred to alternative $y \in X$ by simple majority rule. Before we can conclude whether or not this is the case, we must determine if a set of individuals exists that comprises more than $\frac{n}{2}$ individuals, each of whom strictly prefers $x$ to $y$. The preferences of individuals outside of this set are not relevant to the social preference outcome since it is determined by simple majority rule. Similarly, social strict preference within the framework of the Pareto extension rule requires that every actor strictly prefers one alternative over another. Otherwise, there is no strict social preference for one alternative. We refer to aggregation rules that depend on decisive sets as fuzzy simple rules. We now formally define these concepts in the fuzzy framework.

For any relation $R \subseteq X \times X$, let $\mathrm{Symm}(R)$ denote the symmetric closure of $R$, i.e. the smallest subset of $X \times X$ that contains $R$ and is symmetric.

We will now take family of coalitions, $\mathscr{L}$, and a collection of preference profiles, $\bar{\rho}$, and define an FPAR, $\tilde{g}_{\mathscr{L}}(\bar{\rho})$, which is essentially the coalitions indicated preference.

**Definition 6.11 (coalition preference).** Let $\mathscr{L} \subseteq \mathscr{F}(N)$. Let $\bar{\rho} \in \mathscr{F}\mathscr{R}^n$ and set

$$\tilde{\mathscr{P}}(\bar{\rho}) = \{(x,y) \in X \times X \mid \exists \lambda \in \mathscr{L}, \pi_i(x,y) > 0, \forall i \in \mathrm{Supp}(\lambda)\}.$$

Define $\tilde{g}_{\mathscr{L}} : \mathscr{F}\mathscr{R}^n \to \mathscr{F}\mathscr{R}$ by for all $\bar{\rho} \in \mathscr{F}\mathscr{R}^n$ and all $x, y \in X$,

$$\tilde{g}_{\mathscr{L}}(\bar{\rho})(x,y) = \begin{cases} 1 \text{ if } (x,y) \notin \mathrm{Symm}(\tilde{\mathscr{P}}(\bar{\rho})) \\ \vee\{\wedge\{\rho_i(x,y) \mid i \in \mathrm{Supp}(\lambda)\} \mid \lambda \in \mathscr{L} \text{ and } \pi_i(x,y) > 0, \forall i \in \mathrm{Supp}(\lambda)\} \\ \quad \text{ if } (x,y) \in \tilde{\mathscr{P}}(\bar{\rho}) \\ \wedge\{\wedge\{\rho_i(x,y) \mid i \in \mathrm{Supp}(\lambda)\} \mid \lambda \in \mathscr{L} \text{ and } \pi_i(y,x) > 0, \forall i \in \mathrm{Supp}(\lambda)\} \\ \quad \text{ if } (x,y) \in \mathrm{Symm}(\tilde{\mathscr{P}}(\bar{\rho}))\backslash\tilde{\mathscr{P}}(\bar{\rho}). \end{cases}$$

The connection of the fuzzy aggregation rule $\tilde{g}_{\mathscr{L}}$ in the above definition to the crisp aggregation rule (Austen-Smith and Banks, 1999, p. 59)can be seen from Example 6.13.

**Proposition 6.12.** *If* $(x,y) \in \tilde{\mathscr{P}}(\bar{\rho})$, *then* $g_{\mathscr{L}}(\bar{\rho})(x,y) > g_{\mathscr{L}}(\bar{\rho})(y,x)$. *If strict preferences are partial (regular), then* $\pi_{\mathscr{L}}$ *is partial (regular).*

*Proof.* Since $(x,y) \in \tilde{\mathscr{P}}(\bar{\rho})$.

$$(y,x) \in \mathrm{Symm}(\tilde{\mathscr{P}}(\bar{\rho})) \setminus \tilde{P}(\bar{\rho}) \,.$$

Thus $\exists \lambda \in \mathscr{L}$ such that $\pi_i(x,y) > 0$ for all $i \in \mathrm{Supp}(\lambda)$. For all such $\lambda, \rho_i(x,y) > \rho_i(y,x)$ for all $i \in \mathrm{Supp}(\lambda)$. Hence

$$g_{\mathscr{L}}(\bar{\rho})(x,y) > g_{\mathscr{L}}(\bar{\rho})(y,x)$$

by the definition of $g_{\mathscr{L}}$. Thus if strict preferences are regular, $\pi_{\mathscr{L}}$ is regular. If strict preferences are partial, then $\rho_i(y,x) = 0$ for all $i \in \mathrm{Supp}(\lambda)$ for $\lambda \in \mathscr{L}$. Hence $g_{\mathscr{L}}(\bar{\rho})(y,x) = 0$ and so $\pi_{\mathscr{L}}$ is partial.  □

*Example 6.13.* Let $X = \{x,y,z\}$, $\Delta(X) = \{(x,x),(y,y),(z,z)\}$, and $N = \{1,2,3\}$. Let $\mathscr{L} = \{\{1,2\},\{1,3\},\{2,3\},X\}$. Define the (crisp) relations $R_i$, for $i = 1,2,3$ as follows:

$$R_1 = \Delta X \cup \{(x,y),(y,x)\} \cup \{(x,z),(y,z)\},$$
$$R_2 = \Delta X \cup \{(x,y),(y,x)\} \cup \{(x,z),(z,y)\},$$
$$R_3 = \Delta X \cup \{(x,y),(y,x)\} \cup \{(z,x),(y,z)\}.$$

Then $P_1 = \{(x,z),(y,z)\}$, $P_2 = \{(x,z),(z,y)\}$, and $P_3 = \{(z,x),(y,z)\}$. Let $R = (R_1,R_2,R_3)$ and

$$\widetilde{\mathscr{P}}(R) = \{(u,v) \in X \times X \mid \exists L \in \mathscr{L}, (u,v) \in P_i \forall i \in L\} \,.$$

Then $\widetilde{\mathscr{P}}(R) = \{(x,z),(y,z)\}$, where $L = \{1,2\}$ for $(x,z)$ and $L = \{1,3\}$ for $(y,z)$. Now $\tilde{g}_{\mathscr{L}}(R) = \Delta X \cup \{(x,y),(y,x)\} \cup \{(x,z),(y,z)\}$.

Translating this to the fuzzy case where the mappings are into $\{0,1\}$, we would have

$$\tilde{g}_{\mathscr{L}}(R)(u,v) = \begin{cases} 1 \text{ if } (u,v) \in \Delta X \cup \{(x,y),(y,x)\} \\ 1 \text{ if } (u,v) \in \{(x,z),(y,z)\}, \\ 0 \text{ otherwise.} \end{cases}$$

(In the first case 1 is the indifference part while the second case 1 is the strict part of $\tilde{g}_{\mathscr{L}}(R)$).

Now define $\rho_i(u,v) = 1$ if $(u,v) \in \Delta X \cup \{(x,y),(y,x)\}, i = 1,2,3$ and

$$\rho_1(x,z) = \frac{1}{4}, \rho_1(z,x) = 0, \rho_1(y,z) = \frac{1}{2}, \rho_1(z,y) = 0,$$

$$\rho_2(x,z) = \frac{1}{8}, \rho_2(z,x) = 0, \rho_2(y,z) = 0, \rho_1(z,y) = \frac{3}{8},$$

$$\rho_3(x,z) = 0, \rho_3(z,x) = \frac{1}{2}, \rho_3(y,z) = \frac{7}{8}, \rho_3(z,y) = 0.$$

Let $\mathscr{L} = \{1_{\{1,2\}}, 1_{\{1,3\}}, 1_{\{2,3\}}, 1_X\}$. Let $\rho = (\rho_1, \rho_2, \rho_3)$ and

$$\widetilde{\mathscr{P}}(\rho) = \{(u,v) \in X \times X \mid \exists \lambda \in \mathscr{L}, \pi_i(u,v) > 0 \forall i \in \text{Supp}(\lambda)\}.$$

Then $\widetilde{\mathscr{P}}(\rho) = \{(x,z),(y,z)\}$. Using Definition 6.11, we have

$$\tilde{g}_{\mathscr{L}}(\rho)(u,v) = \begin{cases} 1 \text{ if } (u,v) \in \Delta X \cup \{(x,y),(y,x)\}, \\ \frac{1}{8} \text{ if } (u,v) = (x,z), \frac{1}{2} \text{ if } (u,v) = (y,z), \\ 0 \text{ otherwise.} \end{cases}$$

Note that $(x,x),(y,y),(z,z),(x,y),(y,x)$ are not in $\text{Symm}(\widetilde{\mathscr{P}}(\rho)),(z,x),(z,y)$ are in $\text{Symm}(\widetilde{\mathscr{P}}(\rho))$, but not in $\widetilde{\mathscr{P}}(\rho)$. This works well for strict preferences of type $\pi_{(0)}$. We now consider strict preferences that are regular. Define $\rho_i(u,v) = 1$ if $(u,v) \in \Delta X \cup \{(x,y),(y,x)\}$ for $i = 1,2,3$ and

$$\rho_1(x,z) = \frac{1}{4}, \rho_1(z,x) = \frac{1}{9}, \rho_1(y,z) = \frac{1}{2}, \rho_1(z,y) = \frac{1}{9},$$

$$\rho_2(x,z) = \frac{1}{8}, \rho_2(z,x) = \frac{1}{9}, \rho_2(y,z) = \frac{1}{9}, \rho_1(z,y) = \frac{3}{8},$$

$$\rho_3(x,z) = \frac{1}{9}, \rho_3(z,x) = \frac{1}{2}, \rho_3(y,z) = \frac{7}{8}, \rho_3(z,y) = \frac{1}{9}.$$

Then

$$\tilde{g}_{\mathscr{L}}(\rho)(u,v) = \begin{cases} 1 \text{ if } (u,v) \in \Delta X \cup \{(x,y),(y,x)\}, \\ \frac{1}{8} \text{ if } (u,v) = (x,z), \frac{1}{2} \text{ if } (u,v) = (y,z), \\ \frac{1}{9} \text{ otherwise.} \end{cases}$$

Note that one could make things a little more complicated by having $0 < \rho_i(x,y) = \rho_i(y,x) < 1$, for $i = 1,2,3$. Then Definition 6.11 could be changed to take this into account.

**Definition 6.14 (partial fuzzy simple rule).** Let $\tilde{f}$ be an FPAR. Then $\tilde{f}$ is called a *partial fuzzy simple rule* if for all $\bar{\rho} \in \mathscr{FR}^n$ and all $x,y \in X$,

$$\tilde{f}(\bar{\rho})(x,y) > 0 \iff \tilde{g}_{\mathscr{L}(\tilde{f})}(\bar{\rho})(x,y) > 0.$$

The following are properties of fuzzy aggregation rules. See the Note 6.5 concerning the notation.

**Definition 6.15 (decisive).** Let $\tilde{f}$ be a fuzzy aggregation rule. Then $\tilde{f}$ is called *decisive* if for all $\bar{\rho}, \bar{\rho}' \in \mathscr{FR}^n$ and all $x, y \in X$,

$$[\text{Supp}(\tilde{P}(x,y;\bar{\rho})) = \text{Supp}(\tilde{P}(x,y;\bar{\rho}'))$$
$$\text{and } \pi(x,y) > 0] \implies \pi'(x,y) > 0.$$

In words, an FPAR is decisive when the collection of fuzzy preferences preferring $x$ to $y$ for all preference profiles means that strict preference is greater than zero.

**Definition 6.16 (monotonic).** Let $\tilde{f}$ be a fuzzy aggregation rule. Then $\tilde{f}$ is called *monotonic* if, for all $\bar{\rho}, \bar{\rho}' \in \mathscr{FR}^n$ and all $x, y \in X$,

$$[\text{Supp}(\tilde{P}(x,y;\bar{\rho})) \subseteq \text{Supp}(\tilde{P}(x,y;\bar{\rho}')),$$
$$\text{Supp}(\tilde{R}(x,y;\bar{\rho})) \subseteq \text{Supp}(\tilde{R}(x,y;\bar{\rho}')),$$
$$\text{and } \pi(x,y) > 0] \implies \pi'(x,y) > 0.$$

Since $\text{Supp}(\tilde{P}(x,y;\bar{\rho}))$ and $\text{Supp}(\tilde{R}(x,y;\bar{\rho}))$ is the set of all individuals with a strict preference, $\pi(x,y)$, greater than zero, then an FPAR is monotonic when adding an individual to the support of strict weak preferences yields a new set.

**Definition 6.17 (neutral).** Let $\tilde{f}$ be a fuzzy aggregation rule. Then $\tilde{f}$ is called *neutral* if, for all $\bar{\rho}, \bar{\rho}' \in \mathscr{FR}^n$ and all $w, x, y, z \in X$,

$$[\text{Supp}(\tilde{P}(x,y;\bar{\rho})) = \text{Supp}(\tilde{P}(w,z;\bar{\rho}')) \quad \text{and} \quad \text{Supp}(\tilde{P}(y,x;\bar{\rho})) = \text{Supp}(\tilde{P}(z,w;\bar{\rho}'))]$$
$$\implies$$
$$\pi(x,y) > 0 \iff \pi'(w,z) > 0.$$

Neutral then, is a property of a FPAR where if a group of actors give positive support to both $x$ over $y$ and $w$ to $z$ then the social choice for both pairs of alternatives $(x,y)$ an $(w,z)$ must be zero or positive. Thus the aggregation rule cannot produce significantly different results for similar preference profiles.

**Theorem 6.18.** *Let $\tilde{f}$ be a fuzzy aggregation rule. Assume $\pi$ is partial. Then $\tilde{f}$ is a partial fuzzy simple rule if and only if $\tilde{f}$ is decisive, neutral, and monotonic.*

*Proof.* Let $x, y \in X$ and $\bar{\rho} \in \mathscr{FR}^n$. Suppose $\pi_{\mathscr{L}(\tilde{f})}(x,y) > 0$. Then $\exists \lambda \in \mathscr{L}(\tilde{f})$ such that $\forall i \in \text{Supp}(\lambda), \pi_i(x,y) > 0$ and so $\pi(x,y) > 0$. Suppose $\tilde{f}$ is decisive, neutral, and monotonic. Let $\bar{\rho} \in \mathscr{FR}^n$. Suppose $\pi(x,y) > 0$. In order to show $\tilde{f}$ is a partial fuzzy simple rule, it suffices to show $P(x,y;\bar{\rho}) \in \mathscr{L}(\tilde{f})$ for then $\pi_{\mathscr{L}(\tilde{f})}(x,y) > 0$. Let $a, b \in X$. Let $\bar{\rho}^*$ be any fuzzy preference profile such that $\forall i \in P(x,y;\bar{\rho}), \pi_i^*(a,b) > 0$. Let

$$L^+ = P(a,b;\bar{\rho}^*) \setminus P(x,y;\bar{\rho}).$$

Let $\bar{\rho}^1$ be a fuzzy preference profile such that

$$P(a,b;\bar{\rho}^1) = P(x,y;\bar{\rho})$$

and

$$P(b,a;\bar{\rho}^1) = P(y,x;\bar{\rho}) \,.$$

Since $\tilde{f}$ is neutral, $\pi^1(a,b) > 0$. Let $\bar{\rho}^2 \in \mathscr{F}\mathscr{R}^n$ be defined by

$$\rho_i^2 ]_{\{a,b\}} = \rho_i^* ]_{\{a,b\}} \text{if and only if } i \in L^+ \cup P(x,y;\bar{\rho})$$

and

$$\rho_j^2 ]_{\{a,b\}} = \rho_j^1 ]_{\{a,b\}} \text{ otherwise.}$$

Thus individuals in $\bar{\rho}^2$ and $\bar{\rho}^1$ that differ from 0 on $a,b$ must come from $L^+$. Thus $\pi^2(a,b) > 0$ since $\tilde{f}$ is monotonic. Since $P(a,b;\bar{\rho}^*) = P(a,b;\bar{\rho}^2)$ and $\tilde{f}$ is decisive, $\pi^*(a,b) > 0$. Hence since $\bar{\rho}$ and $a,b$ are arbitrary except for $\pi_i^*(a,b) > 0$ if $i \in P(x,y;\bar{\rho})$ and $\pi^*(a,b) > 0$ it follows that $\lambda \in \mathscr{L}(\tilde{f})$ for any $\lambda$ with $\text{Supp}(\lambda) = P(x,y;\bar{\rho})$.

Conversely, suppose that $\tilde{f}$ is a partial fuzzy simple rule. Then $\tilde{f}$ is neutral since $\mathscr{L}(\tilde{f})$ is defined without regard to alternatives. Monotonicity follows from Proposition 6.10 and the definition of $\tilde{g}_{\mathscr{L}(\tilde{f})}$. That $\tilde{f}$ is decisive follows directly from the definitions of $\mathscr{L}(\tilde{f})$ and $\tilde{g}_{\mathscr{L}(\tilde{f})}$.                                                        □

Theorem 6.18 attains only if $\pi$ is partial. The Theorem does not hold when $\pi$ is regular since monotonicity, neutrality, and decisiveness are defined with respect to strict preference and no relationship necessarily exists between strict preference relationships and regular strict preference relationships. The next three examples illustrate this principle. Furthermore, they highlight the fact that no partial strict preference can exist over two alternatives $x,y \in X$ in a preference relation $\rho \in \mathscr{F}\mathscr{R}$ when $\rho(x,y) > 0$ and $\rho(y,x) > 0$ even though $\rho(x,y) > \rho(y,x)$ or $\rho(y,x) > \rho(x,y)$.

*Example 6.19.* We show that there exists an FPAR that is monotonic and decisive, but not neutral. Let $X = \{w,x,y,z\}$ and $N = \{1,2\}$. Suppose that $\pi$ is regular. Let $\succ$ be the lexicographical order of $X$, i.e. an alphabetical ordering. (Then $\succ$ is irreflexive, complete, and asymmetric.) Define the fuzzy aggregation rule $\tilde{f} : \mathscr{F}\mathscr{R}^n \to \mathscr{F}\mathscr{R}$ by for all $a,b \in X$ and all $\bar{\rho} \in \mathscr{F}\mathscr{R}^n$,

$$\tilde{f}(\bar{\rho})(a,b) = \begin{cases} 1 & \text{if } a = b \text{ or } \pi_i(a,b) > 0, \forall i \in N, \\ \beta & \text{if } a \succ b, \text{ not } [\pi_i(a,b) > 0, \forall i \in N], \\ & \text{and not } [\pi_i(b,a) > 0, \forall i \in N], \\ \gamma & \text{if } a \prec b, \text{ not } [\pi_i(a,b) > 0, \forall i \in N], \\ & \text{and not } [\pi_i(b,a) > 0, \forall i \in N], \\ 0 & \text{if } \pi_i(b,a) > 0, \forall i \in N, \end{cases}$$

where $\beta, \gamma \in (0,1)$ and $\beta > \gamma$. By Definition 6.14, $\tilde{f}$ is a partial fuzzy simple rule, where $\mathscr{L}(\tilde{f}) = \{\lambda \in \mathscr{F}(N) \mid \text{Supp}(\lambda) = N\}$. Consider the following $\bar{\rho}, \bar{\rho}' \in \mathscr{F}\mathscr{R}^2$:

$$\rho_2(x,y) = .4, \ \rho_2'(z,w) = .7,$$
$$\rho_2(y,x) = 0, \ \ \rho_2'(w,z) = 0,$$

where $\rho_i(a,b) = 1$ and $\rho_i'(a,b) = 1$ otherwise, for all $a,b \in X$, $i = 1,2$. Since $\pi$ is regular, $\pi_2(x,y) > 0$, $\pi_2(z,w) > 0$, and $\pi_1(a,b) = 0$ for all $a,b \in X$. In this case,

$$\text{Supp}(\tilde{P}(x,y;\bar{\rho})) = \{2\} = \text{Supp}(\tilde{P}(z,w;\bar{\rho}')),$$

$$\text{Supp}(\tilde{P}(y,x;\bar{\rho})) = \emptyset = \text{Supp}(\tilde{P}(w,z;\bar{\rho}')),$$

and

$$\tilde{f}(\bar{\rho})(x,y) > \tilde{f}(\bar{\rho})(y,x)$$

because $\tilde{f}(\bar{\rho})(x,y) = \beta$ and $\tilde{f}(\bar{\rho})(y,x) = \gamma$. Thus, $\pi(x,y) > 0$. If $\tilde{f}$ is neutral, we would expect $\pi'(z,w) > 0$. However, this is not the case; $\tilde{f}(\bar{\rho}')(w,z) > \tilde{f}(\bar{\rho}')(z,w)$ since $\tilde{f}(\bar{\rho}')(w,z) = \beta$ and $\tilde{f}(\bar{\rho}')(z,w) = \gamma$. Hence, $\tilde{f}$ is not neutral.

We now show that $\tilde{f}$ is monotonic and decisive. To see that $\tilde{f}$ is monotonic, let $a,b \in X$ and $\bar{\rho} \in \mathscr{FR}^n$ be such that $\tilde{f}(\bar{\rho})(a,b) > \tilde{f}(\bar{\rho})(b,a)$. Then there are two cases. Either $a \succ b$ or $\text{Supp}(\tilde{P}(a,b;\bar{\rho})) = N$. Now for any $\bar{\rho}' \in \mathscr{FR}^n$ such that $\text{Supp}(\tilde{P}(a,b;\bar{\rho})) \subseteq \text{Supp}(\tilde{P}(a,b;\bar{\rho}'))$ and $\text{Supp}(\tilde{R}(a,b;\bar{\rho})) \subseteq \text{Supp}(\tilde{R}(a,b;\bar{\rho}'))$, either $a \succ b$ or $\text{Supp}(\tilde{P}(a,b;\bar{\rho}')) = N$. Thus, $\pi'(a,b) > 0$ and $\tilde{f}$ is monotonic. To see that $\tilde{f}$ is decisive, let $a,b \in X$ and $\bar{\rho} \in \mathscr{FR}^n$ be such that $\tilde{f}(\bar{\rho})(a,b) > \tilde{f}(\bar{\rho})(b,a)$. Then either $a \succ b$ or $\text{Supp}(\tilde{P}(a,b;\bar{\rho})) = N$. Hence, for any $\bar{\rho}' \in \mathscr{FR}^n$ such that $\text{Supp}(\tilde{P}(a,b;\bar{\rho})) = \text{Supp}(\tilde{P}(a,b;\bar{\rho}'))$, either $a \succ b$ or $\text{Supp}(\tilde{P}(a,b;\bar{\rho}')) = N$. Thus, $\pi'(a,b) > 0$, and $\tilde{f}$ is decisive.

*Example 6.20.* We show that there exists an FPAR that is neutral and monotonic, but not decisive. Let $X = \{x,y,z\}$ and $N = \{1,2\}$. Suppose that $\pi$ is regular. Define the fuzzy preference aggregation rule $\tilde{f} : \mathscr{FR}^n \to \mathscr{FR}$ by for all $a,b \in X$ and all $\bar{\rho} \in \mathscr{FR}^n$,

$$\tilde{f}(\bar{\rho})(a,b) = \begin{cases} 1 & \text{if } a = b \text{ or } \pi_i(a,b) > 0, \forall i \in N \\ \beta & \text{if } \text{Supp}(\tilde{P}(b,a;\bar{\rho})) = \emptyset, \text{ not } [\pi_i(a,b) > 0, \forall i \in N], \\ & \text{and not } [\pi_i(b,a) > 0, \forall i \in N] \\ \gamma & \text{if } \text{Supp}(\tilde{P}(b,a;\bar{\rho})) \neq \emptyset, \text{ not } [\pi_i(a,b) > 0, \forall i \in N], \\ & \text{and not } [\pi_i(b,a) > 0, \forall i \in N] \\ 0 & \text{if } \pi_i(b,a) > 0, \forall i \in N, \end{cases}$$

where $\beta, \gamma \in (0,1)$ and $\beta > \gamma$. By Definition 6.14, $\tilde{f}$ is a partial fuzzy simple rule, where $\mathscr{L}(\tilde{f}) = \{\lambda \in \mathscr{F}(N) \mid \text{Supp}(\lambda) = N\}$. Consider the following $\bar{\rho}, \bar{\rho}' \in \mathscr{FR}^2$ such that

$$\rho_1(x,y) = .7, \ \rho_1'(x,y) = 1.0,$$
$$\rho_1(y,x) = .3, \ \rho_1'(y,x) = .6,$$
$$\rho_2(x,y) = .6, \ \ \rho_2'(x,y) = 0,$$
$$\rho_2(y,x) = .6, \ \ \rho_2'(y,x) = .2,$$

where $\rho_i(a,b) = 1$ otherwise, for all $a,b \in X$, $i = 1,2$. Then $\mathrm{Supp}(\tilde{P}(x,y;\bar{\rho})) = \{1\}$, $\mathrm{Supp}(\tilde{P}(y,x;\bar{\rho})) = \emptyset$, $\mathrm{Supp}(\tilde{P}(x,y;\bar{\rho}')) = \{1\}$, and $\mathrm{Supp}(\tilde{P}(y,x;\bar{\rho})) = \{2\}$ since $\pi$ is regular. However, $\tilde{f}(\bar{\rho})(x,y) = \beta$ and $\tilde{f}(\bar{\rho})(y,x) = \tilde{f}(\bar{\rho}')(x,y) = \tilde{f}(\bar{\rho}')(y,x) = \gamma$. Thus, $\pi(x,y) > 0$ and $\pi'(x,y) = 0$ even though

$$\mathrm{Supp}(\tilde{P}(x,y;\bar{\rho})) = \{1\}$$
$$= \mathrm{Supp}(\tilde{P}(x,y;\bar{\rho}')).$$

Hence, $\tilde{f}$ is not decisive.

We now show that $\tilde{f}$ is neutral and monotonic. To see that $\tilde{f}$ is neutral, consider any $a,b,c,d \in X$ and $\bar{\rho}, \bar{\rho}' \in \mathcal{FR}^n$ such that $\mathrm{Supp}(\tilde{P}(a,b;\bar{\rho})) = \mathrm{Supp}(\tilde{P}(c,d;\bar{\rho}'))$ and $\mathrm{Supp}(\tilde{P}(b,a;\bar{\rho})) = \mathrm{Supp}(\tilde{P}(d,c;\bar{\rho}'))$. If $\pi(a,b) > 0$, $\mathrm{Supp}(\tilde{P}(a,b;\bar{\rho})) = N$ or $[\mathrm{Supp}(\tilde{P}(b,a;\bar{\rho})) = \emptyset$ and $\mathrm{Supp}(\tilde{P}(a,b;\bar{\rho})) \neq \emptyset]$. Hence, $\mathrm{Supp}(\tilde{P}(c,d;\bar{\rho}')) = N$ or $[\mathrm{Supp}(\tilde{P}(d,c;\bar{\rho})) = \emptyset$ and $\mathrm{Supp}(\tilde{P}(c,d;\bar{\rho})) \neq \emptyset]$. Thus, $\pi'(c,d) > 0$. An identical argument shows that $\pi'(c,d) > 0$ implies $\pi(a,b) > 0$.

We next show that $\tilde{f}$ is monotonic, let $a,b \in X$ and $\bar{\rho} \in \mathcal{FR}^n$ be such that $\pi(a,b) > 0$. Then $\mathrm{Supp}(\tilde{P}(a,b;\bar{\rho})) = N$ or $[\mathrm{Supp}(\tilde{P}(b,a;\bar{\rho})) = \emptyset$ and $\mathrm{Supp}(\tilde{P}(a,b;\bar{\rho})) \neq \emptyset]$. For any $\bar{\rho}' \in \mathcal{FR}^n$ such that $\mathrm{Supp}(\tilde{P}(a,b;\bar{\rho})) \subseteq \mathrm{Supp}(\tilde{P}(a,b;\bar{\rho}'))$ and $\mathrm{Supp}(\tilde{R}(a,b;\bar{\rho})) \subseteq \mathrm{Supp}(\tilde{R}(a,b;\bar{\rho}'))$, $\mathrm{Supp}(\tilde{P}(a,b;\bar{\rho})) = N$ implies $\mathrm{Supp}(\tilde{P}(a,b;\bar{\rho}')) = N$. By the construction of $\bar{\rho}'$, $\mathrm{Supp}(\tilde{P}(b,a;\bar{\rho})) = \emptyset$ and $\mathrm{Supp}(\tilde{P}(a,b;\bar{\rho})) \neq \emptyset$ imply $\mathrm{Supp}(\tilde{R}(a,b;\bar{\rho}')) = N$, which implies $\mathrm{Supp}(\tilde{P}(b,a;\bar{\rho}')) = \emptyset$, and $\mathrm{Supp}(\tilde{P}(a,b;\bar{\rho}')) \neq \emptyset$. Thus, $\pi'(a,b) > 0$, and $\tilde{f}$ is monotonic.

*Example 6.21.* We show that there exists an FPAR that is decisive and neutral, but not monotonic. Let $X = \{x,y,z\}$ and $N = \{1,2,3,4,5\}$. Assume $\pi$ is regular. Define the fuzzy preference aggregation rule $\tilde{f} : \mathcal{FR}^n \to \mathcal{FR}$ by for all $a,b \in X$ and all $\bar{\rho} \in \mathcal{FR}^n$,

$$\tilde{f}(\bar{\rho})(a,b) = \begin{cases} 1 & \text{if } a = b \text{ or } \pi_i(a,b) > 0, \forall i \in N, \\ \beta & \text{if } |\mathrm{Supp}(\tilde{P}(a,b;\bar{\rho}))| \in (\frac{n}{2}, n-1), \text{ not } [\pi_i(a,b) > 0, \forall i \in N] \\ & \text{and not } [\pi_i(b,a) > 0, \forall i \in N], \\ \gamma & \text{if } |\mathrm{Supp}(\tilde{P}(a,b;\bar{\rho}))| \notin (\frac{n}{2}, n-1), \text{ not } [\pi_i(a,b) > 0, \forall i \in N] \\ & \text{and not } [\pi_i(b,a) > 0, \forall i \in N], \\ 0 & \text{if } \pi_i(b,a) > 0, \forall i \in N, \end{cases}$$

where $\beta, \gamma \in (0,1)$ and $\beta > \gamma$. By Definition 6.14, $\tilde{f}$ is a partial fuzzy simple rule, where $\mathcal{L}(\tilde{f}) = \{\lambda \in \mathcal{F}(N) \mid \mathrm{Supp}(\lambda) = N\}$. Consider some $\bar{\rho}, \bar{\rho}' \in \mathcal{FR}^5$ that generate the following individual strict preference relations

$$\forall i \in \{1,2,3\}, \ \pi_i(x,y) > 0 \text{ and } \pi_i'(x,y) > 0,$$
$$\pi_4(x,y) = \pi_4(y,x) = 0 \text{ and } \pi_4'(x,y) > 0,$$
$$\pi_5(y,x) > 0 \text{ and } \pi_5'(y,x) > 0.$$

Then we have that

$$\mathrm{Supp}(\tilde{P}(x,y;\bar{\rho})) = \{1,2,3\},$$
$$\mathrm{Supp}(\tilde{P}(x,y;\bar{\rho}')) = \{1,2,3,4\},$$
$$\mathrm{Supp}(\tilde{R}(x,y;\bar{\rho})) = \{1,2,3,4\}, \text{ and}$$
$$\mathrm{Supp}(\tilde{R}(x,y;\bar{\rho}')) = \{1,2,3,4\}.$$

Thus,

$$\mathrm{Supp}(\tilde{P}(x,y;\bar{\rho})) \subseteq \mathrm{Supp}(\tilde{P}(x,y;\bar{\rho}'))$$

and

$$\mathrm{Supp}(\tilde{R}(x,y;\bar{\rho})) \subseteq \mathrm{Supp}(\tilde{R}(x,y;\bar{\rho}')) .$$

Since $\tilde{f}(\bar{\rho})(x,y) = \beta$ and $\tilde{f}(\bar{\rho})(y,x) = \gamma$, $\pi(x,y) > 0$. If $\tilde{f}$ were monotonic, we would expect $\pi'(x,y) > 0$. However,

$$|\mathrm{Supp}(\tilde{P}(x,y;\bar{\rho}'))| = 4 \notin (\frac{5}{2},4) .$$

Thus, $\tilde{f}(\bar{\rho}')(x,y) = \gamma$, and $\tilde{f}(\bar{\rho}')(y,x) = \gamma$ since $\mathrm{Supp}(\tilde{P}(y,x;\bar{\rho}')) = \{5\}$. Thus, $\pi'(x,y) = 0$, and $\tilde{f}$ is not monotonic.

We now show that $\tilde{f}$ is decisive and neutral. To see that $\tilde{f}$ is decisive, let $a,b \in X$ and $\bar{\rho} \in X$ such that $\pi(a,b) > 0$. Then $\mathrm{Supp}(\tilde{P}(a,b;\bar{\rho})) = N$ or $\mathrm{Supp}(\tilde{P}(a,b;\bar{\rho})) \in (\frac{n}{2},n-1)$, which implies

$$\mathrm{Supp}(\tilde{P}(b,a;\bar{\rho})) \notin (\frac{n}{2},n-1) .$$

Let $\bar{\rho}'$ be such that $\mathrm{Supp}(\tilde{P}(a,b;\bar{\rho})) = \mathrm{Supp}(\tilde{P}(a,b;\bar{\rho}'))$. Then $\mathrm{Supp}(\tilde{P}(a,b;\bar{\rho}')) = N$ or

$$\mathrm{Supp}(\tilde{P}(a,b;\bar{\rho}')) \in (\frac{n}{2},n-1) ,$$

which implies

$$\mathrm{Supp}(\tilde{P}(b,a;\bar{\rho}')) \notin (\frac{n}{2},n-1) ,$$

and $\pi'(a,b) > 0$. Thus $\tilde{f}$ is decisive.

To see that $\tilde{f}$ is neutral, let $a,b,c,d \in X$ and $\bar{\rho},\bar{\rho}' \in X$ be such that $\mathrm{Supp}(\tilde{P}(a,b;\bar{\rho})) = \mathrm{Supp}(\tilde{P}(c,d;\bar{\rho}'))$ and $\mathrm{Supp}(\tilde{P}(b,a;\bar{\rho})) = \mathrm{Supp}(\tilde{P}(d,c;\bar{\rho}'))$. If $\pi(a,b) > 0$, then $\mathrm{Supp}(\tilde{P}(a,b;\bar{\rho})) = N$ or $\mathrm{Supp}(\tilde{P}(a,b;\bar{\rho})) \in (\frac{n}{2},n-1)$. By assumption, $\mathrm{Supp}(\tilde{P}(c,d;\bar{\rho}')) = N$ or $\mathrm{Supp}(\tilde{P}(c,d;\bar{\rho}')) \in (\frac{n}{2},n-1)$. Thus, $\pi'(c,d) > 0$. An identical argument shows that $\pi'(c,d) > 0$ implies $\pi(a,b) > 0$.

The following Theorem demonstrates that Theorem 6.18 holds if and only if $\pi$ is partial.

**Theorem 6.22.** *Let $\tilde{f}$ be a partial fuzzy simple rule. Then $\tilde{f}$ is decisive, monotonic and neutral if and only if the social strict preference relation with respect to $\tilde{f}(\bar{\rho})$ is partial for all $\bar{\rho} \in \mathscr{FR}^n$.*

*Proof.* Suppose $\tilde{f}$ is decisive, neutral and monotonic. The fact that $\tilde{f}(\bar{\rho})(y,x) = 0$ implies $\pi(x,y) > 0$ for all $x,y \in X$ and all $\bar{\rho} \in \mathscr{FR}^n$ is guaranteed by the completeness of $\tilde{g}_{\mathscr{L}(\tilde{f})}(\bar{\rho})$, the definition of a partial fuzzy simple rule, and the regularity of $\pi$. Let $x,y \in X$ and $\bar{\rho} \in \mathscr{FR}^n$ be such that $\pi(x,y) > 0$. The goal of this first part of the proof is to show that $\tilde{P}(x,y;\bar{\rho}) \in \mathscr{L}(\tilde{f})$. In doing so, Definition 6.11 and the properness of $\mathscr{L}(\tilde{f})$ imply

$$\tilde{g}_{\mathscr{L}(\tilde{f})}(\bar{\rho})(x,y) > 0$$

and

$$\tilde{g}_{\mathscr{L}(\tilde{f})}(\bar{\rho})(y,x) = 0\,.$$

Then by definition of a partial fuzzy simple rule, it follows that $\tilde{f}(\bar{\rho})(x,y) > 0$ and $\tilde{f}(\bar{\rho})(y,x) = 0$.

Now consider two arbitrary alternatives $a,b \in X$. Let $\bar{\rho}^*$ be such that $\pi_i^*(a,b) > 0$ for those $i \in N$ such that $\tilde{P}(x,y;\bar{\rho})(i) > 0$. Let $\tilde{L}^* \in \mathscr{F}(N)$ be such that $\tilde{L}^*(i) = \pi_i^*(a,b)$ if

$$i \in \mathrm{Supp}(\tilde{P}(a,b;\bar{\rho}^*))\backslash\mathrm{Supp}(\tilde{P}(x,y;\bar{\rho}))$$

and $\tilde{L}^*(i) = 0$ otherwise. Let $\bar{\rho}^1 \in \mathscr{FR}^n$ be such that

$$\mathrm{Supp}(\tilde{P}(a,b;\bar{\rho}^1)) = \mathrm{Supp}(\tilde{P}(x,y;\bar{\rho}))$$

and

$$\mathrm{Supp}(\tilde{P}(b,a;\bar{\rho}^1)) = \mathrm{Supp}(\tilde{P}(y,x;\bar{\rho}))\,.$$

Because, by assumption, $\pi(x,y) > 0$, neutrality implies $\pi^1(a,b) > 0$. Let $\bar{\rho}^2 \in \mathscr{FR}^n$ be such that

$$\rho_i^2 \rceil_{\{a,b\}} = \rho_i^* \rceil_{\{a,b\}} \text{ if } i \in \mathrm{Supp}(\tilde{L}^*) \cup \mathrm{Supp}(\tilde{P}(x,y;\bar{\rho}))$$

and

$$\rho_j^2 \rceil_{\{a,b\}} = \rho_j^1 \rceil_{\{a,b\}} \text{ otherwise.}$$

Thus, only $i \in \mathrm{Supp}(\tilde{L}^*)$ can have the preference combination of $\pi_i^2(a,b) > 0$ and $\pi_i^1(a,b) = 0$. Since $\tilde{f}$ is monotonic, $\pi^2(a,b) > 0$. Since

$$\mathrm{Supp}(\tilde{P}(a,b;\bar{\rho}^*)) = \mathrm{Supp}(\tilde{P}(a,b;\bar{\rho}^2))$$

and $\tilde{f}$ is decisive, $\pi^*(a,b) > 0$. Since $\bar{\rho}$ and $a,b$ are arbitrary except for $\pi_i^*(a,b) > 0$ if $i \in \mathrm{Supp}(\tilde{P}(x,y;\bar{\rho}))$ and since $\pi^*(a,b) > 0$, it follows that $\lambda \in \mathscr{L}(\tilde{f})$ for any $\lambda$ with $\mathrm{Supp}(\lambda) = P(x,y;\bar{\rho})$.

Conversely, suppose the social strict preference relation with respect to $\tilde{f}(\bar{\rho})$ is partial for all $\bar{\rho} \in \mathscr{FR}^n$. Because $\tilde{f}$ is a partial fuzzy simple rule and $\pi$ is partial, $\tilde{f}$ is decisive, neutral and monotonic by Theorem 6.18.                                                     $\square$

Dutta (1987) demonstrates the equivalence of Definition 6.11 to the definition of a simple rule when $\rho : X \times X \rightarrow \{0,1\}$. In order to model social strict preferences that are regular, we introduce the following definitions.

**Definition 6.23 (regular fuzzy simple rule).** Let $\tilde{f}$ be a fuzzy aggregation rule. Then $\tilde{f}$ is called a *regular fuzzy simple rule* if for all $\bar{\rho} \in \mathscr{FR}^n$ and all $x, y \in X$,

$$\tilde{f}(\bar{\rho})(x,y) > \tilde{f}(\bar{\rho})(y,x) \iff \tilde{g}_{\mathscr{L}(\tilde{f})}(\bar{\rho})(x,y) > \tilde{g}_{\mathscr{L}(\tilde{f})}(\bar{\rho})(y,x).$$

*Example 6.24.* Assume strict preferences are regular. Let $X = \{x, y, z\}$, $N = \{1, 2, 3\}$ and $\bar{\rho} = (\rho_1, \rho_2, \rho_3)$, where $\rho_i(w,w) = 1$ for all $w \in X$ and $i = 1, 2, 3$, and

$$\rho_1(x,y) = \tfrac{3}{4}, \; \rho_1(x,z) = \tfrac{1}{4}, \; \rho_1(y,z) = \tfrac{1}{3},$$
$$\rho_2(x,y) = \tfrac{4}{5}, \; \rho_2(x,z) = \tfrac{2}{3}, \; \rho_2(z,y) = \tfrac{1}{4},$$
$$\rho_3(y,x) = \tfrac{3}{4}, \; \rho_3(z,x) = \tfrac{1}{2}, \; \rho_3(y,z) = \tfrac{2}{3}.$$

Each $\rho_i$ is defined to be $1/2$ otherwise for $i = 1, 2, 3$. Let $\mathscr{L} = \{\lambda \in \mathscr{F}(N) \mid |\text{Supp}(\lambda)| \geq 2\}$. Thus,

$$\pi_1(x,y) > 0, \; \pi_1(z,x) > 0, \; \pi_1(z,y) > 0,$$
$$\pi_2(x,y) > 0, \; \pi_2(x,z) > 0, \; \pi_2(y,z) > 0,$$
$$\pi_3(y,x) > 0, \; \pi_3(z,x) = 0, \; \pi_3(z,y) > 0,$$

where each $\pi_i$ is 0 otherwise for $i = 1, 2, 3$. Using the strict preference relations and definition of $\mathscr{L}$, we can now determine whether or not $\tilde{P}(x,y;\bar{\rho})$ is in $\mathscr{L}$ for all $x, y \in X$. We have that

$$\text{Supp}(\tilde{P}(x,y;\bar{\rho})) = \{1, 2\},$$
$$\text{Supp}(\tilde{P}(y,z;\bar{\rho})) = \{2, 3\}.$$

Since $|\text{Supp}(\tilde{P}(x,y;\bar{\rho}))| \geq 2$ and $|\text{Supp}(\tilde{P}(y,z;\bar{\rho}))| \geq 2$ we know that $\tilde{P}(x,y;\bar{\rho}) \in \mathscr{L}(\tilde{f})$ and $\tilde{P}(y,z;\bar{\rho}) \in \mathscr{L}(\tilde{f})$. Hence, $\tilde{\mathscr{P}}(\bar{\rho}) = \{(x,y), (y,z)\}$ and consequently $\text{Symm}(\tilde{\mathscr{P}}(\bar{\rho})) = \{(x,y), (y,x), (y,z), (z,y)\}$. Thus, it follows that

$$\tilde{g}_{\mathscr{L}}(\bar{\rho})(x,y) = 3/4 \vee 4/5 = 4/5,$$
$$\tilde{g}_{\mathscr{L}}(\bar{\rho})(y,x) = 1/2 \wedge 1/2 = 1/2$$
$$\tilde{g}_{\mathscr{L}}(\bar{\rho})(y,z) = 1/2 \vee 2/3 = 2/3$$
$$\tilde{g}_{\mathscr{L}}(\bar{\rho})(z,y) = 1/4 \wedge 1/2 = 1/4.$$

It also follows that $\tilde{g}_{\mathscr{L}}(\bar{\rho})(x,z) = \tilde{g}_{\mathscr{L}}(\bar{\rho})(z,x) = 1$.

Let $\tilde{f}$ be a fuzzy aggregation rule such that $\mathscr{L}(\tilde{f}) = \mathscr{L}$. Then by definition of a regular fuzzy simple rule, we have that

$$\tilde{f}(\bar{\rho})(x,y) > \tilde{f}(\bar{\rho})(y,x),$$

$$\tilde{f}(\bar{\rho})(y,z) > \tilde{f}(\bar{\rho})(z,y)$$

and

$$\tilde{f}(\bar{\rho})(x,z) = \tilde{f}(\bar{\rho})(z,x).$$

Thus, $\pi(x,y) > 0$, $\pi(y,z) > 0$ and $\pi(x,z) = \pi(z,x) = 0$ by the regularity of $\pi$.

**Definition 6.25 (regularly neutral).** Let $\tilde{f}$ be a fuzzy aggregation rule. Then $\tilde{f}$ is called *regularly neutral* if $\forall \bar{\rho}, \bar{\rho}' \in \mathscr{FR}^n$ and $\forall x, y, z, w \in X$,

$$[P(x,y;\bar{\rho}) = P(z,w;\bar{\rho}') \text{ and } P(y,x;\bar{\rho}) = P(w,z;\bar{\rho}')]$$

imply

$$\tilde{f}(\bar{\rho})(x,y) > \tilde{f}(\bar{\rho})(y,x) \Leftrightarrow \tilde{f}(\bar{\rho}')(z,w) > \tilde{f}(\bar{\rho}')(w,z) .$$

**Theorem 6.26.** *Let $\tilde{f}$ be an FPAR and suppose $\pi$ is regular. Then $\tilde{f}$ is a regular fuzzy simple rule if and only if $\tilde{f}$ is decisive, monotonic, and regularly neutral.*

*Proof.* Let $x, y \in X$ and $\bar{\rho} \in \mathscr{FR}^n$. Suppose $\pi_{\mathscr{L}(\tilde{f})}(x,y) > 0$. Then there exists $\lambda \in \mathscr{L}(\tilde{f})$ such that $\forall i \in \text{Supp}(\lambda), \pi_i(x,y) > 0$, and since $\lambda$ is decisive for $\tilde{f}, \pi(x,y) > 0$.

Suppose $\tilde{f}$ is decisive, neutral, and monotonic. Let $\bar{\rho} \in \mathscr{FR}^n$ and $x, y \in X$. Suppose $\pi(x,y) > 0$. It suffices to show $\pi_{\mathscr{L}(\tilde{f})}(x,y) > 0$. Let $a, b \in X$. Let $\bar{\rho}^*$ be a fuzzy preference profile such that $\forall i \in P(x,y;\bar{\rho}), \pi_i^*(a,b) > 0$. Let

$$L^+ = P(a,b;\bar{\rho}^*) \setminus P(x,y;\bar{\rho}) .$$

Let $\bar{\rho}^1$ be a fuzzy preference profile such that $P(a,b;\bar{\rho}^1) = P(x,y;\bar{\rho})$ and $P(b,a;\bar{\rho}^1) = P(y,x;\bar{\rho})$. Since $\tilde{f}$ is regularly neutral, $\pi(a,b) > 0$. Let $\bar{\rho}^2 \in \mathscr{FR}^n$ be defined by

$$\rho_i^2\rceil_{\{a,b\}} = \rho_i^*\rceil_{\{a,b\}} \text{ if and only if } i \in L^+ \cup P(x,y;\bar{\rho})$$

and

$$\rho_j^2\rceil_{\{a,b\}} = \rho_j^1\rceil_{\{a,b\}} \text{ otherwise.}$$

Then individuals that $\bar{\rho}^2$ and $\bar{\rho}^1$ differ 0 on $a, b$ must come from $L^+$. Hence, we have that

$$P(a,b;\bar{\rho}^1) \subseteq P(a,b;\bar{\rho}^2)$$

and

$$R(a,b;\bar{\rho}^1) \subseteq R(a,b;\bar{\rho}^2) .$$

Thus since $\tilde{f}$ is monotonic and $\pi^1(a,b) > 0, \pi^2(a,b) > 0$. Since

$$P(a,b;\bar{\rho}^*) = P(a,b;\bar{\rho}^2)$$

and $\tilde{f}$ is decisive, $\pi^*(a,b) > 0$.

Since $\bar{\rho}$ and $a, b$ are arbitrary except for $\pi_i^*(a,b) > 0$ if $i \in P(x,y;\bar{\rho})$ and $\pi^*(a,b) > 0$, any $\lambda$ with $\text{Supp}(\lambda) = P(x,y;\bar{\rho})$ is such that $\lambda \in \mathscr{L}(\tilde{f})$. [That is, $\pi_i^*(a,b) > 0 \, \forall i \in P(x,y;\bar{\rho}) \Rightarrow \pi^*(a,b) > 0$.] Hence $(x,y) \in \tilde{\mathscr{P}}(\bar{\rho})$. Thus

$$1 > \tilde{g}_{\mathscr{L}(\tilde{f})}(\bar{\rho})(x,y) > 0$$

and so

$$\pi_{\mathscr{L}(\tilde{f})}(x,y) > 0 .$$

Conversely, suppose $\tilde{f}$ is a regular fuzzy simple rule. Let $\bar{\rho}, \bar{\rho}' \in \mathscr{F}\mathscr{R}^n$ and $x, y, z, w \in X$. Suppose $P(x, y; \bar{\rho}) = P(z, w; \bar{\rho}')$ and $P(y, x; \bar{\rho}) = P(w, z; \bar{\rho}')$. Since $\tilde{f}$ is a regular fuzzy simple rule,

(1)
$$\tilde{f}(\bar{\rho})(x, y) > \tilde{f}(\bar{\rho})(y, x) \Leftrightarrow (2)\tilde{g}_{\mathscr{L}(\tilde{f})}(\bar{\rho})(x, y) > \tilde{g}_{\mathscr{L}(\tilde{f})}(\bar{\rho})(y, x) \text{ and}$$

(2)
$$\tilde{f}(\bar{\rho}')(z, w) > \tilde{f}(\bar{\rho}')(w, z) \Leftrightarrow (4)\tilde{g}_{\mathscr{L}(\tilde{f})}(\bar{\rho}')(z, w) > \tilde{g}_{\mathscr{L}(\tilde{f})}(\bar{\rho}')(w, z).$$

Now $(x, y) \notin \text{Symm}(\mathscr{P}(\bar{\rho}))$ if and only if $(y, x) \notin \text{Symm}(\mathscr{P}(\bar{\rho}))$. Thus if (2) holds, $(x, y) \notin \text{Symm}(\mathscr{P}(\bar{\rho}))$ and $(y, x) \notin \text{Symm}(\mathscr{P}(\bar{\rho}))$. Assuming (2), $(x, y) \in \mathscr{P}(\bar{\rho})$ since $\tilde{f}(\bar{\rho})(x, y) > 0$. Now for $\lambda \in \mathscr{L}(\tilde{f})$, $\text{Supp}(\lambda) \subseteq P(x, y; \bar{\rho})$ when $(x, y) \in \mathscr{P}(\bar{\rho})$. Assuming (2), we have $\pi_{\mathscr{L}(\tilde{f})}(x, y) > 0$ and so $\pi_{\mathscr{L}(\tilde{f})}(y, x) = 0$. Now $\text{Supp}(\lambda) \subseteq P(z, w; \bar{\rho}')$. Since $\lambda$ is decisive, $\pi_{\mathscr{L}(\tilde{f})}(z, w) > 0$ so (4) holds. Hence (2) implies (4). Similarly, (4) implies (2). Thus (1) and (3) are equivalent and so $\tilde{f}$ is regularly neutral.

We next show $\tilde{f}$ is monotonic. Let $\bar{\rho}, \bar{\rho}' \in \mathscr{F}\mathscr{R}^n$ and $x, y \in X$. Suppose

$$P(x, y; \bar{\rho}) \subseteq P(x, y; \bar{\rho}') \text{ and } R(x, y; \bar{\rho}) \subseteq R(x, y; \bar{\rho}') \text{ and } \pi(x, y) > 0.$$

Since $\pi(x, y) > 0, \tilde{f}(\bar{\rho})(x, y) > \tilde{f}(\bar{\rho})(y, x)$. Since $\tilde{f}$ is a regular fuzzy simple rule,

$$\tilde{g}_{\mathscr{L}(\tilde{f})}(\bar{\rho})(x, y) > \tilde{g}_{\mathscr{L}(\tilde{f})}(\bar{\rho})(y, x).$$

Hence $(x, y) \in \mathscr{P}(\bar{\rho})$ so there exists $\lambda \in \mathscr{L}(\tilde{f})$, i.e., a $\lambda$ which is decisive for $\tilde{f}$. Thus

$$\text{Supp}(\lambda) \subseteq P(x, y; \bar{\rho}) \subseteq P(x, y; \bar{\rho}').$$

Thus $\pi_{\mathscr{L}(\tilde{f})}(\bar{\rho}')(x, y) > 0$ and so

$$\tilde{g}_{\mathscr{L}(\tilde{f})}(\bar{\rho}')(x, y) > \tilde{g}_{\mathscr{L}(\tilde{f})}(\bar{\rho}')(y, x).$$

Since $\tilde{f}$ is a regular fuzzy simple rule, $\tilde{f}(\bar{\rho}')(x, y) > \tilde{f}(\bar{\rho}')(y, x)$. Thus $\tilde{f}$ is monotonic.

We next show $\tilde{f}$ is decisive. Let $\bar{\rho}, \bar{\rho}' \in \mathscr{F}\mathscr{R}^n$ and $x, y \in X$. Suppose

$$P(x, y; \bar{\rho}) = P(x, y; \bar{\rho}') \text{and } \pi(x, y) > 0.$$

Since $\pi(x, y) > 0, \tilde{f}(\bar{\rho})(x, y) > \tilde{f}(\bar{\rho})(y, x)$. Since $\tilde{f}$ is a regular fuzzy simple rule,

$$\tilde{g}_{\mathscr{L}(\tilde{f})}(\bar{\rho})(x, y) > \tilde{g}_{\mathscr{L}(\tilde{f})}(\bar{\rho})(y, x).$$

Hence $(x, y) \in \mathscr{P}(\bar{\rho})$ so there exists $\lambda \in \mathscr{L}(\tilde{f})$, i.e., a $\lambda$ which is decisive for $\tilde{f}$. Thus $\text{Supp}(\lambda) \subseteq P(x, y; \bar{\rho}) = P(x, y; \bar{\rho}')$. Hence $\pi_{\mathscr{L}(\tilde{f})}(\bar{\rho}')(x, y) > 0$ and so $\tilde{g}_{\mathscr{L}(\tilde{f})}(\bar{\rho}')(x, y) > \tilde{g}_{\mathscr{L}(\tilde{f})}(\bar{\rho}')(y, x)$. Since $\tilde{f}$ is a regular fuzzy simple rule, $\tilde{f}(\bar{\rho}')(x, y) > \tilde{f}(\bar{\rho}')(y, x)$. Thus $\pi'(x, y) > 0$. Hence $\tilde{f}$ is decisive.                                                        $\square$

Under some aggregation rules, the strict preferences of members in decisive coalitions are not sufficient to determine the social preference. For instance, under plurality rule an alternative $x$ is strictly preferred to alternative $y$ if more individuals strictly prefer $x$ to $y$ than strictly prefer $y$ to $x$. Moreover, under plurality rule weak preferences also help determine the social preference outcome. Under the regular strict preference framework, we refer to these types of aggregation rules as *fuzzy voting rules*. In Mordeson et al. (2010), *partial* fuzzy voting rules were defined and it was illustrated that they are included in the traditional definitions.

We now move on to defining a decisive structure.

**Definition 6.27 (decisive structure).** Let $\tilde{f}$ be a fuzzy aggregation rule. The *decisive structure* of $\tilde{f}$, denoted $\mathscr{D}(\tilde{f})$, is defined to be the set

$$\mathscr{D}(\tilde{f}) = \{(\sigma, \omega) \in \mathscr{F}(N) \times \mathscr{F}(N) \mid \text{Supp}(\sigma) \subseteq \text{Supp}(\omega)$$
$$\text{and } [\forall x, y \in X, \forall \bar{\rho} \in \mathscr{F}\mathscr{R}^n, \pi_i(x,y) > 0 \ \forall i \in \text{Supp}(\sigma) \text{ and } \rho_j(x,y) > 0 \ \forall j \in \text{Supp}(\omega)]$$
$$\text{imply } \pi(x,y) > 0\}.$$

A decisive structure is the set of pairs of coalitions $(\sigma, \omega)$ with the sets of individuals in $\sigma$ and $\omega$ preferring $x$ to $y$ deciding the aggregate preference $\pi$.

**Definition 6.28 (coalition pair preference).** . Let $\mathscr{D} \subseteq \mathscr{F}(N) \times \mathscr{F}(N)$ be such that $\text{Supp}(\sigma) \subseteq \text{Supp}(\omega)$ for all $(\sigma, \omega) \in \mathscr{D}$. Let $\bar{\rho} \in \mathscr{F}\mathscr{R}^n$ and set

$$\tilde{\mathscr{R}}(\bar{\rho}) = \{(x,y) \in X \times X \mid \exists (\sigma, \omega) \in \mathscr{D} \text{ such that}$$
$$\pi_i(x,y) > 0 \ \forall i \in \text{Supp}(\sigma) \text{ and } \pi_j(y,x) = 0 \ \forall j \in \text{Supp}(\omega)\}$$

Define $\tilde{g}_{\mathscr{D}} : \mathscr{F}\mathscr{R}^n \to \mathscr{F}\mathscr{R}$ by for all $\bar{\rho} \in \mathscr{F}\mathscr{R}^n$ and all $x, y \in X$,

$$\tilde{g}_{\mathscr{D}}(\bar{\rho})(x,y) = \begin{cases} 1 & \text{if } (x,y) \notin \text{Symm}(\tilde{\mathscr{R}}(\bar{\rho})) \\ \bigvee\{\bigwedge\{\rho_j(x,y) \mid j \in \text{Supp}(\omega)\} \mid (\sigma, \omega) \in \mathscr{D}, \pi_i(x,y) > 0 \ \forall i \in \text{Supp}(\sigma), \\ \text{and } \pi_j(y,x) = 0 \ \forall j \in \text{Supp}(\omega)\} & \text{if } (x,y) \in \tilde{\mathscr{R}}(\bar{\rho}) \\ \bigwedge\{\bigwedge\{\rho_j(x,y) \mid j \in \text{Supp}(\omega)\} \mid (\sigma, \omega) \in \mathscr{D}, \pi_i(y,x) > 0 \ \forall i \in \text{Supp}(\sigma), \\ \text{and } \pi_j(x,y) = 0 \ \forall j \in \text{Supp}(\omega)\} & \text{if } (x,y) \in \text{Symm}(\tilde{\mathscr{R}}(\bar{\rho})) \backslash \tilde{\mathscr{R}}(\bar{\rho}). \end{cases}$$

The intuition for the following definition is entirely similar to that for Definition 6.11 as illustrated by Example 6.13.

By a proof entirely similar to that of Proposition 6.12, we have the following result.

**Proposition 6.29.** *If* $(x,y) \in \tilde{\mathscr{P}}(\bar{\rho})$, *then* $g_{\mathscr{D}}(\bar{\rho})(x,y) > g_{\mathscr{D}}(\bar{\rho})(y,x)$. *If strict preferences are partial (regular), then* $\pi_{\mathscr{D}}$ *is partial (regular).*

**Definition 6.30.** Let $\tilde{f}$ be a a fuzzy aggregation rule. Then, $\tilde{f}$ is called a *regular fuzzy voting rule* if for all $\bar{\rho} \in \mathscr{F}\mathscr{R}^n$ and all $x, y \in X$,

$$\tilde{f}(\bar{\rho})(x,y) > \tilde{f}(\bar{\rho})(y,x) \iff \tilde{g}_{\mathscr{D}(\tilde{f})}(\bar{\rho})(x,y) > \tilde{g}_{\mathscr{D}(\tilde{f})}(\bar{\rho})(y,x),$$

and $\tilde{f}$ is called a *partial fuzzy voting rule* if for all $\bar{\rho} \in \mathcal{FR}^n$ and all $x, y \in X$,

$$\tilde{f}(\bar{\rho})(x,y) > 0 \iff \tilde{g}_{\mathscr{D}(\tilde{f})}(\bar{\rho})(x,y) > 0.$$

**Theorem 6.31.** *Suppose $\pi$ is partial. An FPAR is a partial fuzzy voting rule if and only if it is neutral and monotonic.*

*Proof.* Suppose $\tilde{f}$ is a partial fuzzy voting rule. We show that $\tilde{f}$ is monotonic. We first prove that $D(\tilde{f})$ is monotonic. Let $(\sigma, \omega) \in D(\tilde{f})$ and $\sigma', \omega' \in \mathcal{F}(N)$ be such that

$$\mathrm{Supp}(\sigma) \subseteq \mathrm{Supp}(\sigma') \subseteq \mathrm{Supp}(\omega')$$

and

$$\mathrm{Supp}(\sigma) \subseteq \mathrm{Supp}(\omega) \subseteq \mathrm{Supp}(\omega') \,.$$

Since $\mathrm{Supp}(\sigma) \subseteq \mathrm{Supp}(\sigma')$ and $\mathrm{Supp}(\omega) \subseteq \mathrm{Supp}(\omega')$, we have $(\sigma', \omega') \in \mathscr{D}(\tilde{f})$. Let $\bar{\rho}, \bar{\rho}' \in \mathcal{FR}^n$ and $x, y \in X$ be such that

$$P(x,y;\bar{\rho}) \subseteq P(x,y;\bar{\rho}') \,,$$

$$R(x,y;\bar{\rho}) \subseteq R(x,y;\bar{\rho}')$$

and

$$\pi(x,y) > 0 \,,$$

By Definitions 6.28 and 6.30, we have that since $\pi(x,y) > 0, \exists (\sigma, \omega) \in \mathscr{D}(\tilde{f})$ such that $\mathrm{Supp}(\sigma) \subseteq \mathrm{Supp}(\omega)$ and $\forall x, y \in X, \forall \bar{\rho} \in \mathcal{FR}^n, \pi_i(x,y) > 0 \forall i \in \mathrm{Supp}(\sigma)$ and $\rho_j(x,y) > 0 \forall j \in \mathrm{Supp}(\omega)$. By hypothesis,

$$\{i \in N \mid \pi_i(x,y) > 0\} \subseteq \{i \in N \mid \pi_i'(x,y) > 0\},$$

$$\{j \in N \mid \rho_j(x,y) > 0\} \subseteq \{j \in N \mid \rho_j'(x,y) > 0\}.$$

Since $(\sigma', \omega') \in \mathscr{D}(\tilde{f})$, we have that $\pi'(x,y) > 0$. Thus $\tilde{f}$ is monotonic. We now show $\tilde{f}$ is neutral. Let $\bar{\rho}, \bar{\rho}' \in \mathcal{FR}^n$ and $x, y, z, w \in X$ be such that

$$\{i \in N \mid \pi_i(x,y) > 0\} = \{i \in N \mid \pi_i'(z,w) > 0\},$$

$$\{i \in N \mid \pi_i(y,x) > 0\} = \{i \in N \mid \pi_i'(w,z) > 0\}.$$

Then it follows that $\pi_i(x,y) > 0 \leftrightarrow \pi_i'(z,w) > 0$ for all $i \in N$. Suppose $\tilde{f}(\bar{\rho})(x,y) > 0$. Then

$$(x,y) \in (X \times X \setminus \mathrm{Symm}(\mathcal{FP}_{\mathscr{D}(\tilde{f})}(\bar{\rho})) \cup \mathcal{FP}_{\mathscr{D}(\tilde{f})}(\bar{\rho})) \,,$$

say $(x,y) \in \mathcal{FP}_{\mathscr{D}(\tilde{f})}(\bar{\rho})$. Thus, $\exists (\sigma, \omega) \in \mathscr{D}(\tilde{f})$ such that $\forall i \in \mathrm{Supp}(\sigma), \pi_i(x,y) > 0, \forall j \in \mathrm{Supp}(\omega), \rho_i(x,y) > 0$. By hypothesis, it follows that $(z,w) \in \mathcal{FP}_{\mathscr{D}(\tilde{f})}(\bar{\rho})$. Hence $\tilde{f}(\bar{\rho}')(z,w) > 0$ and in fact $\pi'(z,w) > 0$. Suppose

$$(x,y) \in X \times X \setminus \text{Symm}(\mathscr{FP}_{\mathscr{D}(\tilde{f})}(\bar{\rho})) \,.$$

By the argument just given, $(z,w) \notin \mathscr{FP}_{\mathscr{D}(\tilde{f})}(\bar{\rho}')$ else $(x,y) \in \mathscr{FP}_{\mathscr{D}(\tilde{f})}(\bar{\rho})$. Thus

$$(z,w) \in X \times X \setminus \mathscr{FP}_{\mathscr{D}()}(\bar{\rho}') \,.$$

Hence $\tilde{f}(\bar{\rho}')(z,w) > 0$. Thus $\tilde{f}$ is neutral.

For the converse, we first show that $\pi_{\mathscr{D}(\tilde{f})}(x,y) > 0$ implies $\pi(x,y) > 0$.

Suppose $\tilde{g}_{\mathscr{D}(\tilde{f})}(\bar{\rho})(x,y) > 0$ for $\bar{\rho} \in \mathscr{FR}^n$ and $x,y \in X$. Then

$$(x,y) \in (X \times X \setminus \text{Symm}(\mathscr{FP}_{\mathscr{D}(\tilde{f})}(\bar{\rho})) \cup \mathscr{FP}_{\mathscr{D}(\tilde{f})}(\bar{\rho})) \,,$$

say $(x,y) \in \mathscr{FP}_{\mathscr{D}(\tilde{f})}(\bar{\rho})$. Thus, $\exists (\sigma,\omega) \in \mathscr{D}(\tilde{f})$ such that $\forall i \in \text{Supp}(\sigma)$, $\pi_i(x,y) > 0$, and $\forall j \in \text{Supp}(\omega)$, $\rho_i(x,y) > 0$. Since $(\sigma,\omega) \in \mathscr{D}(\tilde{f})$, we have $\tilde{f}(\bar{\rho})(x,y) > 0$. Suppose

$$(x,y) \in (X \times X \setminus \text{Symm}(\mathscr{FP}_{\mathscr{D}()}(\bar{\rho}))) \,.$$

Then $(y,x) \in (X \times X) \setminus \text{Symm}(FP_{D()}(\bar{\rho}))$ and so

$$\tilde{g}_{\mathscr{D}(\tilde{f})}(\bar{\rho})(y,x) = \tilde{g}_{\mathscr{D}(\tilde{f})}(\bar{\rho})(x,y) = 1 \,.$$

However, this contradicts the assumption that $\pi_{\mathscr{D}(\tilde{f})}(x,y) > 0$. Consequently,

$$(y,x) \in \text{Symm}(\mathscr{FP}_{\mathscr{D}(\tilde{f})}(\bar{\rho})) \setminus \mathscr{FP}_{\mathscr{D}(\tilde{f})}(\bar{\rho}) \,.$$

Hence it follows that $\tilde{f}(\bar{\rho})(y,x) = 0$. Now suppose that $\tilde{f}$ is neutral and monotonic. Let $x,y \in X$ and $\bar{\rho} \in \mathscr{FR}^n$. Suppose that $\pi(x,y) > 0$. Let $(\sigma,\omega) \in \mathscr{FP}(N) \times FP(N)$, $\text{Supp}(\sigma) \subseteq \text{Supp}(\omega)$, be such that $\text{Supp}(\sigma) = S$ and $\text{Supp}(\omega) = W$, where $S = P(x,y;\bar{\rho})$ and $W = R(x,y;\bar{\rho})$. We wish to show $(\sigma,\omega) \in \mathscr{D}(\tilde{f})$. Let $z,w \in X$. Let $\bar{\rho}' \in \mathscr{FR}(N)$ be such that $\pi_i'(z,w) > 0 \Leftrightarrow i \in S$ and $\pi_i'(w,z) = 0 \Leftrightarrow i \in W$, i.e., $P(z,w;\bar{\rho}') = S$ and $R(z,w;\bar{\rho}') = W$. Since $\tilde{f}$ is neutral, $\pi'(z,w) > 0$. Now let $\bar{\rho}'' \in \mathscr{FR}^n$ be such that

$$P(z,w;\bar{\rho}') \subseteq P(z,w;\bar{\rho}'')$$

and

$$R(z,w;\bar{\rho}') \subseteq R(z,w;\bar{\rho}'') \,.$$

Since $\tilde{f}$ is monotonic, $\pi''(z,w) > 0$. Thus we have that $\pi_i(z,w) > 0 \forall i \in S$ and $\pi_i(w,z) = 0 \ \forall i \in W$ implies $\pi(z,w) > 0$. Hence $(\sigma,\omega) \in \mathscr{D}(\tilde{f})$ and so $\pi_{\mathscr{D}(\tilde{f})}(x,y) > 0$. $\qquad\square$

**Theorem 6.32.** *Let $\tilde{f}$ be an FPAR and suppose $\pi$ is regular. Then $\tilde{f}$ is a regular fuzzy voting rule if and only if $\tilde{f}$ is monotonic and regularly neutral.*

*Proof.* Suppose $\tilde{f}$ is a regal fuzzy voting rule. We show $\tilde{f}$ is monotonic. We fist show that $\mathscr{D}(\tilde{f})$ is monotonic. Let $(\sigma,\omega) \in \mathscr{D}(\tilde{f})$ and $\sigma',\omega' \in \mathscr{FP}(N)$ be such that

$$\text{Supp}(\sigma) \subseteq \text{Supp}(\sigma') \subseteq \text{Supp}(\omega')$$

and

$$\text{Supp}(\sigma) \subseteq \text{Supp}(\omega) \subseteq \text{Supp}(\omega') \, .$$

Since $\text{Supp}(\sigma) \subseteq \text{Supp}(\sigma')$ and $\text{Supp}(\omega) \subseteq \text{Supp}(\omega')$, clearly $(\sigma', \omega') \in \mathscr{D}(\tilde{f})$. Thus $\mathscr{D}(\tilde{f})$ is monotonic. Let $\bar{\rho}, \bar{\rho}' \in \mathscr{F}\mathscr{B}^n$ and $x, y \in X$ be such that

$$P(x, y; \bar{\rho}) \subseteq P(x, y; \bar{\rho}')$$

and

$$R(x, y; \bar{\rho}) \subseteq R(x, y; \bar{\rho}')$$

and $\pi(x, y) > 0$. Since $\pi(x, y) > 0$ and $\tilde{f}$ is a regular fuzzy voting rule, we have that

$$\tilde{g}_{\mathscr{D}(\tilde{f})}(\bar{\rho})(x, y) > \tilde{g}_{\mathscr{D}(\tilde{f})}(\bar{\rho})(y, x) \, .$$

Hence $\pi_{\mathscr{D}(\tilde{f})}(x, y) > 0$ and so by definition of $\tilde{g}_{\mathscr{D}(\tilde{f})}$, $\exists (\sigma, \omega) \in \mathscr{D}(\tilde{f})$ such that $\text{Supp}(\sigma) \subseteq \text{Supp}(\omega)$ and $\forall x, y \in X, \forall \bar{\rho} \in \mathscr{F}\mathscr{R}^n$ we have $\pi_i(x, y) > 0 \ \forall i \in \text{Supp}(\sigma)$ and $\rho_j(x, y) > 0 \ \forall j \in \text{Supp}(\omega)$. Since $(\sigma', \omega') \in \mathscr{D}(\tilde{f})$ and

$$P(x, y; \bar{\rho}) \subseteq P(x, y; \bar{\rho}')$$

and

$$R(x, y; \bar{\rho}) \subseteq R(x, y; \bar{\rho}') \, ,$$

we have that $\pi'(x, y) > 0$. Thus $\tilde{f}$ is monotonic.

We now show $\tilde{f}$ is regularly neutral. Let $\bar{\rho}, \bar{\rho}' \in \mathscr{D}(\tilde{f})$ and $x, y, z, w \in X$ be such that

$$P(x, y; \bar{\rho}) = P(z, w; \bar{\rho}')$$

and

$$P(y, x; \bar{\rho}) = P(w, z; \bar{\rho}') \, .$$

Then it follows easily that $\forall i \in N, \pi_i(x, y) > 0$ if and only if $\pi_i'(z, w) > 0$. Suppose $\tilde{f}(\bar{\rho})(x, y) > \tilde{f}(\bar{\rho})(y, x)$. Then

$$\tilde{g}_{\mathscr{D}(\tilde{f})}(\bar{\rho})(x, y) > \tilde{g}_{\mathscr{D}(\tilde{f})}(\bar{\rho})(y, x)$$

since $\tilde{f}$ is a regular fuzzy voting rule. Thus $(x, y) \in \mathscr{F}\mathscr{P}_{\mathscr{D}(\tilde{f})}(\bar{\rho})$. Hence $\exists (\sigma, \omega) \in \mathscr{D}(\tilde{f})$ such that $\forall i \in \text{Supp}(\sigma), \pi_i(x, y) > 0$ and $\forall j \in \text{Supp}(\omega), \rho_j(x, y) > 0$. By hypothesis, it follows that $(z, w) \in \mathscr{F}\mathscr{P}_{\mathscr{D}(\tilde{f})}(\bar{\rho}')$. Thus $\tilde{f}(\bar{\rho}')(z, w) > \tilde{f}(\bar{\rho}')(w, z)$ and $\pi'(z, w) > 0$. Hence $\tilde{f}$ is regularly neutral.

Let $\bar{\rho} \in \mathscr{F}\mathscr{R}^n$ and $x, y \in X$. We first show that $\pi_{\mathscr{D}(\tilde{f})}(x, y) > 0$ implies $\pi(x, y) > 0$. Suppose $\pi_{\mathscr{D}(\tilde{f})}(x, y) > 0$. Then as previously discussed $(x, y) \in \mathscr{F}\mathscr{P}_{\mathscr{D}(\tilde{f})}(\bar{\rho})$. Thus $\exists (\sigma, \omega) \in \mathscr{D}(\tilde{f})$ such that $\text{Supp}(\sigma) \subseteq \text{Supp}(\omega)$, $\forall x, y \in X, \forall \bar{\rho} \in \mathscr{F}\mathscr{R}^n$, $\pi_i(x, y) > 0 \ \forall i \in \text{Supp}(\sigma)$ and

$$\rho_j(z,x) > 0 \forall j \in \text{Supp}(\omega) \Rightarrow \pi(x,y) > 0.$$

Since $(\sigma,\omega) \in \mathscr{D}(\tilde{f}), \pi(x,y) > 0$. We now show that $\pi(x,y) > 0$ implies $\pi_{\mathscr{D}(\tilde{f})}(x,y) > 0$ under the assumption that $\tilde{f}$ is regularly neutral and monotonic. Suppose that $\pi(x,y) > 0$. Let $(\sigma,\omega) \in \mathscr{F}\mathscr{P}(N) \times FP(N)$ be such that

$$\text{Supp}(\sigma) \subseteq \text{Supp}(\omega),$$

$$\text{Supp}(\sigma) = P(x,y;\bar{\rho})$$

and

$$\text{Supp}(\omega) = R(x,y;\bar{\rho}).$$

It suffices to show that $(\sigma,\omega) \in \mathscr{D}(\tilde{f})$. Let $z,w \in X$. Let $\bar{\rho}' \in \mathscr{F}\mathscr{R}^n$ be such that $\pi_i'(z,w) > 0 \Leftrightarrow i \in \text{Supp}(\sigma)$ and $\rho_i'(z,w) > 0 \Leftrightarrow i \in \text{Supp}(\omega)$. Since $\tilde{f}$ is regularly neutral, $\pi'(z,w) > 0$. Now let $\bar{\rho}'' \in \mathscr{F}\mathscr{B}^n$ be such that $P(z,w;\bar{\rho}') \subseteq P(z,w;\bar{\rho}'')$ and $R(z,w;\bar{\rho}') \subseteq R(z,w;\bar{\rho}'')$. Since $\tilde{f}$ is monotonic, $\pi''(z,w) > 0$. Hence we have that $\pi_i(z,w) > 0 \forall i \in \text{Supp}(\sigma)$ and $\rho_i(z,w) > 0 \forall i \in \text{Supp}(\omega)$ implies that $\pi(z,w) > 0$. Thus $(\sigma,\omega) \in \mathscr{D}(\tilde{f})$ and so $\pi_{\mathscr{D}(\tilde{f})}(x,y) > 0$. $\qquad\square$

## 6.4 Single-Peaked Preferences and the Maximal Set

In this section we consider several combinations of social preferences conditions induced by fuzzy voting rules. We give particular attention to identifying those preference relations that allow for the existence of a non-empty maximal set. While we have yet to consider fuzzy sets of alternatives, it is important to note that both the fuzzy non-dominated set (denoted $ND$ and defined in Def. 3.27 on page 35)and the fuzzy maximal set are determined by the degree to which alternatives exist in the alternative set. Hereafter, we use $\mu \in \mathscr{F}(X)$ to denote an alternative's degree of set inclusion. We now consider the conditions under which fuzzy voting rules or simple rules produce a social preference relation such that there exists a fuzzy non-dominated set. Furthermore, we wish to ensure that such a social preference relation also produces a non-empty fuzzy maximal set as described in Proposition 3.39.

First, however, we present several definitions of different types of preference orderings. These definitions will be used later on to determine the conditions under which a maximum set exists.

**Definition 6.33 (single-peaked).** Let $\mu \in \mathscr{F}(X)$ be such that $|\text{Supp}(\mu)| = r$. Suppose $Q$ is a strict ordering of the elements of $\text{Supp}(\mu)$ and label the elements of $\text{Supp}(\mu)$ so that $a_{t+1}Qa_t$ for all $t = 1,\ldots,r-1$. Let $\rho \in \mathscr{F}\mathscr{R}$ and $\pi$ be the strict preference relation with respect to $\rho$. Then $\rho$ is called *single-peaked* on $\mu$ with respect to $Q$ if and only if there exists some $t \in \{1,\ldots,r\}$ such that

$$\pi(a_t,a_{t+1}) \wedge \pi(a_{t+1},a_{t+2}) \wedge \ldots \wedge \pi(a_{r-1},a_r) > 0$$

and

$$\pi(a_t,a_{t-1}) \wedge \pi(a_{t-1},a_{t-2}) \wedge \ldots \wedge \pi(a_2,a_1) > 0.$$

Let $\mathscr{F}\mathscr{S}$ denote the set of all single-peaked FWPRs. Note that Definition 6.33 ensures that, for some $\mu \in \mathscr{F}(X)$, there exists $x \in \text{Supp}(\mu)$ such that $\pi(x,y) > 0$ for all $y \in \text{Supp}(\mu)$ when $\rho \in \mathscr{F}\mathscr{S}$ is partially quasi-transitive.

**Definition 6.34 (weakly single-peaked).** Let $\mu \in \mathscr{F}(X)$ be such that $|\text{Supp}(\mu)| = r$. Suppose $Q$ is a strict ordering of the elements of $\text{Supp}(\mu)$ and label $\text{Supp}(\mu)$ so that $a_{t+1}Qa_t$ for all $t = 1,\ldots,r-1$. Let $\rho \in \mathscr{F}\mathscr{R}$. Then $\rho$ is called *weakly single-peaked* on $\mu$ with respect to $Q$ if and only if there exists some $t \in \{1,\ldots,r\}$ such that

$$\pi(a_r,a_{r-1}) \vee \pi(a_{r-1},a_{r-2}) \vee \ldots \vee \pi(a_{t+1},a_t) = 0$$

and

$$\pi(a_1,a_2) \vee \pi(a_2,a_3) \vee \ldots \vee \pi(a_{t-1},a_t) = 0.$$

Let $\mathscr{F}\mathscr{W}$ denote the set of all preference relations that are weakly single-peaked, reflexive and complete. In words, Definition 6.34 guarantees that there exists an $a_t \in X$ such that $\pi(y,a_t) = 0$ for all $y \in X$ if $\rho \in \mathscr{F}\mathscr{W}$ is weakly transitive. To see this, note that $\pi(a_{t+1},a_t) = 0$ implies $\rho(a_t,a_{t+1}) \geq \rho(a_{t+1},a_t)$, and $\pi(a_{t+2},a_{t+1}) = 0$ implies $\rho(a_{t+1},a_{t+2}) \geq \rho(a_{t+2},a_{t+1})$. By the weak transitivity of $\rho$, $\rho(a_t,a_{t+2}) \geq \rho(a_{t+2},a_t)$. We can continue this argument until $\rho(a_t,a_s) \geq \rho(a_s,a_t)$ for any $s$ in $\{t+1,\ldots,r\}$. A similar argument holds to show $\rho(a_t,a_s) \geq \rho(a_s,a_t)$ for any $s$ in $\{1,\ldots,t-1\}$. Thus $\pi(y,a_t) = 0$ for all $y$ in $X$.

**Definition 6.35 (single-peaked profile).** A profile $\bar{\rho} \in \mathscr{F}\mathscr{R}^n$ is *(weakly) single-peaked* on $\mu \in \mathscr{F}(X)$ if there is a strict ordering $Q$ of the elements of $\text{Supp}(\mu)$ such that $\rho_i$ is (weakly) single-peaked on $\mu$ with respect to $Q$ for all $i \in N$.

**Theorem 6.36.** *Let $\tilde{f}$ be a fuzzy simple rule such that $\tilde{f}$ is weakly Paretian. Assume $\pi$ is regular. Then, for all $\bar{\rho} \in \mathscr{F}\mathscr{W}$ and all $\mu \in \mathscr{F}(X)$,*

$$Supp(ND(\pi,\mu)) \neq \emptyset,$$

*where $\pi$ is the social strict preference relation with respect to $\tilde{f}(\bar{\rho})$ and $\rho_i$ is weakly transitive for all $i \in N$.*

*Proof.* Let $\mu \in \mathscr{F}(X)$ and let $\bar{\rho} \in \mathscr{F}\mathscr{W}$ be such that $\rho_i$ is weakly transitive for all $i \in N$. Relabel the elements of $\text{Supp}(\mu)$ such that $\text{Supp}(\mu) = \{a_1,\ldots,a_r\}$ and $a_{t+1}Qa_t$ for all $t = 1,\ldots,r-1$, where $Q$ is a strict ordering of $\text{Supp}(\mu)$ such that $\rho_i$ is weakly single-peaked with respect to $Q$ for all $i \in N$. Define $x_i \in \text{Supp}(\mu)$ to be such that $\pi_i(y,x_i) = 0$ for all $y \in \text{Supp}(\mu)$ and there does not exist an $x_i'$ such that $x_i'Qx_i$ and $\pi_i(y,x_i') = 0$ for all $y \in \text{Supp}(\mu)$, for $i = 1,\ldots,n$. Define $\tilde{G}(x) : X \to \mathscr{F}(N)$ by for all $x \in \text{Supp}(\mu)$,

$$\tilde{G}(x)(i) = \begin{cases} \rho_i(x_i,x) & \text{if } xQx_i \text{ or } x = x_i \\ 0 & \text{otherwise.} \end{cases}$$

By definition of $x_i$, $\tilde{G}(x)$ is well-defined for all $x \in \text{Supp}(\mu)$. Note that $\text{Supp}(\tilde{G}(a_t)) \subseteq \text{Supp}(\tilde{G}(a_{t+1}))$ for all $t = 1,\ldots,r-1$. Further, if $\tilde{G}(a_t)(i) > 0$, then by weak transitivity, $\pi_i(a_s,a_t) = 0$ where $s > t$.

Let $x^* \in \{x \in \text{Supp}(\mu) \mid \tilde{G}(x) \in \mathcal{L}(\tilde{f})$ and $\nexists y \in X$ such that $xQy$and $\tilde{G}(y) \in \mathcal{L}(\tilde{f})\}$ .This set is non-empty because $\tilde{G}(a_r) \in \mathcal{L}(\tilde{f})$ by weak Paretianism. We now show that $\pi(y,x^*) = 0$ for all $y \in \text{Supp}(\mu)$ and, accordingly, $ND(\pi,\mu)(x^*) > 0$. To see this, suppose the contrary. Then there exists a $y \in \text{Supp}(\mu)$ such that $\pi(y,x^*) > 0$. There are two cases to consider.

Case 1:  Suppose $x^*Qy$. Then there exists $i \in \text{Supp}(G(x^*))$ such that $x_iQy$. It follows that $\text{Supp}(\tilde{G}(y)) \subset \text{Supp}(\tilde{G}(x^*))$. By definition of $x^*$, $\tilde{G}(y) \notin \mathcal{L}(\tilde{f})$. However $\text{Supp}(\tilde{P}(y,x^*;\bar{\rho})) \subseteq \text{Supp}(\tilde{G}(y))$. By the definition of a fuzzy simple rule, $\pi(y,x^*) = 0$, a contradiction.

Case 2:  Suppose $yQx^*$. Then $\text{Supp}(\tilde{G}(x^*)) \subseteq \text{Supp}(\tilde{G}(y))$. Thus, $\pi_i(y,x^*) = 0$ for those $i \in N$ such that $\tilde{G}(x^*)(i) > 0$ by the previous argument. By the properness of $\mathcal{L}(\tilde{f})$, $\tilde{G}(x^*) \in \mathcal{L}(f)$ implies $\tilde{P}(y,x^*;\bar{\rho}) \notin \mathcal{L}(f)$ since $\text{Supp}(G(x^*)) \cap \text{Supp}(\tilde{P}(y,x^*;\bar{\rho})) = \emptyset$. Hence, $\pi(y,x^*) = 0$, a contradiction. $\square$

**Corollary 6.37.** *Let $\tilde{f}$ be a fuzzy simple rule such that $\tilde{f}$ is weakly Paretian. Assume $\pi$ is regular. Then for all $\mu \in \mathcal{F}(X)$ and $\bar{\rho} \in \mathcal{FW}$ such that $\rho_i$ is weakly transitive for all $i \in N$, $\text{Supp}(M(\tilde{f}(\bar{\rho}),\mu)) \neq \emptyset$.*

To determine under what conditions fuzzy voting rules produce social preference relations with a non-empty non-dominated set, consider the following Lemma.

**Lemma 6.38.** *[Corollary 1, p. 123-124 Montero and Tejada (1988)] Let $\rho \in \mathcal{FR}$. Assume $\pi$ is regular. Then $\text{Supp}(ND(\pi,\mu)) \neq \emptyset$ for all $\mu \in \mathcal{F}(X)$ if and only if $\rho$ is acyclic over $\text{Supp}(\mu)$.*

**Theorem 6.39.** *Let $\tilde{f}$ be a fuzzy voting rule and let $\mu \in \mathcal{F}(X)$. Assume $\pi$ is regular. Let $\bar{\rho} \in \mathcal{FS}^n$ be such that $\rho_i$ is partially quasi-transitive for all $i \in N$. Then $\text{Supp}(ND(\pi,\mu)) \neq \emptyset$, where $\pi$ is the social strict preference relation with respect to $\tilde{f}(\bar{\rho})$.*

*Proof.* Let $\mu \in \mathcal{F}(X)$ be such that $x,y,z \in \text{Supp}(\mu)$, and let $\bar{\rho} \in \mathcal{FS}^n$ be such that $\rho_i$ is partially quasi-transitive for all $i \in N$. Let $Q$ be a strict ordering of the elements of $\text{Supp}(\mu)$ such that $\rho_i$ is single-peaked with respect to $Q$ for all $i \in N$. Suppose, without loss of generality, $zQyQx$. Then the following hold by quasi-transitivity and single-peakedness of $\rho_i$:

$$(1) \quad \text{Supp}(\tilde{P}(x,y;\bar{\rho})) \subseteq \text{Supp}(\tilde{P}(x,z;\bar{\rho}))$$

and

$$\text{Supp}(\tilde{P}(z,x;\bar{\rho})) \subseteq \text{Supp}(\tilde{P}(y,x;\bar{\rho}))$$
$$(2) \quad \text{Supp}(\tilde{P}(z,y;\bar{\rho})) \subseteq \text{Supp}(\tilde{P}(z,x;\bar{\rho}))$$

and

$$\text{Supp}(\tilde{P}(x,z;\bar{\rho})) \subseteq \text{Supp}(\tilde{P}(y,z;\bar{\rho})).$$

There are six cases to consider.

Case 1: Suppose $\pi(x,y) > 0$. Then $(\tilde{P}(x,y;\bar{\rho}),\tilde{R}(x,y;\bar{\rho})) \in \mathscr{D}(\tilde{f})$ by definition of a fuzzy voting rule. Since $\mathscr{D}(\tilde{f})$ is monotonic and $\tilde{f}$ is neutral, (1) implies $(\tilde{P}(x,z;\bar{\rho}),\tilde{R}(x,z;\bar{\rho})) \in \mathscr{D}(\tilde{f})$. Thus, $\pi(x,z) > 0$. Hence, $\pi(x,y) > 0$ and $\pi(y,z) > 0$ imply $\pi(x,z) > 0$.

Case 2: Suppose $\pi(x,z) > 0$. Then $(\tilde{P}(x,z;\bar{\rho}),\tilde{R}(x,z;\bar{\rho})) \in \mathscr{D}(\tilde{f})$. As in case 1, this implies $(\tilde{P}(y,z;\bar{\rho}),\tilde{R}(y,z;\bar{\rho})) \in \mathscr{D}(\tilde{f})$. Thus, $\pi(y,z) > 0$. Hence, $\pi(y,x) > 0$ and $\pi(x,z) > 0$ imply $\pi(y,z) > 0$.

Case 3: Suppose $\pi(z,x) > 0$. Then $(\tilde{P}(z,x;\bar{\rho}),\tilde{R}(z,x;\bar{\rho})) \in \mathscr{D}(\tilde{f})$. By previous argument, $(\tilde{P}(y,x;\bar{\rho}),\tilde{R}(y,x;\bar{\rho})) \in \mathscr{D}(\tilde{f})$. Thus, $\pi(y,x) > 0$. Hence, $\pi(y,z) > 0$ and $\pi(z,x) > 0$ imply $\pi(y,x) > 0$.

Case 4: Suppose $\pi(z,y) > 0$. Then $(\tilde{P}(z,y;\bar{\rho}),\tilde{R}(z,y;\bar{\rho})) \in \mathscr{D}(\tilde{f})$. By previous arguments again, $(\tilde{P}(z,x;\bar{\rho}),\tilde{R}(z,x;\bar{\rho})) \in \mathscr{D}(\tilde{f})$. Thus, $\pi(z,x) > 0$. Hence, $\pi(z,y) > 0$ and $\pi(y,x) > 0$ imply $\pi(z,x) > 0$.

Case 5: Suppose $\pi(x,z) > 0$ and $\pi(z,y) > 0$. Then $\pi(y,z) = 0$. However, this contradicts Case 2 because $\pi(x,z) > 0$ implies $\pi(y,z) > 0$.

Case 6: Suppose $\pi(z,x) > 0$ and $\pi(x,y) > 0$. Then $\pi(y,x) = 0$. However, this contradicts Case 3 because $\pi(z,x) > 0$ implies $\pi(y,x) > 0$.

Thus, $\tilde{f}(\bar{\rho})$ is partially quasi-transitive. By Proposition 6.3, $\tilde{f}(\bar{\rho})$ is acyclic; and by Lemma 6.38, $\mathrm{Supp}(ND(\pi,\mu)) \neq \emptyset$.                                             $\square$

**Corollary 6.40.** *Let $\tilde{f}$ be a fuzzy voting rule and let $\mu \in \mathscr{F}(X)$. Assume $\pi$ is regular. Suppose $\rho_i$ is weakly transitive for all $i \in N$. If $\bar{\rho} \in \mathscr{FS}^n$, then $\mathrm{Supp}(M(\tilde{f}(\bar{\rho}),\mu)) \neq \emptyset$.*

In words, Theorem 6.36 shows that fuzzy simple rules will produce a non-empty maximal set when individuals possess weakly single-peaked, weakly transitive preferences. Theorem 6.39 demonstrates that this result also hold under fuzzy voting rules if individuals possess single-peaked and partially quasi-transitive preferences. However, Theorem 6.36 cannot be extended to fuzzy voting rules and Theorem 6.39 likewise cannot be extended to weakly single-peaked preferences as Example 6.41 demonstrates.

*Example 6.41.* Let $N = \{1,2,3,4,5\}$ and let $\mu \in \mathscr{F}(X)$ be such that $\mathrm{Supp}(\mu) = \{x,y,z\}$. Assume $\pi$ is regular. Let $\tilde{f}$ be a fuzzy voting rule such that $\mathscr{D}(\tilde{f}) = \{(\sigma,\omega) \in \mathscr{F}(N) \times \mathscr{F}(N) \mid \mathrm{Supp}(\sigma) \subseteq \mathrm{Supp}(\omega) \text{ and } |\mathrm{Supp}(\sigma)| > |N \backslash \mathrm{Supp}(\omega)|\}$. In this case, $\tilde{f}$ is said to be a plurality rule. Let $\bar{\rho} \in \mathscr{FR}^5$ be such that for all $i \in \{1,2\}$,

$$\rho_i(x,y) = .9, \rho_i(z,x) = .4, \rho_i(y,z) = .5,$$

for individual 3,

$$\rho_3(x,y) = .3, \rho_3(x,z) = .6, \rho_3(z,y) = 0,$$

for all $j \in \{4,5\}$,

$$\rho_j(y,x) = .8, \rho_j(x,z) = .1, \rho_j(y,z) = .2,$$

where $\rho_k(a,b) = .5$ otherwise, $k = 1,2,3,4,5$ and all $(a,b) \in \text{Supp}(\mu) \times \text{Supp}(\mu)$. Since $\pi$ is regular, we can write the individual strict preference relations as follows:

$$\pi_i(x,y) > 0, \ \pi_i(x,z) > 0, \ \pi_i(y,z) = 0,$$
$$\pi_3(y,x) > 0, \ \pi_3(x,z) > 0, \ \pi_3(y,z) > 0,$$
$$\pi_j(y,x) > 0, \ \pi_j(z,x) > 0, \ \pi_j(z,y) > 0,$$

for all $i \in \{1,2\}$ and $j \in \{4,5\}$, where $\pi_k(a,b) = 0$ otherwise, for $k = 1,2,3,4,5$ and all $(a,b) \in \text{Supp}(\mu) \times \text{Supp}(\mu)$. In this case, $\rho_k$ is weakly transitive for all $k \in N$ by Definition 6.2. Thus, by Proposition 6.3, $\rho_k$ is partially quasi-transitive for all $k \in N$. Relabel $x = a_1$, $y = a_2$ and $z = a_3$, and let $Q$ be a strict ordering on $\mu$ such that $a_3 Q a_2 Q a_1$. Now it is easily verified that $\bar{\rho} \in \mathscr{FW}$, where, according to Definition 6.34, $x = a_t$ for $\rho_i$, $y = a_t$ for $\rho_3$, and $z = a_t$ for $\rho_j$.

Consider some $(\sigma, \omega) \in \mathscr{F}(N) \times \mathscr{F}(N)$ such that

$$(\text{Supp}(\sigma), \text{Supp}(\omega)) = (\{1,2,3\}, \{1,2,3\}).$$

Because $(\sigma, \omega) \in \mathscr{D}(\tilde{f})$, $\pi_k(x,z) > 0$ for all $k \in \text{Supp}(\sigma)$, and $\pi_l(z,x) = 0$ for all $l \in \text{Supp}(\omega)$, $(x,z) \in \tilde{\mathscr{R}}(\bar{\rho})$. Likewise, consider some $(\sigma', \omega') \in \mathscr{F}(N) \times \mathscr{F}(N)$ such that $(\text{Supp}(\sigma'), \text{Supp}(\omega')) = (\{4,5\}, \{1,2,4,5\})$. Since $(\sigma', \omega') \in \mathscr{D}(\tilde{f})$, $\pi_k(z,y) > 0$ for all $k \in \text{Supp}(\sigma')$, and $\pi_l(y,z) = 0$ for all $l \in \text{Supp}(\omega')$, $(z,y) \in \tilde{\mathscr{R}}(\bar{\rho})$. Finally, consider some $(\sigma'', \omega'') \in \mathscr{F}(N) \times \mathscr{F}(N)$ such that $(\text{Supp}(\sigma''), \text{Supp}(\omega'')) = (\{3,4,5\}, \{3,4,5\})$. Since $(\sigma'', \omega'') \in \mathscr{D}(\tilde{f})$, $\pi_k(y,x) > 0$ for all $k \in \text{Supp}(\sigma'')$, and $\pi_l(x,y) = 0$ for all $l \in \text{Supp}(\omega'')$, $(y,x) \in \tilde{\mathscr{R}}(\bar{\rho})$. Using Definition 6.28, we can now determine $\tilde{g}_{\mathscr{D}(\tilde{f})}$ as follows:

$$\tilde{g}_{\mathscr{D}(\tilde{f})}(x,y) = .3 \wedge .5 \wedge .5 = .3,$$
$$\tilde{g}_{\mathscr{D}(\tilde{f})}(y,x) = .5 \vee .8 \vee .8 = .8,$$
$$\tilde{g}_{\mathscr{D}(\tilde{f})}(x,z) = .5 \vee .5 \vee .6 = .6,$$
$$\tilde{g}_{\mathscr{D}(\tilde{f})}(z,x) = .4 \wedge .4 \wedge .5 = .4,$$
$$\tilde{g}_{\mathscr{D}(\tilde{f})}(y,z) = .5 \wedge .5 \wedge .2 \wedge .2 = .2,$$
$$\tilde{g}_{\mathscr{D}(\tilde{f})}(z,y) = .5 \vee .5 \vee .5 \vee .5 = .5,$$

and $\tilde{g}_{\mathscr{D}(\tilde{f})}(a,a) = 1$ for all $a \in X$. By definition of a fuzzy voting rule and the regularity of $\pi$, we now have $\pi(y,x) > 0$, $\pi(x,z) > 0$ and $\pi(z,y) > 0$. Thus $\text{Supp}(ND(\pi, \mu)) = \emptyset$.

## 6.5 Extending Black's Median Voter Theorem

Section 6.4 demonstrated that fuzzy simple rules produce a non-empty fuzzy non-dominated set if individual preferences are weakly single-peaked. Moreover, fuzzy voting rules produce a non-empty fuzzy non-dominated when individual preferences are single-peaked. The fuzzy maximal set is non-empty in both cases. Building on these results, this section concerns itself with the application of Black's Median Voter Theorem to generate maximal elements in social preference relation. We assume that $\rho_i$ is partially quasi-transitive for all $i \in N$ throughout.

**Definition 6.42.** Let $\mu \in \mathscr{F}(X)$ and let $\bar{\rho} \in \mathscr{FS}^n$. Let $Q$ be a strict ordering of the elements of $\text{Supp}(\mu)$ such that $\rho_i$ is single-peaked with respect to $Q$ for all $i \in N$. Define $x_i$ to be that element of $X$ such that $\pi_i(x_i, y) > 0$ for all $y \in \text{Supp}(\mu) \backslash \{x_i\}$, $i = 1, \ldots, n$. Define $\tilde{L}^-, \tilde{L}^+ : \text{Supp}(\mu) \rightarrow \mathscr{F}(N)$ by for all $z \in \text{Supp}(\mu)$ and all $i \in N$,

$$\tilde{L}^-(z)(i) = \begin{cases} \pi_i(x_i, z) & \text{if } zQx_i, \\ 0 & \text{otherwise,} \end{cases}$$

$$\tilde{L}^+(z)(i) = \begin{cases} \pi_i(x_i, z) & \text{if } x_iQz, \\ 0 & \text{otherwise.} \end{cases}$$

**Definition 6.43 (f–median).** Let $\tilde{f}$ be a fuzzy preference aggregation rule and let $\bar{\rho} \in \mathscr{FS}^n$. Let $Q$ be a strict ordering of the elements of $\text{Supp}(\mu)$ such that $\rho_i$ is weakly single-peaked with respect to $Q$ for all $i \in N$. Then for some $\mu \in \mathscr{F}(X)$, an element $z \in \text{Supp}(\mu)$ is called an $\tilde{f}$–median if $\tilde{L}^-(z) \notin \mathscr{L}(\tilde{f})$ and $\tilde{L}^+(z) \notin \mathscr{L}(\tilde{f})$.

**Definition 6.44 (f–median set).** Let $\mu \in \mathscr{F}(X)$ and let $\tilde{f}$ be a fuzzy preference aggregation rule. For all $\bar{\rho} \in \mathscr{FS}$ where $Q$ is some strict ordering of the elements of $\text{Supp}(\mu)$ such that $\rho_i$ is weakly single-peaked with respect to $Q$ for all $i \in N$, define the fuzzy subset $\mu_{\tilde{f}}(\bar{\rho}; Q)$ of $\text{Supp}(\mu)$, by for all $z \in \text{Supp}(\mu)$,

$$\mu_{\tilde{f}}(\bar{\rho}, Q)(z) = \begin{cases} \mu(z) & \text{if } z \text{ is an } \tilde{f}\text{-median,} \\ 0 & \text{otherwise.} \end{cases}$$

We say $\mu_{\tilde{f}}(\bar{\rho}; Q)$ is the *fuzzy subset of $\tilde{f}$-medians* given $\bar{\rho}$. When $Q$ is understood, we simplify the previous notation and write $\mu_{\tilde{f}}(\bar{\rho})$.

**Theorem 6.45.** *Let $\mu \in \mathscr{F}(X)$ and let $\tilde{f}$ be a fuzzy simple rule. If $\bar{\rho} \in \mathscr{FS}^n$, then $\mu_{\tilde{f}}(\bar{\rho})(x) = ND(\pi, \mu)(x)$ for all $x \in \text{Supp}(\mu)$, where $\pi$ is the strict preference relation with respect to $\tilde{f}(\bar{\rho})$.*

*Proof.* Let $\bar{\rho} \in \mathscr{FS}^n$ and let $Q$ be a strict ordering of $\text{Supp}(\mu)$ such that $\rho_i$ is weakly single-peaked with respect to $Q$ for all $i \in N$. Let $x \in \text{Supp}(\mu)$. Further, suppose $x$ is an $\tilde{f}$-median and in $\text{Supp}(ND(\pi, \mu))$. In this case, $\mu_{\tilde{f}}(\bar{\rho})(x) = \mu(x) = ND(\pi, \mu)(x)$. Hence, it suffices to show $\text{Supp}(\mu_{\tilde{f}}(\bar{\rho})) = \text{Supp}(ND(\pi, \mu))$. Since $\text{Supp}(\mu)$ is finite, we relabel the elements of $\text{Supp}(\mu)$ such that $a_{t+1}Qa_t$ for all $t \in \{1, \ldots, |\text{Supp}(\mu)| - 1\}$. Then $x = a_t$ for some $t \geq 1$.

Let $x \in \text{Supp}(ND(\pi, 1_X))$. Suppose $\mu_{\tilde{f}}(\bar{\rho})(x) = 0$, i.e. $x$ is not an $\tilde{f}$-median. Because $x$ is not an $\tilde{f}$-median, either $\tilde{L}^-(x) \in \mathscr{L}(\tilde{f})$ or $\tilde{L}^+(x) \in \mathscr{L}(\tilde{f})$. First, assume the former. Then, for all $u, v \in \text{Supp}(\mu)$, $\pi_i(u, v) > 0$ for all $i \in \text{Supp}(\tilde{L}^-(x))$ implies $\pi(u, v) > 0$. Since $\bar{\rho}$ is single-peaked, $\pi_i(a_{t-1}, a_t) > 0$ for all $i \in \text{Supp}(\tilde{L}^-(x))$.

Since $\check{L}^-(x)$ is decisive, $\pi(a_{t-1},x) > 0$; and $ND(\pi,\mu)(x) = 0$, a contradiction. Second, assume $\check{L}^+(x) \in \mathscr{L}(\tilde{f})$. Then, for all $u,v \in \text{Supp}(\mu)$, $\pi_i(u,v) > 0$ for all $i \in \text{Supp}(\check{L}^+(x))$ implies $\pi(u,v) > 0$. Since $\bar{\rho}$ is single-peaked, $\pi_i(a_{t+1},a_t) > 0$ for all $i \in \text{Supp}(\check{L}^+(x))$. Since $\check{L}^+(x)$ is decisive, $\pi(a_{t+1},x) > 0$; and $ND(\pi,\mu)(x) = 0$. Since both cases establish a contradiction, it follows that $\mu_{\tilde{f}}(\bar{\rho})(x) > 0$. Hence, we have

$$\text{Supp}(ND(\pi,\mu)) \subseteq \text{Supp}(\mu_{\tilde{f}}(\bar{\rho})).$$

Now let $x \in \text{Supp}(\mu(\bar{\rho}))$ and suppose $x \notin \text{Supp}(ND(\pi,\mu))$. Then there exists some $a_\tau \in \text{Supp}(\mu)$ such that $\pi(a_\tau,x) > 0$. Thus, $\tilde{P}(a_\tau,a_t;\bar{\rho}) \in \mathscr{L}(\tilde{f})$. Since $Q$ is a strict ordering on $X$, either $a_t Q a_\tau$ or $a_\tau Q a_t$.

First, assume the former. Then $\text{Supp}(\tilde{P}(a_\tau,a_t;\bar{\rho})) \subseteq \text{Supp}(\check{L}^-(a_t))$. However, $\mu_{\tilde{f}}(\bar{\rho})(a_t) > 0$ implies $\check{L}^-(a_t) \notin \mathscr{L}(\tilde{f})$. This contradicts the monotonicity of $\mathscr{L}(\tilde{f})$.

Second, assume $a_\tau Q a_t$. Then $\text{Supp}(\tilde{P}(a_\tau,a_t;\bar{\rho})) \subseteq \text{Supp}(\check{L}^+(a_t))$. However, $\mu_{\tilde{f}}(\bar{\rho})(a_t) > 0$ implies $\check{L}^+(a_t) \notin \mathscr{L}(\tilde{f})$. Likewise, this contradicts the monotonicity of $\mathscr{L}(\tilde{f})$.

Hence, $a_t \in \text{Supp}(ND(\pi,\mu))$ and $x \in \text{Supp}(ND(\pi,\mu))$. Thus,

$$\text{Supp}(ND(\pi,\mu)) = \text{Supp}(\mu_{\tilde{f}}(\bar{\rho})). \qquad \square$$

**Corollary 6.46.** *Let* $\mu \in \mathscr{F}(X)$ *and let* $\tilde{f}$ *be a fuzzy simple rule. Assume* $\pi$ *is partial. If* $\bar{\rho} \in \mathscr{FS}^n$, *then*

$$Supp(\mu_{\tilde{f}}(\bar{\rho})) = Supp(ND(\pi,\mu)) = Supp(M(\tilde{f}(\bar{\rho}),\mu)),$$

*where* $\pi$ *is the strict preference relation with respect to* $\tilde{f}(\bar{\rho})$.

**Corollary 6.47.** *Let* $\mu \in \mathscr{F}(X)$ *and let* $\tilde{f}$ *be a fuzzy simple rule. Assume* $\pi$ *is regular. If* $\bar{\rho} \in \mathscr{FS}^n$, *then*

$$Supp(\mu_{\tilde{f}}(\bar{\rho})) = Supp(ND(\pi,\mu)) \subseteq Supp(M(\tilde{f}(\bar{\rho}),\mu)),$$

*where* $\pi$ *is the strict preference relation with respect to* $\tilde{f}(\bar{\rho})$.

**Corollary 6.48.** *Let* $\mu \in \mathscr{F}(X)$ *and let* $\tilde{f}$ *be a fuzzy simple rule. Assume* $\pi$ *is partial and regular. If* $\bar{\rho} \in \mathscr{FS}^n$, *then for all* $x \in X$,

$$\mu_{\tilde{f}}(\bar{\rho})(x) = ND(\pi,\mu)(x) = M(\tilde{f}(\bar{\rho}),\mu)(x),$$

*where* $\pi$ *is the strict preference relation with respect to* $\tilde{f}(\bar{\rho})$.

In words, Theorem 6.45 proves that Black's Median Voter Theorem holds with respect to the fuzzy non-dominated set regardless of any restrictions placed on the strict preference relation. Corollary 6.46 demonstrates that these results also hold with respect to the the fuzzy maximal set if the strict preference relation is partial. As Corollaries 6.46 and 6.48 demonstrate, these results do not hold if strict preferences are regular but not partial. Example 6.49 below further illustrates this.

## 6.6   An Application

The following example presents an application of Black's Median Voter Theorem
with a "majority rules" voting rule and regular strict preferences.

*Example 6.49.* Let $N = \{1,2,3\}$ and $X = \{x,y,z\}$. Let $\mu \in \mathcal{F}(X)$ be such that
$\mu(x) = .9$, $\mu(y) = .3$, $\mu(z) = .6$. Assume $\pi$ is regular. Let $\bar{\rho} \in \mathcal{F}\mathcal{R}^3$ be defined
as

$$\rho_1(x,y) = .7, \ \rho_1(x,z) = 1.0, \ \rho_1(z,y) = .3,$$
$$\rho_2(x,y) = 0, \ \ \rho_2(z,x) = .6, \ \rho_2(y,z) = .7,$$
$$\rho_3(y,x) = .8, \ \ \rho_3(x,z) = .2, \ \rho_3(y,z) = .1,$$

where $\rho_i(a,b) = .5$ otherwise, for $i = 1,2,3$ and all $(a,b) \in \text{Supp}(\mu) \times \text{Supp}(\mu)$.
Because $\pi$ is regular, we can write the individual strict preference relations as follows:

$$\pi_1(x,y) > 0, \ \pi_1(x,z) > 0, \ \pi_1(y,z) > 0,$$
$$\pi_2(y,x) > 0, \ \pi_2(z,x) > 0, \ \pi_2(y,z) > 0,$$
$$\pi_3(y,x) > 0, \ \pi_3(z,x) > 0, \ \pi_3(z,y) > 0.$$

Using Definition 6.33, it can be verified that $\rho_i$ is single-peaked with respect to some
ordering of $\text{Supp}(\mu)$. In this case, $zQyQx$ or $xQyQz$, where $a_t = x$ for $\rho_1$, $a_t = y$ for
$\rho_2$, and $a_t = z$ for $\rho_3$. Suppose $zQyQx$. Let $\tilde{f} : \mathcal{F}\mathcal{R}^n \to \mathcal{F}\mathcal{R}$ be defined as follows:
for all $\bar{\rho} \in \mathcal{F}\mathcal{R}^n$ and all $a,b \in X$,

$$\tilde{f}(\bar{\rho})(a,b) = \begin{cases} 1 & \text{if } a = b, \\ 1 & \text{if } |\text{Supp}(\tilde{P}(a,b;\bar{\rho}))| > \frac{n}{2}, \\ \beta & \text{otherwise}, \end{cases}$$

where $\beta \in (0,1)$. In the case of $N = \{1,2,3\}$,

$$\{\text{Supp}(\lambda) \mid \lambda \in \mathcal{L}(\tilde{f})\} = \{\{1,2\}, \{1,3\}, \{2,3\}, \{1,2,3\}\}.$$

From the definition of $\tilde{f}$, it follows that $\tilde{f}$ is monotonic, neutral and decisive. Thus,
$\tilde{f}$ is a fuzzy simple rule. Now, $\tilde{L}^-$ and $\tilde{L}^+$ can be written as

$$\tilde{L}^-(x)(1) = 0, \ \tilde{L}^-(x)(2) = 0, \ \tilde{L}^-(x)(3) = 0,$$
$$\tilde{L}^-(y)(1) > 0, \ \tilde{L}^-(y)(2) = 0, \ \tilde{L}^-(y)(3) = 0,$$
$$\tilde{L}^-(z)(1) > 0, \ \tilde{L}^-(z)(2) > 0, \ \tilde{L}^-(z)(3) = 0,$$

and

$$\tilde{L}^+(x)(1) = 0, \ \tilde{L}^+(x)(2) > 0, \ \tilde{L}^+(x)(3) > 0,$$
$$\tilde{L}^+(y)(1) = 0, \ \tilde{L}^+(y)(2) = 0, \ \tilde{L}^+(y)(3) > 0,$$
$$\tilde{L}^+(z)(1) = 0, \ \tilde{L}^+(y)(2) = 0, \ \tilde{L}^+(y)(3) = 0.$$

By definition of $\tilde{f}$ and $\mathcal{L}(\tilde{f})$, $\tilde{L}^-(z)$ and $\tilde{L}^+(x)$ are decisive. Hence, $x$ and
$z$ are not $\tilde{f}$-medians. Also, $\tilde{L}^-(y)$ and $\tilde{L}^+(y)$ are not decisive since

$|\text{Supp}(\tilde{L}^-(y))| = |\text{Supp}(\tilde{L}^+(y))| = 1$. Thus, $y$ is an $\tilde{f}$-median. It follows that $\mu_{\tilde{f}}(\bar{\rho}, Q)(x) = 0$, $\mu_{\tilde{f}}(\bar{\rho}, Q)(y) = \mu(y) = .3$, and $\mu_{\tilde{f}}(\bar{\rho}, Q)(z) = 0$.

The social preference relation given $\bar{\rho}$ defined above is

$$\tilde{f}(\bar{\rho})(x,y) = \beta, \ \tilde{f}(\bar{\rho})(x,z) = \beta,$$
$$\tilde{f}(\bar{\rho})(y,x) = 1, \ \tilde{f}(\bar{\rho})(y,z) = 1,$$
$$\tilde{f}(\bar{\rho})(z,x) = 1, \ \tilde{f}(\bar{\rho})(z,y) = \beta,$$

where $\tilde{f}(\bar{\rho})(a,a) = 1$ for all $a \in X$. Since $\pi$ is regular and $\beta < 1$, $\pi(y,x) > 0$, $\pi(y,z) > 0$ and $\pi(z,x) > 0$. Since there does not exists an $a \in \text{Supp}(\mu)$ such that $\pi(a,y) > 0$, then $nd(\tilde{f}(\bar{\rho})(y) = 0$ and $ND(\pi,\mu)(y) = \mu(y) = .3$. Because $\pi(y,x) > 0$ and $\pi(y,z) > 0$, $ND(\pi,\mu)(x) = ND(\pi,\mu)(z) = 0$. Thus, $ND(\pi,\mu) = \mu_{\tilde{f}}(\bar{\rho}, Q)$. However, $M(\tilde{f}(\bar{\rho}), \mu)$ is calculated as follows. Here we use the residuum operator (see Sec. 1.2.5) where $a, b \in [0,1]$ and the standard residuum is defined as $a \to b = \bigvee \{t \in [0,1] \mid a \wedge t \leq b\}$.

$$M(\tilde{f}(\bar{\rho}), \mu)(x) = \mu(x) \wedge (\bigwedge \{\tilde{f}(\bar{\rho})(a,x) \to \tilde{f}(\bar{\rho})(x,a)\} \mid a \in \text{Supp}(\mu)\})$$
$$= .9 \wedge (\{\tilde{f}(\bar{\rho})(y,x) \to \tilde{f}(\bar{\rho})(x,y)\} \wedge$$
$$\{\tilde{f}(\bar{\rho})(z,x) \to \tilde{f}(\bar{\rho})(x,z)\})$$
$$= .9 \wedge (\{1 \to \beta\}\} \wedge \{1 \to \beta\}\})$$
$$= .9 \wedge (\beta \wedge \beta)$$
$$> 0.$$

$$M(\tilde{f}(\bar{\rho}), \mu)(y) = \mu(y) \wedge (\bigwedge \{\tilde{f}(\bar{\rho})(a,y) \to \tilde{f}(\bar{\rho})(y,a)\} \mid a \in \text{Supp}(\mu)\})$$
$$= .3 \wedge (\{\tilde{f}(\bar{\rho})(x,y) \to \tilde{f}(\bar{\rho})(y,x)\} \wedge$$
$$\{\tilde{f}(\bar{\rho})(z,y) \to \tilde{f}(\bar{\rho})(y,z)\})$$
$$= .3 \wedge (\{\beta \to 1\} \wedge \{\beta \to 1\})$$
$$= .3 \wedge (1 \wedge 1)$$
$$= .3.$$

$$M(\tilde{f}(\bar{\rho}), \mu)(z) = \mu(z) \wedge (\bigwedge \{\tilde{f}(\bar{\rho})(a,z) \to \tilde{f}(\bar{\rho})(z,a)\} \mid a \in \text{Supp}(\mu)\}\})$$
$$= .3 \wedge (\{\tilde{f}(\bar{\rho})(y,z) \to \tilde{f}(\bar{\rho})(z,y)\} \wedge$$
$$\{\tilde{f}(\bar{\rho})(x,z) \to \tilde{f}(\bar{\rho})(z,x)\})$$
$$= .3 \wedge (\{1 \to \beta\}\} \wedge \{\beta \to 1\})$$
$$= .3 \wedge (\beta \wedge 1)$$
$$> 0.$$

Hence, $M(\tilde{f}(\bar{\rho}), \mu) \neq \mu_{\tilde{f}}(\bar{\rho}, Q)$ and $\text{Supp}(M(\tilde{f}(\bar{\rho}), \mu)) \neq \text{Supp}(\mu_{\tilde{f}}(\bar{\rho}, Q))$.

## 6.7   Conclusions and Spatial Models

Our work in this chapter demonstrates that the definitions of the fuzzy maximal set and fuzzy strict preference relations are critical in determining whether or not Black's Median Voter Theorem holds in the fuzzy framework. These results beg the question as to whether fuzzy social choice models behave better or worse than their conventional crisp set counterparts in multi-dimensional space. Specifically, we wish to know the conditions under which the fuzzy spatial model produces a non-empty maximal set. This question is the focus of the next chapter.

Before turning to this question in the next chapter, we present two final results.

**Definition 6.50 (Property M).** Let $\rho$ be a fuzzy function on $X$. Then $\rho$ is said to have *Property M* if $\forall x \in X, \wedge\{\rho(x,y)|y \in X\} = 0$ implies $\exists y \in X$ such that $\rho(x,y) = 0$.

**Definition 6.51 (fuzzy core).** Let $\tilde{f}$ be a fuzzy aggregation rule. $\forall \bar{\rho} \in \mathscr{FR}^n$, define $C_{\tilde{f}}(\bar{\rho}) : X \to [0,1]$ by $\forall x \in X, C_{\tilde{f}}(\bar{\rho})(x) = M(\tilde{f}(\bar{\rho}), 1_x)(x)$. Then $C_{\tilde{f}}(\bar{\rho})$ is called the *fuzzy core* of $\tilde{f}$ at $\bar{\rho}$.

**Theorem 6.52.** *Suppose strict preferences are partial. Let $\tilde{f}$ be a partial fuzzy voting rule. Let $\bar{\rho} \in \mathscr{FR}^n$. Suppose $\tilde{f}(\bar{\rho})$ and $\tilde{g}_{\mathscr{L}(\tilde{f})}(\bar{\rho})$ have property M. If $\bar{\rho}$ is strictly convex, then $Supp(C_{\tilde{f}}(\bar{\rho})) = Supp(C_{\tilde{g}_{\mathscr{L}(\tilde{f})}}(\bar{\rho}))$.*

*Proof.* We first note that

$$Supp(C_{\tilde{f}}(\bar{\rho})) \subseteq Supp(C_{\tilde{g}_{\mathscr{L}(\tilde{f})}}(\bar{\rho})).$$

Let $x \in Supp(C_{\tilde{f}}(\bar{\rho}))$ and suppose $x \notin Supp(C_{\tilde{g}_{\mathscr{L}(\tilde{f})}}(\bar{\rho}))$. Then $\exists y \in X$ such that $\tilde{g}_{\mathscr{L}(\tilde{f})}(\bar{\rho})(x,y) = 0$ since $\tilde{g}_{\mathscr{L}(\tilde{f})}$ has property M and so $\pi_{\mathscr{L}(\tilde{f})}(x,y) > 0$. Thus $(y,x) \in \mathscr{P}(\bar{\rho})$ and $(x,y) \in Symm(\mathscr{P}(\bar{\rho})) \setminus \mathscr{P}(\bar{\rho})$. Hence $\pi(y,x) > 0$. $(\pi_i(y,x) > 0 \,\forall i \in Supp(\lambda) \Rightarrow \pi(y,x) > 0.)$ Since $x \in Supp(C_{\tilde{f}}(\bar{\rho}))$, we have that $\tilde{f}(\bar{\rho})(x,y) > 0$, a contradiction. Thus $x \in Supp(C_{\tilde{g}_{\mathscr{L}(\tilde{f})}}(\bar{\rho}))$.

We now show

$$Supp(C_{\tilde{g}_{\mathscr{L}(\tilde{f})}}(\bar{\rho})) \subseteq Supp(C_{\tilde{f}}(\bar{\rho})).$$

Let $x \in Supp(C_{\tilde{g}_{\mathscr{L}(\tilde{f})}}(\bar{\rho}))$ and suppose $x \notin Supp(C_{\tilde{f}}(\bar{\rho}))$. Then $\exists y \in X$ such that $\tilde{f}(\bar{\rho})(x,y) = 0$ since $\tilde{f}(\bar{\rho})$ has property M and so $\pi(x,y) > 0$. Hence by the proof of Theorem 6.31, $(\sigma, \omega) \in \mathscr{D}(\tilde{f})$, for any $\sigma, \omega \in \mathscr{FP}(N)$ such that $Supp(\sigma) = P(y,x;\bar{\rho})$ and $Supp(\omega) = R(y,x;\bar{\rho})$. For all $i \in R(y,x;\bar{\rho}), \pi_i(z,x) > 0$, where $z = ax + (1-a)y$ for some $a \in (0,1)$ by the strict convexity of $\bar{\rho}$. Thus $\omega \in \mathscr{L}(\tilde{f})$, where $Supp(\omega) = R(y,x;\bar{\rho})$. Hence $\pi_{\mathscr{L}(\tilde{f})}(z,x) > 0$ and so $\tilde{g}_{\mathscr{L}(\tilde{f})}(\bar{\rho})(x,z) = 0$. Thus $x \notin Supp(C_{\tilde{g}_{\mathscr{L}(\tilde{f})}}(\bar{\rho}))$. $\square$

We note that if a fuzzy relation $\rho$ on $X$ is such that $|Im(\rho)| < \infty$, then $\rho$ has property M. Fuzzy relations with property M may be of importance when considering problems of thick indifference, an issue we consider in the next chapter.

**Theorem 6.53.** *Suppose strict preferences are regular. Let $\tilde{f}$ be a regular fuzzy voting rule. Let $\bar{\rho} \in \mathscr{FR}^n$. Suppose $\tilde{f}(\bar{\rho})$ and $\tilde{g}_{\mathscr{L}(\tilde{f})}(\bar{\rho})$ have property M. If $\bar{\rho}$ is strictly convex, then $Supp(C_{\tilde{f}}(\bar{\rho})) = Supp(C_{\tilde{g}_{\mathscr{L}(\tilde{f})}}(\bar{\rho}))$.*

*Proof.* We first note that

$$Supp(C_{\tilde{f}}(\bar{\rho})) \subseteq Supp(C_{\tilde{g}_{\mathscr{L}(\tilde{f})}}(\bar{\rho})) .$$

Let $x \in Supp(C_{\tilde{f}}(\bar{\rho}))$ and suppose $x \notin Supp(C_{\tilde{g}_{\mathscr{L}(\tilde{f})}}(\bar{\rho}))$. Then $\exists y \in X$ such that

$$\tilde{g}_{\mathscr{L}(\tilde{f})}(\bar{\rho})(x,y) = 0$$

since $\tilde{g}_{\mathscr{L}(\tilde{f})}$ has property M and so $\pi_{\mathscr{L}(\tilde{f})}(y,x) > 0$. Thus $(y,x) \in \mathscr{P}(\bar{\rho})$ and $(x,y) \in Symm(\mathscr{P}(\bar{\rho})) \setminus \mathscr{P}(\bar{\rho})$. Hence $\pi(y,x) > 0$. $(\pi_i(y,x) > 0 \ \forall i \in Supp(\lambda) \Rightarrow \pi(y,x) > 0.)$ Since $x \in Supp(C_{\tilde{f}}(\bar{\rho})), \tilde{f}(\bar{\rho})(x,y) > 0$, a contradiction. Thus $x \in Supp(C_{\tilde{g}_{\mathscr{L}(\tilde{f})}}(\bar{\rho}))$.

We now show

$$Supp(C_{\tilde{g}_{\mathscr{L}(\tilde{f})}}(\bar{\rho})) \subseteq Supp(C_{\tilde{f}}(\bar{\rho})) .$$

Let $x \in Supp(C_{\tilde{g}_{\mathscr{L}(\tilde{f})}}(\bar{\rho}))$ and suppose $x \notin Supp(C_{\tilde{f}}(\bar{\rho}))$. Then $\exists y \in X$ such that $\tilde{f}(\bar{\rho})(x,y) = 0$ since $\tilde{f}(\bar{\rho})$ has property M and so $\pi(y,x) > 0$. Since $\tilde{f}$ is a regular fuzzy voting rule, $\tilde{f}$ is neutral and monotonic by Theorem 6.32. Thus by the proof of Theorem 6.32, $(\sigma, \omega) \in \mathscr{D}(\tilde{f})$, for any $\sigma, \omega \in \mathscr{FP}(N)$ such that $Supp(\sigma) = P(y,x;\bar{\rho})$ and $Supp(\omega) = R(y,x;\bar{\rho})$. For all $i \in R(y,x;\bar{\rho})$, $\pi_i(z,x) > 0$, where

$$z = ax + (1-a)y$$

or some $a \in (0,1)$ by the strict convexity of $\bar{\rho}$. Hence $\omega \in \mathscr{L}(\tilde{f})$, where $Supp(\omega) = R(y,x;\bar{\rho})$. Thus $\pi_{\mathscr{L}(\tilde{f})}(z,x) > 0$ and so $\tilde{g}_{\mathscr{L}(\tilde{f})}(\bar{\rho})(x,z) = 0$. Hence $x \notin Supp(C_{\tilde{g}_{\mathscr{L}(\tilde{f})}}(\bar{\rho}))$, a contradiction. □

# References

Arrow, K.: Social Choice and Individual Values. Wiley, New York (1951)

Austen-Smith, D., Banks, J.S.: Positive Political Theory I: Collective Preference. University of Michigan Press, Ann Arbor (1999)

Black, D.: On Arrow's impossibility theorem. Journal of Law and Economics 12(2), 227–248 (1969), http://EconPapers.repec.org/RePEc:ucp:jlawec:v:12:y:1969:i:2:p:227-48

Dutta, B.: Fuzzy preferences and social choice. Mathematical Social Sciences 13(3), 215–229 (1987)

Fono, L.A., Andjiga, N.G.: Fuzzy strict preference and social choice. Fuzzy Sets Syst. 155, 372–389 (2005), http://dx.doi.org/10.1016/j.fss.2005.05.001

Gibbard, A.: Manipulation of voting schemes: A general result. Econometrica 41(4), 587–601 (1973)

Gibilisco, M.B., Mordeson, J.N., Clark, T.D.: Fuzzy Black's median voter theorem: Examining the structure of fuzzy rules and strict preference. New Mathematics and Natural Computation (NMNC) 8(2), 195–217 (2012)

Kiewiet, D.R., McCubbins, M.D.: Presidential influence on congressional appropriations decisions. American Journal of Political Science 32(3), 713–736 (1988)

Montero, F., Tejada, J.: A necessary and sufficient condition for the existence of orlovsky's choice set. Fuzzy Sets and Systems 26(1), 121–125 (1988),
http://www.sciencedirect.com/science/article/pii/
0165011488900103

Mordeson, J.N., Clark, T.D.: Fuzzy Arrow's theorem. New Mathematics and Natural Computation 5(2), 371–383 (2009),
http://www.worldscientific.com/doi/abs/10.1142/
S1793005709001362

Mordeson, J.N., Nielsen, L., Clark, T.D.: Single peaked fuzzy preferences in one-dimensional models: Does Black's median voter theorem hold? New Mathematics and Natural Computation 6(1), 1–16 (2010)

Romer, T., Rosenthal, H.: The elusive median voter. Journal of Public Economics 12(2), 143–170 (1979), http://www.sciencedirect.com/science/article/pii/
0047272779900100

Satterthwaite, M.A.: Strategy-proofness and Arrow's conditions: Existence and correspondence theorems for voting procedures and social welfare functions. Journal of Economic Theory 10(2), 187–217 (1975)

# Chapter 7
# Representing Thick Indifference in Spatial Models

**Abstract.** This chapter demonstrates that a fuzzy approach to modeling thick indifference can accommodate highly irregularly shaped indifference curves, even those that are concave or multi-modal. Moreover, it permits the calculation of a majority rule maximal set with relative ease under assumptions of non-separability. This approach relies on a homomorphism that permits a region of interest to be mapped to a simpler region with a suitable and natural partial ordering where the results are determined and then faithfully transferred back to the original region of interest.

## 7.1 Stability and Thick Indifference in Individual Preferences

It has been long known that the probability of a majority rule maximal set increases in spatial models when actors possess thick indifference over individual preferences (Bräuninger, 2007; Balke et al., 2006; Barberà and Ehlers, 2011; Gehrlein and Valognes, 2001; Skog, 1994; Sloss, 1973; Tovey, 1991). Many of the studies in this genre make use of Tovey's (1991; 2010) concept of an epsilon-core ($\varepsilon$−core), a threshold distance in Euclidean space that must be exceeded before players distinguish between alternatives (Bräuninger, 2007; Koehler, 2001). Unless an alternative lies outside of the region defined by the $\varepsilon$−core, a player is indifferent between it and the core's center. Essentially, actors have thick indifference curves Sloss (1973). Unfortunately, applying the approach in empirical analyses is hampered by the complexity of calculating the existence of a majority rule maximal set. It is even more problematic when thick indifference introduces irregularly shaped preference curves.

This chapter demonstrates that a fuzzy approach to modeling thick indifference can accommodate highly irregularly shaped indifference curves, even those that are concave or multi-modal. Moreover, it permits the calculation of a majority rule maximal set with relative ease under assumptions of non-separability. Section 7.2 develops the approach, which relies on a homomorphism that permits a region of interest (spatial model) to be mapped to a simpler region with a suitable and natural partial ordering where the results are determined and then faithfully transferred back to the original region of interest. Section 7.3 provides an empirical application of the approach. Section 7.4 then presents a proof of the homomorphism. Section 7.5 presents

M.B. Gibilisco et al., *Fuzzy Social Choice Theory*, 149
Studies in Fuzziness and Soft Computing 315,
DOI: 10.1007/978-3-319-05176-5_7, © Springer International Publishing Switzerland 2014

a formal that in all but a limited number of cases, spatial models of individuals with thick indifference curves result in an empty majority rule maximal set *if and only if* the Pareto set contains a union of cycles. The section also completely characterizes the elements that constitute the exception for a three-person game based on the general definition for *n* players. The substantive interpretation is that if the degree to which a majority find a given alternative acceptable is relatively high, then a stable outcome is assured under majority rule. Section 7.6 concludes with a consideration of the theoretical implications of the approach and observations on its utility for empirical studies.

## 7.2   Modeling Thick Indifference in Individual Preferences

The conventional approach to fuzzy spatial modeling, where $X$ is the set of alternatives, makes use of fuzzy preference relations Bezdek et al. (1978, 1979); Blin (1974); Kacprzyk and Fedrizzi (1988); Kacprzyk et al. (1992); Nurmi (1981a); Orlovsky (1978). Arguing that most data available to social sciences do not measure preference relations, Clark, Larson, Mordeson, Potter, and Wierman (2008) follow the lead of Nurmi (1981b) and use fuzzy sets to denote individual preferences. We gave consideration to individual preferences in Chapter 3.

Let $N$ be the set of political actors and $\mathscr{A}$ be the set of alternatives. We assume that $\mathscr{A}$ is a subset of an arbitrary universe of interest. Applied to spatial models, $\mathscr{A} \subseteq \mathbb{R}^k$, where $\mathbb{R}$ is the set of real numbers and $k$ is the number of dimensions in Euclidean space. Let a function, $\sigma_i$, indicate the degree to which political actor $i \in N$ views a particular alternative in the policy space as more or less ideal. Thus, $\sigma_i$ is a function mapping $\mathscr{A}$ onto the closed interval $[0, 1]$, where $\sigma_i(x) = 1$ represents all ideal policies and $\sigma_i(x) = 0$ represents all policies that are totally unacceptable to player $i$. If $\sigma_i$ is restricted to a discrete set, actors possess thick indifference. For example, $\mathrm{Im}(\sigma_i) \subseteq T = \{0, .25, .5, .75, 1\}$ would impose preferences similar to a Likert scale, where $\mathrm{Im}(\sigma_i)$ denotes the image of $\sigma_i$. $T$ denotes the granularity of individual preferences, how discerning players are over alternatives. We can set $T$ to any finite scale. In essence, political actors partition $\mathscr{A}$ into a finite number of classes, each class comprising an indifference set. While the boundaries between each indifference set may be rather sharp, this problem can be resolved by increasing the granularity in the region of a boundary. Doing so does not effect our results. For ease of presentation and without loss of generalization, we consider coarse granularity at the boundaries of indifference sets in our examples.

Both the geometry representing spatial preferences and its corresponding relation space can be mapped into a simpler, more appropriate set $U$. We assume that $U$ is an arbitrary set with a partial order, making it a lattice and allowing for a simpler analysis of the relation space. Nonetheless, we can specify $U$ to be more intuitive. We specify $U = T^n$, where $T^n = \{(a_1, ..., a_n) \mid a_i \in T, i = 1, ..., n\}$ and $n = |N|$. The mapping of the relation space into $T^n$ permits the characterization of any policy space in its entirety with $n$−tuples, $(a_1, a_2, ..., a_n)$, which represent the specific $\sigma$−values of a policy space. Essentially, $T^n$ is a lattice of $n$−tuples with entries from $T$ under this construction.

Let $\mathscr{R}$ denote the set of all binary relations on $U$ that are reflexive, complete, and transitive and $\mathscr{B}$ the set of all reflexive and complete binary relations on $U$. Let $R_i \in \mathscr{R}, i \in N$. Then $xP_iy$ if and only if $xR_iy$ and not $yR_ix$. In such case, we say that $x$ is *strictly preferred* to $y$ by player $i$. Let $\mathscr{R}^n = \{\overline{\rho} \mid \overline{\rho} = (R_1, ..., R_n), R_i \in \mathscr{R}, i = 1, ..., n\}$, where $|N| = n$.

**Definition 7.1 (simple majority rule).** Let $\overline{\rho} \in \mathscr{R}^n$. Let $f$ be an aggregation rule, that is, a function from $\mathscr{R}^n$ into $\mathscr{B}$. Let $X \subseteq U$. Define simple majority rule as follows. For all $(x,y) \in X^2$,

$$(x,y) \in f(\overline{\rho}) \text{ if and only if } |\{i \in N \mid xR_iy\}| > \frac{n}{2}.$$

Then $(x,y) \in f(\overline{\rho})$ and not $(y,x) \in f(\overline{\rho})$ if and only if $|\{i \in N \mid xR_iy\}| > \frac{n}{2}$ and $|\{i \in N \mid yR_ix\}| \leq \frac{n}{2}$ if and only if $|\{i \in N \mid xP_iy\}| > \frac{n}{2}$.

**Definition 7.2 (simple majority relation).** Let $\overline{\rho} \in \mathscr{R}^n$. Define the binary relation $R$ on $X$ by for all $x, y \in X, (x,y) \in R$ if and only if

$$|\{i \in N \mid xR_iy\}| \geq \frac{n}{2}.$$

Define $P \subseteq X \times X$ by for all $x, y \in X, (x,y) \in P$ if and only if $(x,y) \in R$ and $(y,x) \notin R$. Let $R(x,y;\overline{\rho}) = \{i \in N \mid xR_iy\}$ and $P(x,y;\overline{\rho}) = \{i \in N \mid xP_iy\}$.

Note that $R$ is a social preference, and $\overline{\rho}$ is an n-tuple of individual preference relations.

**Proposition 7.3.** *Let $\overline{\rho}$ and $R$ be defined as in Definition 7.2. Let $x, y \in X$. Then $(x,y) \in P$ if and only if*

$$|P(x,y;\overline{\rho})| > \frac{n}{2}.$$

*Proof.* $xPy$ if and only if $xRy$ and not $yRx$ if and only if

$$|\{i \in N \mid xR_iy\}| \geq \frac{n}{2}$$

and

$$|\{j \in N \mid yR_jx\}| < \frac{n}{2}.$$

Since each $R_i$ is complete, $R$ is complete. Hence $xPy$ if and only if

$$|\{i \in N \mid xP_iy\}| > \frac{n}{2}$$

by a simple counting procedure. Thus $xPy$ if and only if

$$|P(x,y;\overline{\rho})| > \frac{n}{2}. \qquad \square$$

**Definition 7.4 (maximal set).** (Austen-Smith and Banks (1999), p. 3) Let

$$M(R,X) = \{x \in X \mid xRy \text{ for all } y \in X\}.$$

Then $M(R,X)$ is called the *maximal set* of $R$.

## 7.3 An Empirical Application

Figure 7.1 is a spatial model using fuzzy sets to define the individual preferences of three political actors, $A$, $B$, and $C$, where $T = \{0, .25, .5, .75, 1\}$. If $\widetilde{X}$ is a fuzzy subset of a set $S$, i.e., $\widetilde{X} : S \rightarrow [0, 1]$, the $t$−level set of $\widetilde{X}$ is the set $\widetilde{X}^t = \{s \in S \mid \widetilde{X}(s) \geq t\}$, where $t \in [0, 1]$. The inner-most region is the $t = 1$ level for each player.

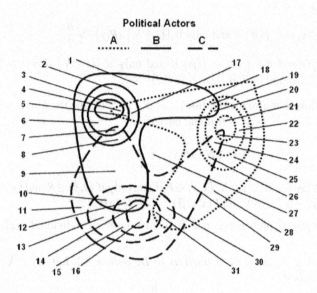

| Point ID | Three-Tuple | Winset |
|---|---|---|
| 1 | { 0.0, 0.25, 0.0 } | [ 4, 5, 7, 8, 17, 23, 24, 26, 29, 30, 31 ] |
| 2 | { 0.0, 0.5, 0.0 } | [ 4, 5, 7, 23, 24, 26, 29, 30, 31 ] |
| 3 | { 0.0, 1.0, 0.0 } | [ 23, 24, 26, 29, 30, 31 ] |
| 4 | { 0.25, 1.0, 0.0 } | [ 23, 24, 26 ] |
| 5 | { 0.25, 0.75, 0.0 } | [ 23, 24, 26 ] |
| 6 | { 0.0, 0.75, 0.0 } | [ 4, 23, 24, 26, 29, 30, 31 ] |
| 7 | { 0.0, 0.75, 0.25 } | [ 4, 30, 31 ] |
| 8 | { 0.0, 0.5, 0.25 } | [ 4, 5, 30, 31 ] |
| 9 | { 0.0, 0.25, 0.25 } | [ 4, 5, 17, 30, 31 ] |
| 10 | { 0.0, 0.25, 0.5 } | [ 4, 5, 17, 31 ] |
| 11 | { 0.0, 0.0, 0.25 } | [ 4, 5, 10, 12, 14, 17, 18, 19, 20, 30, 31 ] |
| 12 | { 0.0, 0.25, 0.75 } | [ 4, 5, 17 ] |
| 13 | { 0.0, 0.0, 0.5 } | [ 4, 5, 12, 14, 17, 18, 19, 20, 31 ] |
| 14 | { 0.0, 0.25, 1.0 } | [ 4, 5, 17 ] |
| 15 | { 0.0, 0.0, 0.75 } | [ 4, 5, 14, 17, 18, 19, 20 ] |
| 16 | { 0.0, 0.0, 1.0 } | [ 4, 5, 17, 18, 19, 20 ] |
| 17 | { 0.25, 0.5, 0.0 } | [ 7, 23, 24, 26 ] |
| 18 | { 0.25, 0.25, 0.0 } | [ 7, 8, 23, 24, 26 ] |
| 19 | { 0.5, 0.25, 0.0 } | [ 7, 8, 23, 24 ] |
| 20 | { 0.75, 0.25, 0.0 } | [ 7, 8, 23 ] |
| 21 | { 1.0, 0.0, 0.0 } | [ 7, 8, 9, 10, 12, 14 ] |
| 22 | { 0.75, 0.0, 0.0 } | [ 7, 8, 9, 10, 12, 14, 23 ] |
| 23 | { 1.0, 0.0, 0.25 } | [ 10, 12, 14 ] |
| 24 | { 0.75, 0.0, 0.25 } | [ 10, 12, 14 ] |
| 25 | { 0.5, 0.0, 0.0 } | [ 7, 8, 9, 10, 12, 14, 20, 23, 24 ] |
| 26 | { 0.5, 0.0, 0.25 } | [ 10, 12, 14, 20 ] |
| 27 | { 0.0, 0.0, 0.0 } | [ 4, 5, 7, 8, 9, 10, 12, 14, 17, 18, 19, 20, 23, 24, 26, 29, 30, 31 ] |
| 28 | { 0.25, 0.0, 0.0 } | [ 7, 8, 9, 10, 12, 14, 19, 20, 23, 24, 26 ] |
| 29 | { 0.25, 0.0, 0.25 } | [ 10, 12, 14, 19, 20 ] |
| 30 | { 0.25, 0.0, 0.5 } | [ 12, 14, 19, 20 ] |
| 31 | { 0.25, 0.0, 0.75 } | [ 14, 19, 20 ] |

**Fig. 7.1** A Three-Player Fuzzy Spatial Model

The complete set of alternatives is numbered, and the corresponding three-tuple $t$-levels $(a_A, a_B, a_C)$, the intersection of the $t$-levels for the three players, are noted in braces.

We call a pair $(\mathscr{A}, R)$ a relation space if $R$ is a relation on the set $\mathscr{A}$. The transformation of the universe from a relation space into another relation space is a homomorphism, or a function that maps an arbitrary relation $R$ from one set into another set while faithfully reproducing $R$. As a consequence, the calculation of the majority rule maximal set using the partial ordering can be faithfully transferred back to the spatial diagram. The results obtained can be applied to $T^n$. The existence of the homomorphism linking the spatial model to a natural partial ordering, where the results are determined, makes it possible for these models to deal with highly complex preferences that would be difficult, if not impossible, for step-wise utility functions of the sort proposed by Sloss (1973) and others to resolve. This greatly simplifies the task of empirical testing of spatial models of thick indifference.

The set of options that are majority preferred (the winset) to each numbered alternative are noted in brackets. In this case, there is no majority rule maximal set. All points are majority preferred by at least one other point. However, alternatives $4, 16$, and $27$ comprise an externally stable cycle. Alternative $4$ is majority preferred to alternative $16$, which is majority preferred to alternative $27$, which is majority preferred to alternative $4$. No alternative outside of this set is majority preferred to any alternative within it, and all alternatives outside of it are majority preferred by at least one point within it.

## 7.4 Proof of the Homomorphism

We now provide a proof of the homomorphism. We define an appropriate function $f^*$ from $\mathscr{A}$ onto $X$, where $\mathscr{A}$ denotes the region of interest (for example, a spatial representation of fuzzy preferences) and $X$ the region onto which $\mathscr{A}$ is mapped by the homomorphism $f^*$. We work in general in our theories. We let $\mathscr{A}$ be an arbitrary set mapped onto $X$, where $f^* : \mathscr{A} \to X \subseteq U$. In our application, we work in $T^n$, the lattice, where $\mathscr{A} = \mathbb{R}_+^2, U = T^n$, and $\mathbb{R}_+$ denotes the set of nonnegative real numbers.. The following formal discussion shows that the results determined in $X$ concerning the maximal set and Pareto set can be transferred faithfully back to $\mathscr{A}$.

**Definition 7.5 (homomorphism).** Let $(\mathscr{A}, \widetilde{R})$ and $(X, R)$ be relation spaces. Let $f^*$ be a function of $\mathscr{A}$ onto $X$. Then $f^*$ is called a *homomorphism* of $(\mathscr{A}, \widetilde{R})$ into $(X, R)$ if for all $a, b \in \mathscr{A}, (a, b) \in \widetilde{R}$ if and only if $(f^*(a), f^*(b)) \in R$. If $f^*$ maps $\mathscr{A}$ onto $X$, we say $f^*$ maps $(\mathscr{A}, \widetilde{R})$ onto $(X, R)$. For all $(a, b) \in \widetilde{R}$, we write $f^*((a, b)) = (f^*(a), f^*(b))$ and $f^*(\widetilde{R}) = \{(f^*((a, b)) \mid (a, b) \in \widetilde{R}\}$.

Thus if $a, a', b, b' \in \mathscr{A}$ and $f^*(a) = f^*(a'), f^*(b) = f^*(b')$, it is not possible that $(a, b) \in \widetilde{R}$ and $(a', b') \notin \widetilde{R}$.

**Proposition 7.6.** *Let $f^*$ be a homomorphism of $(\mathscr{A}, \widetilde{R})$ onto $(X, R)$. Then $f^*(\widetilde{R}) = R$.*

*Proof.* Clearly, $f^*(\widetilde{R}) \subseteq R$. Let $(x,y) \in R$. Since $f^*$ maps $\mathscr{A}$ onto $X$, there exists $a,b \in \mathscr{A}$ such that $f^*(a) = x$ and $f^*(b) = y$. Thus $(x,y) = (f^*(a),f^*(b)) = f^*((a,b)) \in f^*(\widetilde{R})$. □

**Proposition 7.7.** *Let $f^*$ be a homomorphism of $(\mathscr{A},\widetilde{R})$ onto $(X,R)$. Then for all $a,b \in \mathscr{A}, (a,b) \in \widetilde{P}$ if and only if $(f^*(a),f^*(b)) \in P$.*

*Proof.* Let $a,b \in \mathscr{A}$. Then

$$(a,b) \in \widetilde{P} \iff (a,b) \in \widetilde{R}, (b,a) \notin \widetilde{R}$$
$$\iff (f^*(a),f^*(b)) \in R, (f^*(a),f^*(b)) \notin R$$
$$\iff (f^*(a),f^*(b)) \in P.$$

In the next result we show that not only does a homomorphism preserve the notion of a maximal set, but furthermore the preimage of the maximal set is exactly the maximal set in the domain. □

**Theorem 7.8.** *Let $f^*$ be a homomorphism of $(\mathscr{A},\widetilde{R})$ onto $(X,R)$. Then*

$$f^*(M(\widetilde{R},\mathscr{A})) = M(R,X),$$

*where $M(\widetilde{R},\mathscr{A})$ denotes the maximal set of $\widetilde{R}$ in $\mathscr{A}$ and $M(R,X)$ denotes the maximal set of $R$ in $X$. Furthermore,*

$$f^{*-1}(M(R,X)) = M(\widetilde{R},\mathscr{A}),$$

*where $f^{*-1}(M(R,X))$ denotes the preimage of $M(R,X)$ in $\mathscr{A}$.*

*Proof.* $a \in M(\widetilde{R},\mathscr{A}) \iff$ for all $b \in \mathscr{A}, a\widetilde{R}b \iff$ for all $f^*(b) \in X, f^*(a)Rf^*(b) \iff f^*(a) \in M(R,X)$, where the latter equivalence holds since $f^*$ maps $\mathscr{A}$ onto $X$. Thus if $f^*(a) \in f^*(M(\widetilde{R},\mathscr{A}))$, then $a \in M(\widetilde{R},\mathscr{A})$. Hence $f^*(a) \in M(R,X)$. Thus $f^*(M(\widetilde{R},\mathscr{A})) \subseteq M(R,X)$. Let $x \in M(R,X)$. Then $\forall y \in X, xRy$. Let $a \in \mathscr{A}$ be such that $f^*(a) = x$. Let $b \in \mathscr{A}$. Then $f^*(a)Rf^*(b)$ since $x = f^*(a)$ and $x \in M(R,X)$. Hence $a\widetilde{R}b$ by Definition 7.5. Thus $a \in M(\widetilde{R},\mathscr{A})$ and so $x = f^*(a) \in f^*(M(\widetilde{R},\mathscr{A}))$. Thus, $M(R,X) \subseteq f^*(M(\widetilde{R},\mathscr{A})$.

Clearly, $f^{*-1}(M(R,X)) \supseteq M(\widetilde{R},\mathscr{A})$. Let $a \in f^{*-1}(M(R,X))$. Suppose there exists $b \in \mathscr{A}$ such that $(a,b) \notin \widetilde{R}$. Then $(f^*(a),\rho^*(b)) \notin R$ since $f^*$ is a homomorphism. Thus, $f^*(a) \notin M(R,X)$, a contradiction of $a \in f^{*-1}(M(R,X))$. Hence, $(a,b) \in \widetilde{R} \forall b \in \mathscr{A}$. Thus, $a \in M(\widetilde{R},\mathscr{A})$. Hence, $f^{*-1}(M(R,X)) \subseteq M(\widetilde{R},\mathscr{A})$.

Let $(\mathscr{A},\widetilde{R}_i)$ be a relation space, $i = 1,...,n$. Let $f_i^*$ be a homomorphism of $(\mathscr{A},\widetilde{R}_i)$ onto $(X,R_i), i = 1,...,n$. Then $R_i = f_i^*(\widetilde{R}_i), i = 1,...,n$ by Proposition 7.6. □

**Definition 7.9 (preserve the pair).** Let $\widetilde{f}$ be an aggregation rule on

$$(\mathscr{A},(\widetilde{R}_1,...,\widetilde{R}_n))$$

and let $f$ be an aggregation rule on

$$(X, (R_1, ..., R_n)).$$

Let $f_i^*$ be a homomorphism of $(\mathscr{A}, \widetilde{R_i})$ onto $(X, R_i)$, for $i = 1, ..., n$. Let $f^*$ be a homomorphism of $(\mathscr{A}, \widetilde{f}((\widetilde{R}_1, ..., \widetilde{R}_n)))$ onto $(X, f((R_1, ..., R_n)))$. Then $f^*$ is said to *preserve the pair* $(\widetilde{f}, f)$ *with respect to* $(f_1^*, ..., f_n^*)$ if

$$f^*(\widetilde{f}(\widetilde{R}_1, ..., \widetilde{R}_n)) = f((R_1, ..., R_n)).$$

In Definition 7.9, it is understood that if $f^*$ preserves $(\widetilde{f}, f)$ with respect to $(f_1^*, ..., f_n^*)$, then $f_i^*(\widetilde{R}_i) = R_i, i = 1, ..., n$.

**Definition 7.10 (Pareto set).** (Austen-Smith and Banks (2005), p. 7) Let $\overline{\rho} \in \mathscr{R}^n$. Define the *Pareto set* at $\overline{\rho}, PS_N(\overline{\rho})$, to be the set $PS_N(\overline{\rho}) = \{x \in X \mid$ for all $y \in X$ where $y \neq x$ (there exists $i \in N, yP_ix$ implies there exists $j \in N, xP_jy)\}$.

An alternative $x$ is in the Pareto set if whenever a player strictly prefers an alternative $w$ to $x$, then there is a player who strictly prefers $x$ to $w$. Any effort by the group to choose other than an alternative in the Pareto set will leave at least one player worse off. Note that the Pareto set is not determined by majority rule but rather by unanimity.

In Definition 7.9, let $\widetilde{\rho} = (\widetilde{R}_1, ..., \widetilde{R}_n)$ and $\rho = (R_1, ..., R_n)$.

In the following result, we show that a homomorphism preserves the notion of a Pareto set and that the preimage of the Pareto set is exactly the Pareto set.

**Theorem 7.11.** *Let* $\widetilde{f}$ *be an aggregation rule on* $(\mathscr{A}, (\widetilde{R}_1, ..., \widetilde{R}_n))$ *and let* $f$ *be an aggregation rule on* $(X, (R_1, ..., R_n))$. *Let* $f_i^*$ *be a homomorphism of* $(\mathscr{A}, \widetilde{R_i})$ *onto* $(X, R_i)$, $i = 1, ..., n$. *Let* $f^*$ *be a homomorphism of* $(\mathscr{A}, \widetilde{f}((\widetilde{R}_1, ..., \widetilde{R}_n)))$ *onto* $(X, f((R_1, ..., R_n)))$ *such that* $f^*$ *preserves* $(\widetilde{f}, f)$ *w.r.t.* $(f_1^*, ..., f_n^*)$. *Then* $f^*(PS_N(\widetilde{\rho})) = PS_N(\rho)$. *Furthermore,* $f^{*-1}(PS_N(\rho)) = PS_N(\widetilde{\rho})$.

*Proof.* Recall that $a \in PS_N(\widetilde{\rho}) \Leftrightarrow \forall b \in \mathscr{A}, $ (there exists $i \in N, b\widetilde{P}_ia$ implies there exists $j \in N, a\widetilde{P}_jb)$. Let $a \in \mathscr{A}$. Suppose $f^*(a) \in f^*(PS_N(\widetilde{\rho}))$. Then $a \in PS_N(\widetilde{\rho})$ by definition of a homomorphism and Propositions 7.6 and 7.7. Hence $f^*(a) \in PS_N(\rho)$. Thus

$$f^*(PS_N(\widetilde{\rho})) \subseteq PS_N(\rho).$$

Let $x \in PS_N(\rho)$. Let $y \in X$. If there exists $i \in N$ such that $yP_ix$, then there exists $j \in N$ such that $xP_jy$. Let $a \in \mathscr{A}$ be such that $f^*(a) = x$. Let $b \in \mathscr{A}$. Then $f^*(b)P_if^*(a)$ if and only if $b\widetilde{P}_ia$ and $f^*(a)P_jf^*(b)$ if and only if $a\widetilde{P}_jb$. Thus if there exists $i \in N$ such that $b\widetilde{P}_ia$, then there exists $j \in N$ such that $a\widetilde{P}_jb$. Thus $a \in PS_N(\widetilde{\rho})$ and so

$$x = f^*(a) \in f^*(PS_N(\widetilde{\rho}).$$

Hence

$$PS_N(\rho) \subseteq f^*(PS_N(\widetilde{\rho}).$$

Clearly $f^{*-1}(PS_N(\rho)) \supseteq PS_N(\widetilde{\rho})$. Let $a \in f^{*-1}(PS_N(\rho))$. Suppose $a \notin PS_N(\widetilde{\rho})$. Then $\sim$ (for all $b \in \mathscr{A}$, there exists $i \in N, b\widetilde{P}_ia$ implies there exists $j \in N, a\widetilde{P}_jb)$.

Then there exists $b \in \mathscr{A}$ such that $\sim$ (there exists $i \in N, b\widetilde{P}_i a$ implies there exists $j \in N, a\widetilde{P}_j b$). Thus if there exists $i \in N, b\widetilde{P}_i a$, then there does not exist $j \in N, a\widetilde{P}_j b$ and so $b\widetilde{R}_j a$ for all $j \in N$. Hence $(b,a) \in \widetilde{R}i$ for all $i \in N$ and so $(f^*(a), f^*(b)) \in R_i$ for all $i \in N$. Thus $(b,a) \in \widetilde{R}_i$ for all $i \in N$ and so $(f^*(b), f^*(a)) \in R_i$ for all $i \in N$. Hence $f^*(a) \notin PS_N(\rho)$ which contradicts the fact that $a \in f^{*-1}(PS_N(\rho))$. Thus $a \in PS_N(\widetilde{\rho})$ and so $f^{*-1}(PS_N(\rho)) \subseteq PS_N(\widetilde{\rho})$. $\qquad \square$

Let $f^*$ be a homomorphism of the relation space $(\mathscr{A}, \widetilde{R})$ onto the relation space $(X, R)$. Define the relation $\sim$ on $\mathscr{A}$ by $\forall (x,y) \in \mathscr{A} \times \mathscr{A}, (x,y) \in \sim$ if and only if $f^*(x) = f^*(y)$. Then $\sim$ is an equivalence relation on $\mathscr{A}$. For all $x \in A$, let $[x] = \{y \in A \mid y \sim x\}$. Then $[x]$ is the equivalence class of $x$ with respect to $\sim$.

**Definition 7.12 (quotient top cycle set).** Let $f^*$ be a homomorphism of the relation space $(\mathscr{A}, \widetilde{R})$ onto the relation space $(X, R)$. Define the *quotient top cycle set* of $\widetilde{R}$ with respect to $\sim$, denoted $\overline{T}(\widetilde{R})$, to be the set $\{x \in \mathscr{A} \mid \forall y \in \mathscr{A} \setminus [x], \exists a_0, a_1, ..., a_r \in \mathscr{A}$ such that $a_t \widetilde{P} a_{t+1}, i = 0, 1, ..., r-1, a_0 = x, a_r = y\}$.

It is clear that $T(\widetilde{R}) \subseteq \overline{T}(\widetilde{R})$. Thus $f^*(T(\widetilde{R})) \subseteq f^*(\overline{T}(\widetilde{R}))$, where $T(\widetilde{R})$ is the top cycle set of $\widetilde{R}$. Let $T(R)$ denote the top cycle set of $R$.

**Theorem 7.13.** *Let $f^*$ be a homomorphism of the relation space $(\mathscr{A}, \widetilde{R})$ onto the relation space $(X, R)$. Then $f^*(\overline{T}(\widetilde{R})) = T(R)$. Furthermore, $f^{*-1}(T(R)) = \overline{T}(\widetilde{R})$.*

*Proof.* Let $x \in \overline{T}(\widetilde{R})$. Let $z \in X \setminus \{f^*(x)\}$. Let $y \in \mathscr{A} \setminus [x]$ be such that $f^*(y) = z$. Hence $\exists a_0, a_1, ..., a_r \in A$ such that $a_t \widetilde{P} a_{t+1}, i = 0, 1, ..., r-1, a_0 = x, a_r = y$. By Proposition 7.7, $f^*(a_t) P f^*(a_{t+1}), t = 0, 1, ..., r-1$. Now $f^*(a_0) = f^*(x)$ and $f^*(a_r) = f^*(y) = z$. Thus $f^*(x) \in T(R)$. Hence $f^*(\overline{T}(\widetilde{R})) \subseteq T(R)$. Let $w \in T(R)$. Let $x \in \mathscr{A}$ be such that $f^*(x) = w$. Let $y \in \mathscr{A} \setminus [x]$. Then $f^*(y) \neq f^*(x)$ and in fact $f^*(y)$ can be considered an arbitrary element of $X \setminus \{f^*(x)\}$. Thus since $f^*(x) \in T(R), \exists b_0, b_1, ..., b_r \in X$ such that $b_0 = f^*(x), b_r = f^*(y)$ and $b_t P b_{t+1}$ for $t = 0, 1, ..., r-1$. Let $a_t \in \mathscr{A}$ be such that $f^*(a_t) = b_t, t = 0, 1, ..., r-1$. Then $a_t \widetilde{P} a_{t+1}, t = 0, 1, ..., r-1$ and we can take $a_0 = x, a_r = y$. Hence $x \in \overline{T}(\widetilde{R})$ and $f^*(x) = w$. Thus $T(R) \subseteq f^*(\overline{T}(\widetilde{R}))$.

Clearly, $f^{*-1}(T(R)) \supseteq \overline{T}(\widetilde{R})$. Let $x \in f^{*-1}(T(R))$. Then $f^*(x) \in T(R)$. Let $y \in \mathscr{A} \setminus [x]$. Then $\exists b_0, b_1, ..., b_r \in X$ such that $b_0 = f^*(x), b_r = f^*(y)$ and $b_t P b_{t+1}, t = 0, 1, ..., r-1$. Let $a_t \in \mathscr{A}$ be such that $f^*(a_t) = b_t, t = 0, 1, ..., r-1$. Then $a_t \widetilde{P} a_{t+1}, t = 0, 1, ..., r-1$ by Proposition 7.7 and we can take $a_0 = x, a_r = y$. Thus $x \in \overline{T}(\widetilde{R})$. Hence $f^{*-1}(T(R)) \subseteq \overline{T}(\widetilde{R})$. $\qquad \square$

**Theorem 7.14.** *Let $\widetilde{f}$ be an aggregation rule on $(\mathscr{A}, (\widetilde{R}_1, ..., \widetilde{R}_n))$ and let $f$ be an aggregation rule on $(X, (R_1, ..., R_n))$. Let $f_i^*$ be a homomorphism of $(\mathscr{A}, \widetilde{R}_i)$ onto $(X, R_i)$, $i = 1, ..., n$. Let $f^*$ be a homomorphism of $(\mathscr{A}, \widetilde{f}((\widetilde{R}_1, ..., \widetilde{R}_n)))$ onto $(X, f((R_1, ..., R_n)))$ such that $f^*$ preserves $(\widetilde{f}, f)$ w.r.t. $(f_1^*, ..., f_n^*)$. Then $\widetilde{f}$ is a simple majority rule if and only if $f$ is a simple majority rule.*

*Proof.* Since by Proposition 7.7, for all $a, b \in \mathscr{A}, (a,b) \in \widetilde{P}_i$ if and only if $(f^*(a), f^*(b)) \in P_i, i = 1, ..., n$, it follows that $|\widetilde{P}(a, b; \widetilde{f}((\widetilde{R}_1, ..., \widetilde{R}_n)))| = |P(f^*(a), f^*(b)); f((R_1, ..., R_n)))|$. The desired result now follows. $\qquad \square$

**Proposition 7.15.** *Let $f^*$ be a homomorphism of the relation space $(\mathscr{A}, \tilde{R} = (\tilde{R}_1, ...,$ $\tilde{R}_n))$ onto the relation space $(X, R = (R_1, ..., R_n))$. Let $\tilde{f}$ and $f$ be fuzzy aggregation rules for $(\mathscr{A}, \tilde{R})$ and $(X, R)$, respectively. Let $f_i^*$ be a homomorphism of $(A, \tilde{R}_i)$ onto $(X, R_i), i = 1, ..., n$. Suppose $f^*$ preserves $(\tilde{f}, f)$ with respect to $(f_1^*, ..., f_n^*)$. Then $f$ is a Pareto extension rule if and only if $\tilde{f}$ is a Pareto extension rule.*

*Proof.* Since $\forall a, b \in \mathscr{A}, a\tilde{R}_i b \Leftrightarrow f^*(a)R_i f^*(b), i = 1, ..., n$, it follows that

$$\left| \tilde{R}(a, b; (R_1, ..., R_n)) \right| = |R(f^*(a), f^*(b); (R_1, ..., R_n))| \; .$$

Thus

$$\tilde{R}(a, b; (R_1, ..., R_n)) = N$$
$$\Leftrightarrow R(f^*(a), f^*(b); (R_1, ..., R_n)) = N.$$

Also $a\tilde{P}b \Leftrightarrow f^*(a)Pf^*(b)$. Consider the statements.

(1) $a\tilde{P}b \Leftrightarrow \tilde{R}(a, b; (R_1, ..., R_n)) = N$,
(2) $f^*(a)Pf^*(b) \Leftrightarrow R(f^*(a), f^*(b); (R_1, ..., R_n)) = N$.

Now $\tilde{f}$ is a Pareto extension rule if and only if (1) holds and $f$ is a Pareto extension rule if and only if (2) holds. The desired result is now immediate. □

**Proposition 7.16.** *Let $f^*$ be a homomorphism of the relation space $(\mathscr{A}, \tilde{R} = (\tilde{R}_1, ...,$ $\tilde{R}_n))$ onto the relation space $(X, R = (R_1, ..., R_n))$. Let $\tilde{f}$ and $f$ be fuzzy aggregation rules for $(\mathscr{A}, \tilde{R})$ and $(X, R)$, respectively. Let $f_i^*$ be a homomorphism of $(A, \tilde{R}_i)$ onto $(X, R_i), i = 1, ..., n$. Suppose $f^*$ preserves $(\tilde{f}, f)$ with respect to $(f_1^*, ..., f_n^*)$. Then $f$ is dictatorial if and only if $\tilde{f}$ is dictatorial.*

*Proof.* We have $\forall i \in N, \forall a, b \in \mathscr{A}, a\tilde{P}_i b \Leftrightarrow f^*(a)P_i f^*(b)$ and $a\tilde{P}b \Leftrightarrow f^*(a)Pf^*(b)$. Consider the statements

(1) $\exists i \in N, \forall a, b \in \mathscr{A}, a\tilde{P}_i b \Rightarrow a\tilde{P}b$,
(2) $\exists i \in N, \forall a, b \in \mathscr{A}, f^*(a)P_i f^*(b) \Rightarrow f^*(a)Pf^*(b)$.

Now $\tilde{f}$ is dictatorial if and only if (1) holds and $f$ is dictatorial if and only (2) holds since $f^*$ maps $\mathscr{A}$ onto $X$. The desired result is now immediate. □

**Proposition 7.17.** *Let $f^*$ be a homomorphism of the relation space $(\mathscr{A}, \tilde{R} = (\tilde{R}_1, ...,$ $\tilde{R}_n))$ onto the relation space $(X, R = (R_1, ..., R_n))$. Let $\tilde{f}$ and $f$ be fuzzy aggregation rules for $(\mathscr{A}, \tilde{R})$ and $(X, R)$, respectively. Let $f_i^*$ be a homomorphism of $(A, \tilde{R}_i)$ onto $(X, R_i), i = 1, ..., n$. Suppose $f^*$ preserves $(\tilde{f}, f)$ with respect to $(f_1^*, ..., f_n^*)$. Then $f$ is weakly Paretian if and only if $\tilde{f}$ is weakly Paretian.*

*Proof.* We have $\forall i \in N, \forall a, b \in \mathscr{A}, a\tilde{P}_i b \Leftrightarrow f^*(a)P_i f^*(b)$ and $a\tilde{P}b \Leftrightarrow f^*(a)Pf^*(b)$. Consider the statements

(1) $\exists i \in N, \forall a, b \in \mathscr{A}, a\widetilde{P_i}b \Rightarrow a\widetilde{P}b,$
(2) $\exists i \in N, \forall a, b \in \mathscr{A}, f^*(a)P_i f^*(b) \Rightarrow f^*(a)P f^*(b).$

Now $\widetilde{f}$ is weakly Paretian if and only if (1) holds and $f$ is weakly Paretian if and only if (2) holds since $f^*$ maps $\mathscr{A}$ onto $X$. The desired result is now immediate.

## 7.5   The Existence of a Majority Rule Maximal Set

We have demonstrated the equivalence between $R$ on $X$ and $\widetilde{R}$ on $\mathscr{A}$, as a result of which spatial models making use of fuzzy sets representing thick indifference can easily solve for the existence of a majority rule maximal set. The reader should also be able to see for herself that the approach can be extended with relative ease to extraordinarily complex representations, to include multi-modal individual preferences. However, as the example at Figure 6.1 makes clear, such a set may not always exist. We now consider the conditions under which a majority rule maximal set exists.

### 7.5.1   Conditions for the Existence of a Majority Rule Maximal Set

The results derived here are for arbitrary finite $T \subseteq [0, 1]$ with $0, 1 \in T$ and with $U$ arbitrary. In the ensuing subsection we demonstrate our results by applying them to the case $U = T^3$, where $T^3 = \{(a_1, a_2, a_3) \mid a_i \in T, i = 1, 2, 3\}$, $T = \{0, .25, .5, .75, 1\}$. $T^3$ is the set of all ordered 3−tuples with the entries from $T$.

We will be interested in fuzzy subsets of $\mathscr{A}$. When $\mathscr{A}$ represents spatial alternatives, let $\mathbb{R}_+$ denote the nonnegative real numbers and let $\mathbb{R}_+^2$ denote the set of ordered pairs $\mathbb{R}_+ \times \mathbb{R}_+$. Of primary interest to us in our application and subsequent demonstration are fuzzy subsets in $\mathscr{C} = \{ \widetilde{X} : \mathbb{R}_+^2 \to T \mid 0, 1 \in \mathrm{Im}(\widetilde{X}),$ the t-level set $\widetilde{X}^t$ is the interior and boundary of a simple closed curve for all $t \in T \setminus \{0\}\}$, where $\mathrm{Im}(\widetilde{X})$ denotes the image of $\widetilde{X}$. A simple closed curve is a curve for which there is a one-to-one open continuous function of the unit circle onto it. A simple closed curve has an interior that is bounded and an exterior. Individual preferences over alternatives in space need not be convex.

We also have a particular interest in those fuzzy subsets $\widetilde{X}$ in $\mathscr{C}$ for which $\widetilde{X}^t$ is a compact set for all $t \in T \setminus \{0\}$. A compact set is one that is closed and bounded.

If we are to avoid cycling, then under majority rule, our models must predict a maximal set. Let $N$ denote a set of players and $\mathscr{A}$ denote an arbitrary set of interest. We consider $\mathscr{A}$ as a set of alternatives. Our goal is to characterize the maximal set in $\mathscr{A}$ with respect to a binary relation $\widetilde{R}$ on $\mathscr{A}$. We first characterize the maximal set for a set of alternatives $X$ in a universe $U$ for which there is a special function $f^*$ of $\mathscr{A}$ onto $X$ such that the results determined in $X$ can be carried back to corresponding results in $\mathscr{A}$. We first characterize the maximal set in $X$ with respect to a relation $R$ such that $f^*(\widetilde{R}) = R$. In Theorem 7.8 we proved that the characterized maximal set in $X$ characterizes the maximal set in $\mathscr{A}$.

We demonstrated earlier in this chapter that a maximal set does not always exist in fuzzy set spatial models. In what follows, we characterize the conditions under which that is the case. Our analysis begins with the Pareto set. As we will demonstrate, the characteristics of the elements in the Pareto set determine whether a maximal exists in fuzzy set spatial models.

Let $\preceq$ be any partial order on $U$. Recall that $\forall x, y \in U, x \prec y$ if and only if $x \preceq y$ and $x \neq y$. For $\overline{\rho} = (R_1, ..., R_n)$, we assume in subsequent sections of this paper that $\preceq$ satisfies the following properties:

(1) for all $x, y \in U, x \preceq y$ implies for all $i \in N, yR_ix$;
(2) for all $x, y, z \in U$, for all $i \in N, x \preceq y$ and $xR_iz$ implies $yR_iz$;
(3) for all $x, y, z \in U$, for all $i \in N, x \preceq y$ and $xP_iz$ implies $yP_iz$;
(4) for all $x, y \in U, x \prec y$ implies there exists $i \in N$ such that $yP_ix$;
(5) for all $x, y, z \in U$, for all $i \in N, x \preceq y$ and $zR_iy$ implies $zR_ix$;
(6) for all $x, y \in U, x$ and $y$ incomparable under $\preceq$ and there exists $i \in N$ such that $xP_iy$ implies there exists $j \in N$ such that $yP_jx$.

The above properties give the relationship of $\preceq$ to $R$ of Definition 7.2. An example of a partial order $\preceq$ that satisfies properties (1)-(6) is the one given in our three-player example in the ensuing section.

**Definition 7.18 (largest element).** Let $X \subseteq U$ and $L_R = \{x \in X \mid$ there does not exist $y \in X, x \prec y\}$. An element of $L_R$ is called a *largest element* of $X$ with respect to $\preceq$.

Mathematics refers to an element in $L_R$, which in defined in terms of $\preceq$, as a maximal element. We use "largest element" to avoid confusion with the elements comprising the maximal set given in Definition 7.4. In what follows, we demonstrate the relationship of the set of maximal elements to the Pareto set (Definition 7.10).

The relation between $R$ and $\overline{\rho}$ in the following results and those to follow is given by Definition 7.2.

**Proposition 7.19.** $L_R = PS_N(\overline{\rho})$.

*Proof.* Suppose $x \in L_R$. Let $y \in X$. Suppose there exists $i \in N$ such that $yP_ix$. Now there does not exist $y \in X$ such that $x \prec y$. Thus for all $y \in X$, either $y \preceq x$ or $x$ and $y$ are not comparable. Since $yP_ix, y \preceq x$ is impossible else $xR_iy$ for all $i \in N$ by (1). Hence $x$ and $y$ are incomparable under $\preceq$. Thus there exists $j \in N$ such that $xP_jy$ by property (6). Hence $x \in PS_N(\overline{\rho})$. Thus $L_R \subseteq PS_N(\overline{\rho})$.

Suppose $x \in PS_N(\overline{\rho})$. Suppose there exists $y \in X$ such that $x \prec y$. Then there exists $i \in N$ such that $yP_ix$. Since $x \in PS_N(\overline{\rho})$, there exists $j \in N$ such that $xP_jy$. Thus $x \prec y$ is impossible by (1). Hence $x \in L_R$. Therefore $PS_N(\overline{\rho}) \subseteq L_R$. $\blacksquare$

**Corollary 7.20.** *Let* $x \in X$.

*(1) Suppose for all* $y \in X, x \preceq y$ *implies* $x = y$. *Then* $x \in PS_N(\overline{\rho})$.
*(2) If* $x \notin PS_N(\overline{\rho})$, *then there exists* $y \in PS_N(\overline{\rho})$ *such that* $x \prec y$.

*Proof.* (1) Clearly $x \in L_R$, but $L_R = PS_N(\overline{\rho})$.

(2) Since $x \notin PS_N(\overline{p}), x \notin L_R$. Thus there exists $y \in X$ such that $x \prec y$. Let $y$ be the largest such element. Then $y \in L_R = PS_N(\overline{p})$.                                                     □

**Definition 7.21 (closure operator).** Define $\langle\rangle : \mathscr{P}(U) \to \mathscr{P}(U)$ by $\forall S \in \mathscr{P}(U), \langle S \rangle = \{x \in U \mid \exists s \in S, x \preceq s\}$.

The notation $\langle\rangle$ in the foregoing definition is standard mathematical notation used to denote a closure operator. In our case, it yields the smallest set containing $S$ that is closed under $\preceq$.

**Proposition 7.22.** *Let* $\langle\rangle : \mathscr{P}(U) \to \mathscr{P}(U)$ *be defined as above. Then the following conditions hold.*

*(1) for all $S \in \mathscr{P}(U), S \subseteq \langle S \rangle$;*
*(2) for all $S_1, S_2 \in \mathscr{P}(U), S_1 \subseteq S_2$ implies $\langle S_1 \rangle \subseteq \langle S_2 \rangle$;*
*(3) for all $S \in \mathscr{P}(U), \langle S \rangle = \langle \langle S \rangle \rangle$;*
*(4) for all $S \in \mathscr{P}(U), \langle S \rangle = \cup_{s \in S} \langle \{s\} \rangle$;*
*(5) for all $S \in \mathscr{P}(U)$, for all $x, y \in U, x \in \langle S \cup \{y\} \rangle$ and $x \notin \langle S \rangle$ implies $x \in \langle \{y\} \rangle$.*

*Proof.* (1) Let $s \in S$. Then $s \preceq s$ and so $s \in \langle S \rangle$. Thus $S \subseteq \langle S \rangle$.
(2) Let $x \in \langle S_1 \rangle$. Then there exists $s \in S_1$ such that $x \preceq s$. Since $s \in S_2, x \in \langle S_2 \rangle$.
(3) By (1), $\langle S \rangle \subseteq \langle \langle S \rangle \rangle$. Let $x \in \langle \langle S \rangle \rangle$. Then there exists $y \in \langle S \rangle$ such that $x \preceq y$. There exists $s \in S$ such that $y \preceq s$. Since $\preceq$ is transitive, $x \preceq s$. Thus $x \in \langle S \rangle$. Hence $\langle \langle S \rangle \rangle \subseteq \langle S \rangle$.
(4) For all $s \in S, \langle \{s\} \rangle \subseteq \langle S \rangle$ by (2). Thus $\cup_{s \in S} \langle \{s\} \rangle \subseteq \langle S \rangle$. Let $x \in \langle S \rangle$. Then there exists $s \in S$ such that $x \preceq s$. Thus $x \in \langle \{s\} \rangle$ and so $x \in \cup_{s \in S} \langle \{s\} \rangle$. Hence $\langle S \rangle \subseteq \cup_{s \in S} \langle \{s\} \rangle$.
(5) Suppose $x \in \langle S \cup \{y\} \rangle$ and $x \notin \langle S \rangle$. Then there does not exist $s \in S$ such that $x \preceq s$. Hence $x \preceq y$. Thus $x \in \langle \{y\} \rangle$.                                                     □

The function $\langle\rangle$ is similar to that used to obtain structure results for (fuzzy) directed graphs, (fuzzy) finite state machines, and approximation spaces Kuroki and Mordeson (1997); Malik and Mordeson (2002); Mordeson (1999); Mordeson and Nair (1996). It may be possible to apply these structure results to $PS_N(R)$.

The following result is the *gateway* to our main conclusion. The result is critical in a series of lemmas that lead to our main conclusion (Theorem 7.27).

**Theorem 7.23.** $\langle X \rangle = \langle PS_N(\overline{p}) \rangle$.

*Proof.* Clearly, $PS_N(\overline{p}) \subseteq X$. Thus $\langle PS_N(\overline{p}) \rangle \subseteq \langle X \rangle$. Let $x \in X$. If $x \notin PS_N(\overline{p})$, then by (2) of Corollary 15, there exists $y \in PS_N(\overline{p})$ such that $x \prec y$. Thus $x \in \langle \{y\} \rangle \subseteq \langle PS_N(\overline{p}) \rangle$. If $x \in PS_N(\overline{p})$ clearly $x \in \langle PS_N(\overline{p}) \rangle$. Hence $X \subseteq \langle PS_N(\overline{p}) \rangle$ and so $\langle X \rangle \subseteq \langle PS_N(\overline{p}) \rangle$.                                                     □

The following lemma shows that if a maximal set exists, then at least one element of the maximal set is in the Pareto set. Recall that the relationship between $R$ and $\overline{p}$ is given in Definition 7.2.

**Lemma 7.24.** $M(R,X) \cap PS_N(\overline{p}) = \emptyset$ *if and only if* $M(R,X) = \emptyset$.

*Proof.* Suppose $M(R,X) \neq \emptyset$. Let $x \in M(R,X)$. By Theorem 7.23, there exists $y \in PS_N(\overline{\rho})$ such that $y \succeq x$. Since $x \in M(R,X)$, $xRz$ for all $z \in X$. Since $y \succeq x, yRz$ for all $z \in X$. Thus $y \in M(R,X)$. Hence $M(R,X) \cap PS_N(\overline{\rho}) \neq \emptyset$.                                      $\square$

The next lemma shows that an element in the Pareto set is in the maximal set of $R$ if and only if it is not strictly preferred to (by majority rule) by another element in the Pareto set. Hence, the search for an element strictly preferred to another can be confined to the Pareto set.

**Lemma 7.25.** *Let $s \in PS_N(\overline{\rho})$. Then there does not exist $c \in PS_N(\overline{\rho})$ such that $cPs$ if and only if $s \in M(R,X)$.*

*Proof.* Since $R$ is complete and not $cPs$ for all $c \in PS_N(\overline{\rho})$, it follows that $sRc$ for all $c \in PS_N(\overline{\rho})$. Let $x \in X$. By Theorem 7.23, there exists $c \in PS_N(\overline{\rho})$ such that $c \succeq x$. Thus $sRx$ by property (5). Hence $s \in M(R,X)$. The converse is immediate.

The next lemma establishes that if an element in the Pareto set can be majority defeated by any other element in $X$, it can be defeated by at least one element in the Pareto set.

**Lemma 7.26.** *(1) Let $s \in PS_N(\overline{\rho})$. If there exists $x \in X$ such that $xPs$, then there exists $c \in PS_N(\overline{\rho})$ such that $cPs$.*
*(2) $M(R,X) = \emptyset$ if and only if $\forall s \in PS_N(\overline{\rho})$, there exists $c \in PS_N(\overline{\rho})$ such that $cPs$.*

*Proof.* (1) By Theorem 7.23, there exists $c \in PS_N(\overline{\rho})$ such that $c \succeq x$. Hence $cPs$ by property (3).
(2) Suppose $M(R,X) = \emptyset$. Then the result holds by Lemma 7.25. Conversely, suppose $M(R,X) \neq \emptyset$. By Lemma 7.24, $M(R,X) \cap PS_N(\overline{\rho}) \neq \emptyset$ and so there exists $s \in M(R,X) \cap PS_N(\overline{\rho})$. Hence there does not exist $c \in PS_N(\overline{\rho})$ such that $cPs$.  $\square$

We can now state our main conclusion, which lays out the conditions under which a maximal set is empty in fuzzy set spatial models.

Let $C$ be a nonempty subset of $X$. Let $R$ be a binary relation on $X$ and $P$ the strict binary relation associated with $R$. Then $C$ is a cycle with respect to $R$ if there exists an ordering of the elements of $C$, say $c_1,...,c_k$, such that $c_1Pc_2, c_2Pc_3,...,$ $c_{k-1}Pc_k, c_kPc_1$.

Let $V = \{v \in U \mid v$ is not in a cycle$\}$. Let $N_1 = V \backslash N_2$, where $N_2 = \{w \in V \mid \forall R \in \mathscr{R}, w \in PS_N(R) \Rightarrow M(R,X) \neq \emptyset\}$. Let $M_1 = \{w \in V \mid \forall R \in \mathscr{R}^n, w \notin PS_N(R)\}$. Assume $M_1 \subseteq N_1$. Let $N_1' = N_1 \backslash M_1$. Suppose $N_1$ is such that none of its elements are strictly preferred to one of $U \backslash V$.

**Theorem 7.27.** *$M(R,X) = \emptyset$ if and only if*

$$PS_N(\overline{\rho}) = \left( \bigcup_{k=1}^{n} C_k \right) \cup \left( \bigcup_{j=1}^{m} C_j' \right) \cup N_1'',$$

*where $N_1'' \subseteq N_1', C_k$ are cycles, $k = 1,...,n, C_j'$ are subsets of cycles which are not themselves cycles, $j = 1,...,m$, and*

*(1)* $\forall s \in \bigcup_{j=1}^{m} C_j'$, there exists $c \in (\bigcup_{k=1}^{n} C_k) \cup \left(\bigcup_{j=1}^{m} C_j'\right)$ such that $cPs$,

*(2)* $\forall s \in N_1''$ there exists $c \in (\bigcup_{k=1}^{n} C_k) \cup \left(\bigcup_{j=1}^{m} C_j'\right)$ such that $cPs$.

*Proof.* It follows that

$$PS_N(\overline{p}) \subseteq \left(\bigcup_{k=1}^{n} C_k\right) \cup \left(\bigcup_{j=1}^{m} C_j'\right) \cup V.$$

Since no element of $M_1$ can be in $PS_N(\overline{p})$

$$PS_N(\overline{p}) \subseteq \left(\bigcup_{k=1}^{n} C_k\right) \cup \left(\bigcup_{j=1}^{m} C_j'\right) \cup (N_1 \backslash M_1) \cup N_2.$$

Hence it follows that

$$PS_N(\overline{p}) = \left(\bigcup_{k=1}^{n} C_k\right) \cup \left(\bigcup_{j=1}^{m} C_j'\right) \cup N_1'' \cup N_2'$$

for certain cycles $C_k, k = 1, ..., n, C_j'$ subsets of cycles which are not themselves cycles, $j = 1, ..., m$, and for some $N_1'' \subseteq N_1'$, and $N_2' \subseteq N_2$.

Suppose $M(R,X) = \emptyset$. Since $N_2 \cap PS_N(\overline{p}) \neq \emptyset$ implies $M(R,X) \neq \emptyset$,

$$PS_N(\overline{p}) = \left(\bigcup_{k=1}^{n} C_k\right) \cup \left(\bigcup_{j=1}^{m} C_j'\right) \cup N_1'',$$

i. e., $N_2' = \emptyset$. Since no element of $N_1 \backslash M_1$ is preferred to one of $U \backslash V$, no element of $N_1 \backslash M_1$ is preferred to one of $PS_N(\overline{p})$. Hence for all $s \in \bigcup_{j=1}^{m} C_j'$, there exists

$$c \in \left(\bigcup_{k=1}^{n} C_k\right) \cup \left(\bigcup_{j=1}^{m} C_j'\right)$$

such that $cPs$ by Lemma 7.25, else $M(R,X) \neq \emptyset$. By Lemma 7.25, $\forall s \in N_1''$, there exists

$$c \in \left(\bigcup_{k=1}^{n} C_k\right) \cup \left(\bigcup_{j=1}^{m} C_j'\right)$$

such that $cPs$.

For the converse, the conditions imply for all $s \in PS_N(\overline{p})$, there exists $c \in PS_N(\overline{p})$ such that $cPs$. Hence no element of $PS_N(\overline{p})$ is in $M(R,X)$. Thus by Lemma 7.24, $M(R,X) = \emptyset$.                                     □

**Theorem 7.28.** *If $PS_N(\overline{p})$ has no cycles, then $M(X,R) \neq \emptyset$.*

*Proof.* Suppose that there are no strict preferences among the elements of $PS_N(R)$. Then by Lemma 7.25, it follows that

$$PS_N(\overline{\rho}) \subseteq M(R,X) .$$

Hence $M(R,X) \neq \emptyset$ in this case. Let $\overline{a}_1, ..., \overline{a}_n$ be distinct elements of $PS_N(R)$ such that $\overline{a}_1 P \overline{a}_2, \overline{a}_2 P \overline{a}_3, ..., \overline{a}_{n-1} P \overline{a}_n$ is of maximal length. If $n = 2$, then clearly it is not the case that $\overline{a}_1 P \overline{a}_1$ or $\overline{a}_2 P \overline{a}_1$. If $n \geq 3$, there does not exist $\overline{a}_i$ such that $\overline{a}_i P \overline{a}_1$ else $\overline{a}_1 P \overline{a}_2, ..., \overline{a}_{i-1} P \overline{a}_i$ is a cycle but $PS_N(\overline{\rho})$ has no cycles. By the maximality of $n$, there does not exist $\overline{b} \in PS_N(\overline{\rho})$ distinct from $\overline{a}_i (i = 1, ..., n)$ such that $\overline{b} P \overline{a}_1$. Hence no element of $PS_N(\overline{\rho})$ is strictly preferred to $\overline{a}_1$. Thus $\overline{a}_1$ is a maximal element by Lemma 7.25. Hence $M(R,X) \neq \emptyset$.                                   $\square$

### 7.5.2 The Three-Player Case

Our main conclusion is presented in Theorem 7.27. In all but a limited number of cases the maximal set is empty *if and only if* the Pareto set is a union of cycles or a subset of a union of cycles. In this section, we completely characterize the set of elements that constitute the exceptions to Theorem 7.27 in a three-player game. The characterized set is presented in Theorem 7.35.

Let the set of political players be $N = \{1, 2, 3\}$ and $i = N$. Let $\sigma_i : \mathscr{A} \to T$. Define the binary relation $R_i$ on $\mathscr{A}$ by for all $x, y \in \mathscr{A}, x R_i y$ if and only if $\sigma_i(x) \geq \sigma_i(y)$. In such a case, we say that $x$ is at least as good as $y$ for player $i$. Clearly, $R_i$ is reflexive, complete, and transitive. Now $x P_i y$ if and only if $x R_i y$ and not $y R_i x$ if and only if $\sigma_i(x) > \sigma_i(y)$. In such case, we say that $x$ is *strictly preferred* to $y$ by player $i$.

We now consider our application area. Let $R_i$ be defined in terms of $\sigma_i$ as above, $i = 1, 2, 3$. Let $\mathscr{R}$ denote the set of all reflexive, complete, and transitive binary relations on $\mathscr{A}$; $\mathscr{R}^3$ the set of all ordered triples of elements of $\mathscr{R}$; and $\mathscr{B}$ the set of all reflexive and complete binary relations on $\mathscr{A}$. Define simple majority rule $f : \mathscr{R}^3 \to \mathscr{B}$ as follows: for all $\rho = (R_1, R_2, R_3)$, for all $(x, y) \in \mathscr{A} \times \mathscr{A}, (x, y) \in f(\rho)$ if and only if $|\{i \in N \mid x R_i y\}| \geq 2$. Then $(x, y) \in f(\rho)$ and not

$$(y, x) \in f(\rho) \Leftrightarrow |\{i \in N \mid x R_i y\}| \geq 2$$

and

$$|\{i \in N \mid y R_i x\}| \leq 1 \Leftrightarrow |\{i \in N \mid x P_i y\}| \geq 2 .$$

Recall the definition of the Pareto set (Definition 7.10). An alternative $x$ is in the Pareto set if whenever a player strictly prefers an alternative $w$ to $x$, then there is a player who strictly prefers $x$ to $w$. Any effort by the group to choose an alternative not in the Pareto set will leave at least one player worse off. Note that the Pareto set is not determined by majority rule but rather by unanimity.

Let $T^3$ denote the set of all ordered triples of elements from $T$.

**Definition 7.29 (assignment).** Define $f^* : \mathscr{A} \to T^3$ by $\forall x \in \mathscr{A}, f^*(x) = (r, s, t)$, where

$$r = \bigvee \{a \in T \mid x \in \sigma_1^a\},$$

$$s = \bigvee \left\{b \in T \mid x \in \sigma_1^b\right\},$$

$$t = \bigvee \{c \in T \mid x \in \sigma_3^c\}.$$

Then $(r,s,t)$ is called the *assignment* of $x$ with respect to $\sigma_1, \sigma_2, \sigma_3$. An element $(a,b,c)$ of $T^3$ is called *allowable* with respect to $\sigma_1, \sigma_2, \sigma_3$ if there exists $x \in \mathscr{A}$ such that $f^*(x) = (a,b,c)$. Let $X$ denote the set of all elements of $T^3$ that are allowable with respect to $\sigma_1, \sigma_2, \sigma_3$.

In Definition 7.29, it is clear that $x \in \sigma_1^r \cap \sigma_2^s \cap \sigma_3^t$. It is also clear that for $r', s', t' \in T$ with $r \le r', s \le s', t \le t'$, it is not the case that $x \in \sigma_1^{r'} \cap \sigma_2^{s'} \cap \sigma_3^{t'}$ if any of the inequalities are strict.

The following results show that the main results, which are determined in $T^3$, can be transferred faithfully to $\mathscr{A}$ via the function $f^*$ of $\mathscr{A}$ into $T^3$.

**Definition 7.30** ($RT^3$)**.** Define the binary relation $R$ on $T^3$ as follows: for all

$$(a,b,c), (d,e,f) \in T^3, (a,b,c)R(d,e,f)$$

if and only if either $a \ge d, b \ge e$ or $a \ge d, c \ge f$ or $b \ge e, c \ge f$. Define the binary relation $P$ on $T^3$ by for all $(a,b,c), (d,e,f) \in T^3$,

$$(a,b,c)P(d,e,f) \text{ if and only if } (a,b,c)R(d,e,f) \text{ and not } (d,e,f)R(a,b,c).$$

**Proposition 7.31.** *Let $R$ and $P$ be defined as in Definition 7.30. Then for all $(a,b,c)$, $(d,e,f) \in T^3$,*

$$(a,b,c)P(d,e,f) \text{ if and only if } a > d, b > e \text{ or } a > d, c > f \text{ or } b > e, c > f.$$

*Proof.* It follows easily that $R$ is complete. Thus for all

$$(a,b,c), (d,e,f) \in T^3, (a,b,c)P(d,e,f)$$

if and only if not $(d,e,f)R(a,b,c)$. Now not

$$
\begin{aligned}
(d,e,f)R(a,b,c) &\Leftrightarrow \text{not}\,(d \ge a, e \ge b \text{ or } d \ge a, f \ge c \text{ or } e \ge b, f \ge c)\\
&\Leftrightarrow \text{not}\,(d \ge a, e \ge b) \text{ and not}\,(d \ge a, f \ge c) \text{ and not}\,(e \ge b, f \ge c))\\
&\Leftrightarrow (\text{not}\, d \ge a \text{ or not}\, e \ge b)\\
&\quad \text{and } (\text{not}\, d \ge a \text{ or not}\, f \ge c)\\
&\quad \text{and } (\text{not}\, e \ge b \text{ or not}\, f \ge c)\\
&\Leftrightarrow (a > d \text{ or } b > e) \text{ and } (a > d \text{ or } c > f) \text{ and } (b > e \text{ or } c > f)\\
&\Leftrightarrow a > d, b > e \text{ or } a > d, c > f \text{ or } b > e, c > f.
\end{aligned}
$$

Hence the desired result holds.                                                                        □

**Definition 7.32 (strict preference in $T^3$).** Define the binary relation $\widehat{R}$ on $\mathscr{A}$ as follows: for all $x, y \in \mathscr{A}, x\widehat{R}y$ if and only if $(a,b,c)R(d,e,f)$, where $(a,b,c)$ and $(d,e,f)$ are the assignments of $x$ and $y$, respectively, with respect to $\sigma_1, \sigma_2, \sigma_3$. Define the binary relation $\widehat{P}$ on $\mathscr{A}$ by for all $x, y \in \mathscr{A}, x\widehat{P}y$ if and only if $x\widehat{R}y$ and not $y\widehat{R}x$.

The following result is an easy consequence of the definitions. It makes the connection between strict preference in $\mathscr{A}$ and strict preference in $T^3$.

**Proposition 7.33.** *For all $x, y \in \mathscr{A}$, $x\widehat{P}y$ if and only if $(a,b,c)P(d,e,f)$, where $(a,b,c)$ and $(d,e,f)$ are the assignments of $x$ and $y$, respectively, with respect to $\sigma_1, \sigma_2, \sigma_3$.*

Let $x, y \in \mathscr{A}$. Let $(a,b,c) = f^*(x)$ and $(d,e,f) = f^*(y)$. Then $x\widehat{R}y$ if and only if $f^*(x)Rf^*(y)$ if and only if $(a,b,c)R(d,e,f)$ if and only if $\sigma_1(x) \geq \sigma_1(y), \sigma_2(x) \geq \sigma_2(y)$ or $\sigma_1(x) \geq \sigma_1(y), \sigma_3(x) \geq \sigma_3(y)$ or $\sigma_2(x) \geq \sigma_2(y), \sigma_3(x) \geq \sigma_3(y)$ if and only if $xR_1y, xR_2y$ or $xR_1y, xR_3y$ or $xR_2y, xR_3y$ if and only if $x\widehat{R}y$. Thus $\widehat{R}$ and $\widetilde{R}$ are the equivalent. (This argument shows that $f^*$ is a homomorphism as defined in section three.) It also follows that

$$PS_N(\overline{\rho}) = \{(a,b,c) \in f^*(\mathscr{A}) \mid \forall(d,e,f) \in f^*(\mathscr{A}), a < d \Rightarrow b > e \text{ or } c > f$$
$$\text{and } b < e \Rightarrow a > d \text{ or } c > f \text{ and } c < f \Rightarrow a > d \text{ or } b > e\}.$$

Thus we see that the preimage of $PS_N(R)$ under $f^*$ equals $PS_N(\widetilde{R})$.

Let $U = T^3$, where $T = \{0, .25, .5, .75, 1\}$. The partial order $\preceq$ on $T^3$ is defined by $\forall(a_1, a_2, a_3), (b_1, b_2, b_3) \in R^n, (a_1, a_2, a_3) \preceq (b_1, b_2, b_3)$ if and only if $a_i \leq b_i$ for $i = 1, 2, 3$. Then $(a_1, a_2, a_3) \prec (b_1, b_2, b_3)$ if and only if $(a_1, a_2, a_3) \preceq (b_1, b_2, b_3)$ and $a_i < b_i$ for some $i = 1, 2, 3$. For properties $(1) - (6)$ in Section 7.5.1, the relations $R_i$ are defined as follows: for all $(a_1, a_2, a_3), (b_1, b_2, b_3) \in T^3, (a_1, a_2, a_3)R_i(b_1, b_2, b_3)$ if and only if $a_i \geq b_i, i = 1, 2, 3$. It follows easily that properties $(1) - (6)$ hold. (Suppose $f^* : \mathscr{A} \to T^3$ is such that for $x, y \in \mathscr{A}$, $f_i^*(x) = (a_1, a_2, a_3)$ and $f_i^*(y) = (b_1, b_2, b_3)$. Then for $i = 1, 2, 3$,

$$f_i^*(x)R_if_i^*(y) \Leftrightarrow (a_1, a_2, a_3)R_i(b_1, b_2, b_3)$$
$$\Leftrightarrow a_i \geq b_i$$
$$\Leftrightarrow \widetilde{A}_i(x) \geq \widetilde{A}_i(y)$$
$$\Leftrightarrow x\widetilde{R}_iy.$$

Thus $f_i^*$ satisfies the properties in Theorem 7.14.

**Lemma 7.34.** *Let*

$$N_1 = \{(a,b,c) \mid a,b,c \in \{0, 0.25\}\}$$
$$\cup \{(0,0,a) \mid a \in T\}$$
$$\cup \{(0,a,0) \mid a \in T\}$$
$$\cup \{(a,0,0) \mid a \in T\}$$

*and*

$$N_2 = \{(d,e,f) \mid d,e,f \in \{.75,1\}\}$$
$$\cup\{(1,1,a) \mid a \in T\}$$
$$\cup\{(1,a,1) \mid a \in T\}$$
$$\cup\{(a,1,1) \mid a \in T\}\,.$$

*Let $(x,y,z) \in T^3$. Then there does not exist a cycle in $T^3$ containing $(x,y,z)$ if and only if $(x,y,z) \in N_1 \cup N_2$.*

*Proof.* Let $\bar{x} \in N_2$. If there does not exist $\bar{y} \in T^3$ such that $\bar{y}P\bar{x}$, then $\bar{x}$ is not in a cycle. If there exists $\bar{y} \in T^3$ such that $\bar{y}P\bar{x}$, then two of the components of $\bar{y}$ equal 1. Thus there does not exist $\bar{z} \in T^3$ such that $\bar{z}P\bar{y}$. Hence $\bar{x}$ is not in a cycle. Let $\bar{x} \in N_1$. If there does not exist $\bar{y} \in T^3$ such that $\bar{x}P\bar{y}$, then $\bar{x}$ is not in a cycle. If there exists $\bar{y} \in T^3$ such that $\bar{x}P\bar{y}$, then two components of $\bar{y}$ equal 0. Hence there does not exist $\bar{z} \in T^3$ such that $\bar{y}P\bar{z}$. Thus $\bar{x}$ is not in a cycle.

We complete the proof by showing any other element of $T^3$ is in a cycle. Consider $(a,b,c)$, where $a,b,c$ are pairwise distinct, say $a < b < c$. Then $\{(c,a,b),(b,c,a),(a,b,c)\}$ is a cycle. Suppose $a = b = c$. Then the previous paragraph shows that $a = b = c = .5$. Clearly

$$\{(.75,0,.75),(.5,.5,.5),(.25,.25,1)\}$$

is a cycle. Suppose exactly two of $a,b,c$ are equal, say $a = b$. There are 10 remaining elements $(a,a,c)$ and their permutations $(a,c,a),(c,a,a)$. It suffices to consider the 10 elements,

$$(.5,.5,0),(.5,.5,.25),(.5,.5,.75),(.5,.5,1),$$
$$(.25,.25,.5),(.25,.25,.75),(.25,.25,1),$$
$$(.75,.75,0),(.75,.75,.25),(.75,.75,.5).$$

The following are cycles involving the 10 elements or permutations of the elements:

$$\{(0,.75,.75),(.5,.5,0),(.25,.25,1)\},\{(0,.75,.5),(.5,.5,.25),(,25,.25,.75)\},$$
$$\{(0,.75,1),(.5,.5,.75),(.25,1,.25)\},\{(.75,.75,c),(.5,.5,1),(.25,1,.75)\},$$
$$\{(.5,.5,0),(.25,.25,.5),(.75,0,.25)\},$$

where $c = 0,.25$ or $.5$.                                                                              $\square$

Let

$$N_1' = \{(.25,.25,.25),(.25,.25,0),(.25,0,.25),(0,.25,.25)\}$$

and

$$I_1 = \{(1,0,0),(0,0,1),(0,1,0)\}\,.$$

We can now state our main conclusion, which lays out the conditions under which a maximal set is empty in fuzzy three-player spatial models.

**Theorem 7.35.** $M(R,X) = \emptyset$ *if and only if* $PS_N(\overline{\rho}) = (\cup_{k=1}^n C_k) \cup (\cup_{j=1}^m C_j') \cup N_1'' \cup I_1'$,
*where* $N_1'' \subseteq N_1', I_1' \subseteq I_1, C_k$ *are cycles,* $k = 1,..,n, C_j'$ *are subsets of cycles which are
not themselves cycles,* $j = 1,...m$, *and*

*(1) for all* $\overline{s} \in \cup_{j=1}^m C_j'$, *there exists* $\overline{c} \in (\cup_{k=1}^n C_k) \cup (\cup_{j=1}^m C_j')$ *such that* $\overline{c} P \overline{s}$,
*(2) for all* $\overline{s} \in N_1''$ *there exists* $\overline{c} \in (\cup_{k=1}^n C_k) \cup (\cup_{j=1}^m C_j')$ *such that* $\overline{c} P \overline{s}$ *and*
*(3) for all* $\overline{i} \in I_1'$, *there exists* $\overline{d} \in (\cup_{k=1}^n C_k) \cup (\cup_{j=1}^m C_j') \cup N_1''$ *such that* $\overline{d} P \overline{i}$.

*Proof.* By Lemma 7.34, it follows that

$$PS_N(\overline{\rho}) \subseteq \left( \bigcup_{k=1}^n C_k \right) \cup \left( \bigcup_{j=1}^m C_j' \right) \cup N_1 \cup N_2.$$

Since no element of $(\{(0,0,a) \mid a \in T\} \cup \{(0,a,0) \mid a \in T\} \cup \{(a,0,0) \mid a \in T\}) \backslash I_1$
can be in $PS_N(\overline{\rho})$ and since no element of $N_1 \backslash (N_1' \cup I_1)$ can be in

$$PS_N(\overline{\rho}), PS_N(\overline{\rho}) \subseteq \left( \bigcup_{k=1}^n C_k \right) \cup \left( \bigcup_{j=1}^m C_j' \right) \cup N_1' \cup I_1 \cup N_2.$$

Hence it follows that

$$PS_N(\overline{\rho}) = \left( \bigcup_{k=1}^n C_k \right) \cup \left( \bigcup_{j=1}^m C_j' \right) \cup N_1'' \cup I_1' \cup N_2'$$

for certain cycles $C_k, k = 1,...,n, C_j'$ subsets of cycles which are not themselves cycles, $j = 1,...,m$, and for some $N_1'' \subseteq N_1', I_1' \subseteq I_1$, and $N_2' \subseteq N_2$.
    Suppose $M(R,X) = \emptyset$. By Lemma 7.24,

$$PS_N(\overline{\rho}) = \left( \bigcup_{k=1}^n C_k \right) \cup \left( \bigcup_{j=1}^m C_j' \right) \cup N_1'' \cup I_1',$$

i. e., $N_2' = \emptyset$. If $\overline{s} \in N_1$ is such that $\overline{s} P \overline{c}$ for some $\overline{c} \in PS_N(\overline{\rho})$, then two of the components of $\overline{c}$ equal 0 and so $\overline{c} \in I_1'$. Thus no element of $N_1$ is strictly preferred to one
of

$$\left( \bigcup_{k=1}^n C_k \right) \cup \left( \bigcup_{j=1}^m C_j' \right).$$

No element of $I_1'$ is strictly preferred to any element of $PS_N(\overline{\rho})$. Hence for all $\overline{s} \in \cup_{j=1}^m C_j'$, there exists

$$\overline{c} \in \left( \bigcup_{k=1}^n C_k \right) \cup \left( \bigcup_{j=1}^m C_j' \right)$$

such that $\overline{c} P \overline{s}$ by Lemma 7.25 else $M(R,X) \neq \emptyset$. By Lemma 7.25, for all $\overline{s} \in N_1''$
there exists

$$\overline{c} \in \left( \bigcup_{k=1}^{n} C_k \right) \cup \left( \bigcup_{j=1}^{m} C_j' \right)$$

such that $\overline{c}P\overline{s}$ and for all $\overline{i} \in I_1'$, there exists

$$\overline{d} \in \left( \bigcup_{k=1}^{n} C_k \right) \cup \left( \bigcup_{j=1}^{m} C_j' \right) \cup N_1''$$

such that $\overline{d}P\overline{i}$.

For the converse, the conditions imply $\forall \overline{s} \in PS_N(\overline{p})$, $\exists \overline{c} \in PS_N(\overline{p})$ such that $\overline{c}P\overline{s}$. Hence no element of $PS_N(\overline{p})$ is in $M(R,X)$. Thus by Lemma 7.26, $M(R,X) = \emptyset$.  □

We can have

$$\left( \bigcup_{k=1}^{n} C_k \right) \cap \left( \bigcup_{j=1}^{m} C_j' \right) = \emptyset$$

in Theorem 7.35 by deleting any element

$$\overline{s} \in \left( \bigcup_{k=1}^{n} C_k \right) \cap \left( \bigcup_{j=1}^{m} C_j' \right)$$

from $\bigcup_{j=1}^{m} C_j'$ since $\overline{s}$ is in some $C_k$. Also any elements of $\bigcup_{j=1}^{m} C_j'$ that form a cycle can be removed from $\bigcup_{j=1}^{m} C_j'$ and moved to $\bigcup_{k=1}^{n} C_k$ since they form a cycle.

*Example 7.36.* Let $PS_N(\overline{p}) = C \cup \{(.75,.75,.75)\}$, where

$$C = \{(1,.5,0),(.5,0,1),(0,1,.5)\}.$$

Then $C$ is a cycle. Thus $M(R,X) \cap PS_N(R) = \{(.75,.75,.75)\}$. It is easily verified that $M(R,X) = \{(.75,.75,.75,(.75,.75,.5),(.75,.5,.75),(.5,.75,.75)\}$. Here $M(R,X) \nsubseteq PS_N(R)$.

*Example 7.37.* Let $PS_N(\overline{p}) = C \cup \{(.75,.75,.75)\}$, where

$$C = \{(1,.75,0),(.75,0,1),(0,1,.75)\}.$$

Then $C$ is a cycle. Thus $M(R,X) \cap PS_N(R) = \{(.75,.75,.75)\}$. It is easily verified that $M(R,X) = \{(.75,.75,.75\}$. In this example, $M(R,X) \subseteq PS_N(\overline{p})$.

## 7.6  Implications

Our main argument is contained in Theorem 7.27, which demonstrates that spatial models of fuzzy individual preferences can predict a majority rule maximal set under far less restrictive conditions than the conventional approach. Theorem 7.27 can be applied to any *n*-player game ($n > 2$). Moreover, Mordeson and Clark (2010) demonstrates similar results in greater dimensionality spatial models. Our results in this chapter establish that in all but a limited number of cases, the majority rule maximal set is empty *if and only if* the Pareto set is a union of cycles or a subset of a union of cycles. Hence, if the elements in the Pareto set cycle under majority rule or if they constitute a subset of a cycle under majority rule, the maximal set is empty, and vice versa. Moreover, Theorem 7.28 establishes that for the majority rule maximal set to be empty, the Pareto set must contain at least one undefeated alternative.

Furthermore, we have demonstrated that if the maximal set exists, at least one of its elements must be contained in the Pareto set (Lemma 7.24). If an element in the Pareto set can not be majority defeated by any other element in the Pareto set, then it is an element in the maximal set. However, if every element in the Pareto set can be majority defeated by at least one other element in the Pareto set, then the maximal set is empty. Moreover, if an element in the Pareto set can be majority defeated by any element in the total set of available alternatives then at least one of the elements that defeats it must be in the Pareto set (Lemma 7.26). The implication is that we may confine our initial search for the existence of a majority rule maximal set to the Pareto set.

The substantive interpretation of our argument is straightforward and intuitive. If the degree to which a majority find a given alternative acceptable is relatively high, then a stable outcome is assured under majority rule. This is born out by Theorem 7.35, which completely characterizes the limited cases that constitute the exceptions for a three-player game. While none of the following seven distinct elements are ever part of any cycle, if they are part of the Pareto set together with a cycle or a subset of a cycle, and they are defeated by at least one other element in the Pareto set, then the maximal set is empty.

$$\{(.25,.25,.25),(.25,.25,0),(.25,0,.25),(0,.25,.25),$$
$$(1,0,0),(0,0,1),(0,1,0)\}.$$

If they are not defeated by at least one other element in the Pareto set or if they uniquely comprise the Pareto set, then they are elements in the maximal set. The first four elements represent a situation in which the players' preferences intersect at the lowest $t$−level possible. In the most trivial case, represented by $\{(1,0,0),(0,0,1),(0,1,0)\}$, the preferences of three players do not intersect at any $t$−level. In this case, the Pareto set and maximal set comprise the same three alternatives (the "ideal" points of the players).

Furthermore, if any of the following elements are in the Pareto set, a maximal set always exists:

$$\{(1,1,a),(1,a,1),(a,1,1),(1,.75,.75),(.75,1,.75),(.75,.75,1),$$
$$(.75,.75,.75)\}, \text{ where } a \text{ is any element of } T.$$

In essence, if the preferences for all three players intersect at relatively high $t$−levels or at $t$−level $= 1$ (the ideal region of alternatives) for any two of them, a maximal set exists.

We would be remiss if we did highlight the ability of the approach considered in this paper to deal with highly irregular preferences, which standard mathematical approaches can only tackle with substantial difficulty. If spatial models are to be tested empirically, they will need to be able to accommodate any shape imposed upon them by estimates derived from data sets. As of yet, no convention has emerged on how best to estimate preferences, but judging by the efforts to do so using Nominate Bianco et al. (2004) and the Comparative Manifesto Project Laver (2001), preferences are likely to take any number of shapes. The oddity of those shapes is far likelier in the case of collective institutions.

We should also note the relative ease with which the approach deals with non-separable issues. The conventional approach relying on the Euclidean utility function $u_i(y) = -[(y-x)A_i(y-x)^T]$ requires constructing a $k \times k$ positive definite matrix $A_i$ to weight a player's preference calculations, where $k$ denotes the number of dimensions in Euclidean space and $T$ denotes transpose. In the case of non-separable preferences, the model must specify the off-diagonal entries of $A_i$ to numbers other than zero. The approach put forth by Clark, Larson, Mordeson, and Wierman (2008) does not require this degree of specificity. The homomorphism, the proof of which is provided in section three, permits the model to solve for solution sets when preferences are non-separable, even though $\sigma$ can be represented by any geometric shape. In fact, Clark, Larson, Mordeson, and Wierman (2008) demonstrate that separable fuzzy preferences in two-dimensional space are rectangular. Fuzzy preferences represented by other shapes, to include circles, are non-separable.

We conclude with a brief comment on our major theorem (Theorem 7.27). The theorem makes it clear that it is neither the shape of players' preferences nor the positioning of players in space that matters when determining the maximal set. In fact, convex ellipses might result in an empty maximal set, and non-convex preferences can result in a maximal set. What is important are the intersections of the players' preferences. If the preferences for all three players intersect at relatively high $t$−levels or at $t$−level $= 1$ (the ideal region of alternatives) for any two of them, a maximal set always exists. Furthermore, our theory demonstrates that fuzzy set spatial models can map any geometric shape representing the preferences of players to a region with suitable and natural partial ordering.

# References

Austen-Smith, D., Banks, J.S.: Positive Political Theory I: Collective Preference. University of Michigan Press, Ann Arbor (1999)

Austen-Smith, D., Banks, J.S.: Positive Political Theory II: Strategy and Structure. University of Michigan Press, Ann Arbor (2005)

Balke, W.T., Guntzer, U., Siberski, W.: Exploiting indifference for customization of partial order skylines. In: 10th International Database Engineering and Applications Symposium, IDEAS 2006, pp. 80–88 (2006)

Barberà, S., Ehlers, L.: Free triples, large indifference classes and the majority rule. Social Choice and Welfare 37(4), 559–574 (2011),
http://dx.doi.org/10.1007/s00355-011-0584-8

Bezdek, J.C., Spillman, B., Spillman, R.: A fuzzy relation space for group decision theory. Fuzzy Sets and Systems 1, 255–268 (1978)

Bezdek, J.C., Spillman, B., Spillman, R.: A fuzzy relation space for group decision theory. Fuzzy Sets and Systems 2, 5–14 (1979)

Bianco, W.T., Jeliazkov, I., Sened, I.: The Uncovered Set and the Limits of Legislative Action. Political Analysis 12, 256–276 (2004)

Blin, J.M.: Fuzzy relations in group decision theory. Journal of Cybernetics 4(2), 17–22 (1974)

Bräuninger, T.: Stability in spatial voting games with restricted preference maximizing. Journal of Theoretical Politics 19(2), 173–191 (2007),
http://jtp.sagepub.com/content/19/2/173.abstract

Clark, T.D., Larson, J.M., Mordeson, J.N., Potter, J.D., Wierman, M.J. (eds.): Applying Fuzzy Mathematics to Formal Models in Comparative Politics. STUDFUZZ, vol. 225. Springer, Heidelberg (2008)

Gehrlein, W.V., Valognes, F.: Condorcet efficiency: A preference for indifference. Social Choice and Welfare 18(1), 193–205 (2001),
http://dx.doi.org/10.1007/s003550000071

Kacprzyk, J., Fedrizzi, M.: A 'soft' measure of consensus in the setting of partial (fuzzy) preferences. European Journal of Operational Research 34(3), 316–325 (1988)

Kacprzyk, J., Fedrizzi, M., Nurmi, H.: Group decision making and consensus under fuzzy preferences and fuzzy majority. Fuzzy Sets and Systems 49, 21–31 (1992)

Koehler, D.H.: Convergence and restricted preference maximizing under simple majority rule: Results from a computer simulation of committee choice in two-dimensional space. American Political Science Review, 155–167 (March 2001),
http://journals.cambridge.org/article_S0003055401000065

Kuroki, N., Mordeson, J.N.: Structures of rough sets and rough groups. Journal of Fuzzy Mathematics 5, 183–191 (1997)

Laver, M. (ed.): Estimating the Policy Position of Political Actors. Routledge, London (2001)

Malik, D.S., Mordeson, J.N.: Structure of upper and lower approximation spaces of infinite sets. In: Lin, T.Y., Yao, Y.Y., Zadeh, L.A. (eds.) Data Mining, Rough Sets and Granular Computing, ch. 5.2, pp. 461–473. Physica-Verlag GmbH, Heidelberg (2002),
http://dl.acm.org/citation.cfm?id=783032.783055

Mordeson, J.N.: Algebraic properties of lower approximation spaces. Journal of Fuzzy Mathematics 7, 631–637 (1999)

Mordeson, J.N., Clark, T.D.: The existence of a majority rule maximal set in arbitrary n-dimensional spatial models. New Mathematics and Natural Computation 06(03), 261–274 (2010)

Mordeson, J.N., Nair, P.S.: Successor and source of (fuzzy) finite state machines and (fuzzy) directed graphs. Inf. Sci. 95(1), 113–124 (1996)

Nurmi, H.: Approaches to collective decision making with fuzzy preference relations. Fuzzy Sets and Systems 6, 249–259 (1981a)

Nurmi, H.: A fuzzy solution to a majority voting game. Fuzzy Sets and Systems 5, 187–198 (1981b)

Orlovsky, S.: Decision-making with a fuzzy preference relation. Fuzzy Sets and Systems 1, 155–167 (1978)

Skog, O.J.: "volontè generale" and the instability of spatial voting games. Rationality and
    Society 6(2), 271–285 (1994)

Sloss, J.: Stable outcomes in majority rule voting games. Public Choice 15(1), 19–48 (1973),
    http://dx.doi.org/10.1007/BF01718841

Tovey, C.A.: The instability of instability. Tech. Rep. NPSOR-91-15, Department of Oper-
    ations Research, Department of Operations Research, Naval Postgraduate School, Mon-
    terey, CA (1991)

Tovey, C.A.: The instability of instability of centered distributions. Mathematical Social Sci-
    ences 59(1), 53–73 (2010)

# Chapter 8
# Conclusion

**Abstract.** This concluding chapter provides a summary of the findings from this book. After showing that a fuzzy maximal set exists, a fuzzy aggregation rule was shown to exist which satisfies all five Arrowian conditions including non-dictatorship. Although the Gibbard-Satterthwaite Theorem has considered individual fuzzy preferences, this book shows that both individuals and groups can choose alternatives to varying degrees resulting in a social choice that can be both strategy-proof and non-dictatorial. Under strict fuzzy preferences, the Median Voter Theorem is shown to hold; however, this is not found under weak fuzzy preferences.

## Introduction

In the preceding chapters, we described the set-theoretic work produced to date on social choice theory and reanalyzed its implications in the fuzzy framework. Though we presented many previously-obtained results regarding the basic behavior of fuzzy preferences (Chapters 1 and 3), our primary focus was on the novel application of fuzzy set theory to the normatively-problematic results in the crisp social choice literature. Specifically, we generalized the well-known works of Arrow (Chapter 4), Gibbard and Satterthwaite (Chapter 5), Black (Chapter 6), and KcKelvey (Chapter 7). We extended these theorems in the fuzzy framework, with novel results that differ substantially from the crisp framework.

We began to discuss our novel results in Chapter 4, wherein we generalized Arrow's Impossibility Theorem in the fuzzy framework. Table 8.1 summarizes our results, as well as the results of previous attempts by others in the social choice literature. Specifically, Table 8.1 classifies generalized versions of Arrow's Theorems by the type of strict preference relation used. Recall from Corollary 3.23 that $\pi_{(n)}$ is given as follows.

**Proposition 8.1.** *Let $\rho$ be an FWPR on $X$ such that $\rho = \iota \cup \pi$.*
*Then we list the following results.*

*(1) If $\cup = \cup_1$, where Gödel union $\cup_1$ is given by*

$$(A \cup_1 B)(x) = max\{A(x), B(x)\},$$

then $\pi = \pi_{(1)}$ where

$$\pi_{(1)}(x,y) = \begin{cases} \rho(x,y) & \text{if } \rho(x,y) > \rho(y,x), \\ 0 & \text{otherwise.} \end{cases}$$

(2) If $\cup = \cup_2$, where Łukasiewicz union $\cup_2$ is given by

$$(A \cup_2 B)(x) = min\{1, A(x) + B(x)\},$$

and conditions (iv) and (v) from Proposition 3.18 hold, then $\pi = \pi_{(2)}$ where

$$\pi_{(2)}(x,y) = 1 - \rho(y,x).$$

(3) If $\cup = \cup_2$, then $\pi = \pi_{(3)}$ where

$$\pi_{(3)}(x,y) = max\{0, \rho(x,y) - \rho(y,x)\}.$$

(4) If $\cup = \cup_3$, where strict union $\cup_3$ is given by

$$(A \cup_3 B)(x) = \begin{cases} B(x) & \text{if } A(x) = 0 \\ A(x) & \text{if } B(x) = 0 \\ 1 & \text{otherwise,} \end{cases}$$

then $\pi = \pi_{(4)}$ where

$$\pi_{(4)}(x,y) = \begin{cases} \rho(x,y) & \text{if } \rho(y,x) = 0, \\ 0 & \text{otherwise.} \end{cases}$$

(5) If $\cup = \cup_4$, where algebraic union $\cup_4$ is given by

$$(A \cup_4 B)(x) = A(x) + B(x) - A(x)B(x)$$

then $\pi = \pi_{(5)}$ where

$$\pi_{(5)}(x,y) = \begin{cases} \frac{\rho(x,y) - \rho(y,x)}{1 - \rho(y,x)} & \text{if } \rho(x,y) > \rho(y,x), \\ 0 & \text{otherwise.} \end{cases}$$

**Table 8.1** Summary of Results From Chapter 4

| Strict Preference Relation Type | Assumptions on Individual Preferences | Assumptions on Social Preferences | Properties of the FPAR | Social Choice Results | Arrow's Theorem | Reference |
|---|---|---|---|---|---|---|
| $\pi$-1 | Reflexive, connected, and max-min-transitive | Reflexive, connected and max-min-transitive | IIA-1 and PC | Nondictatorial (Oligarchy) | No | Dutt Props 3.7 & 3.9 |
| $\pi$-1 | Reflexive, connected, and max-min-transitive | Reflexive, connected and max-min-transitive | IIA-1, PC, and positive responsiveness | Dictatorial | Holds | Dutt Prop 3.12 |
| $\pi$-1 | Reflexive, connected, and Ł-transitive | Reflexive, connected, and Ł-transitive | IIA-1 and PC | Nondictatorial | No | Dutt Prop 4.14 |
| $\pi$-1 | Reflexive, complete, max-* transitive | Reflexive, complete, and max-* transitive | IIA-1, IIA-3, PC, weakly-Paretian | Nondictatorial | No | Thrm 4.50 |
| $\pi$-regular | Reflexive, complete, transitive | Reflexive, complete, and max-* transitive | IIA-1, positive responsiveness, weakly-Paretian | Nondictatorial | No | Thrm 4.43 |
| $\pi$-1 or $\pi$-3 | Reflexive, complete, max-* transitive | Reflexive, complete, max-* transitive | IIA-1, PC, positive responsiveness, weakly-Paretian | Nondictatorial | No | Thrm 4.44 |
| $\pi$-2 | Reflexive, s-connected and max-min and Ł-transitive | Reflexive, s-connected and max-min and Ł-transitive | IIA-1 and PC | Dictatorial | Holds | Bann Prop 3.1 |
| $\pi$-regular (of sorts) | Connected and weakly transitive | Connected and weakly transitive | IIA-1 and PC | Dictatorial | Holds | Bill Thrm 4 |
| $\pi$-regular | Reflexive, s-connected and m-transitive | Reflexive, s-connected and m-transitive | IIA-1 and PC | Dictatorial | Holds | Rich Prop 2.1 |
| $\pi$-3 | Reflexive, connected, and Ł-transitive | Reflexive, connected, and Ł-transitive | IIA-1, PC, and PR | Nondictatorial | No | Rich Prop 2.1 |

**Table 8.1** (*continued*)

| Strict Preference Relation Type | Assumptions on Individual Preferences | Assumptions on Social Preferences | Properties of the FPAR | Social Choice Results | Arrow's Theorem | Reference |
|---|---|---|---|---|---|---|
| $\pi$-3 | Reflexive, connected, and weakly max-min transitive | Reflexive, connected, and weakly max-min transitive | IIA-1 and PC | Weakly dictatorial | No | DaDe |
| $\pi$-regular | Reflexive, connected, max-$*$ transitive and partially quasi-transitive | Reflexive, connected, max-$*$ transitive and partially quasi-transitive | IIA-1 and PC | Nondictatorial (Oligarchy) | No | FoAd Thrm 9 & Prop 11 |
| $\pi$-regular | Reflexive, connected, max-$*$ transitive and partially quasi-transitive | Reflexive, connected, max-$*$ transitive and partially quasi-transitive | IIA-1, PC and positive-responsiveness | Dictatorial | Holds | FA Thrm 4.12 |
| $\pi$-4 | Reflexive, complete, and partially transitive | Reflexive, complete, and partially transitive | IIA-2 and WP | Dictatorial | Holds | Thrm 4.24 & MC Thrm 3.12 |
| $\pi$-regular | Reflexive, connected, max-$*$ transitive and partially quasi-transitive | Reflexive, connected, max-$*$ transitive and negative transitive | IIA-1 and NI | Null, dictatorial or antidictatorial | No | FDKA Theorem 4.2 |
| none | Reflexive, w-connected, max-$*$ transitive, and no zero divisor | Reflexive, w-connected, max-$*$ transitive, and no zero divisor | IIA-1 and Unanimity | Nondictatorial | No | DPPP Thrm 1: Sufficiency |
| none | Reflexive, w-connected, max-$*$ transitive & zero divisor | Reflexive, w-connected, max-$*$ transitive & zero divisor | IIA-1 and Unanimity | Dictatorial | Holds | DPPP Theorem 1: Necessity |

Table 8.1 then specifies the necessary assumptions that each generalized theorem makes regarding individual and social preferences. The definitions of each of these assumptions (reflexiveness, completeness, max-∗ transitivity, and so on) can be found in Chapter 3, while the properties of each Fuzzy Preference Aggregation Rule (IIA-$n$, weak-paretianism, etc.) are defined in Chapters 3 and 4. The social choice results of these FPARs and their properties are then summarized at the end of the table.

In the Assumptions on Individual/Social Preferences columns we use the following conventions. Max-min and max-∗ transitivity are from Definitions 1.19 and 3.32 respectively. Preference is Ł-transitive if it is max-∗ transitive with ∗ being the Łukasiewicz t–norm, so that $\forall x, y, z \in X$,

$$\rho(x,z) \geq \rho(x,y) + \rho(y,z) - 1.$$

Preference is m-transitive (minimal transitive) if $\forall x, y, z \in X$,

$$(\rho(x,y) = 1 \text{ and } \rho(y,z) = 1) \text{ imply } \rho(x,z) = 1.$$

Finally, $w$-connected stands for weakly connected, i.e., $\rho(x,y) = 0$ implies $\rho(y,x) = 1$ which is a slightly stronger requirement than the completeness condition.

In the Properties of the FPAR column PC stands for Pareto Condition, see Definition 4.5 and NI stands for non-imposition where $\pi_i(x,y)$ for all $i \in N$ then $n\pi(y,x) = 0$.

In the Social Choice Results column, a weak dictator implies there exists a dictator over $(x,y) \in X \times X$ for all $(x \neq y, y \neq x) \notin X \times X$.

The key to the Reference column is

| | |
|---|---|
| Bill | Billot (1992), |
| Bann | Banerjee (1994), |
| Dutt | Dutta (1987), |
| DPPP | Duddy, Perote-Peña, and Piggins (2011), |
| DaDe | Dasgupta and Deb (1999), |
| FoAn | Fono and Andjiga (2005), |
| FDKA | Fono, Donfack-Kommogne, and Andjiga (2009), |
| MoCl | Mordeson and Clark (2009) and |
| Rich | Richardson (1998). |

Theorem and proposition numbers refer to the numbering in the associated paper. If no paper is specified, the number references a Theorem in this book.

We encountered many difficulties in formulating fuzzy definitions of an ideal aggregation rule – particularly with respect to the IIA criterion – but we found that there exist specific combinations of conditions that allow for a fuzzy aggregation rule that satisfies all of the fuzzy counterparts of Arrow's conditions. This result and its assumptions are listed in Table 8.1 below, under Theorem 4.24. Essentially, Theorem 4.24 indicates that, given an FPAR which satisfies fuzzy definitions of transitivity, weak Paretianism and independence of irrelevant alternatives, the FPAR must be dictatorial. However, the implication of dictatorship cannot be generalized over all derivations of fuzzy Arrowian conditions, as the rest of Table 8.1 demonstrates.

Specifically, as we show in Sections 4.1.3 and 4.1.4, the fuzzy Arrowian framework allows for a non-dictatorial aggregation of fuzzy preferences that satisfies the five Arrowian criteria laid out in Section 4.1. Theorems 4.43, 4.44, and 4.50 (also listed in Table 8.1) illustrate this result. Theorem 4.50, for instance, satisfies the fuzzy Pareto rule and is reflexive, complete, weakly Paretian, IIA-1, IIA-3 and max-$*$ transitive, without placing excessive restrictions on $\pi$.

The results obtained in Theorems 4.43, 4.44, and 4.50 of Chapter 4 are distinct from those previously obtained by approaches that use exact preferences. Most importantly, our results suggest that the normatively-undesirable results of classic social choice theory no longer hold when groups possess fuzzy preferences.

Chapter 5 parallels Chapter 4 with a generalization of the Gibbard-Satterthwaite Theorem, which states that voters, if they are able, can manipulate voting procedures to obtain more preferred social outcomes by reporting insincere preferences. We begin by considering three definitions of strategic manipulation of fuzzy social choice functions wherein actors can choose alternatives to varying degrees. We then demonstrate that with minimal assumptions on individual preferences, strategy-proof fuzzy social choice functions satisfy fuzzy versions of peak-only, weak Paretianism, and monotonicity. Furthermore, a fuzzy social choice function is strategy-proof if and only if it is a form of the fuzzy augmented median. These results, outlined in Table 8.2, provide a new dimension to the strategic manipulation literature, which is currently divided as to whether choice functions can be both non-manipulable and nondictatorial when individual preferences are restricted to a single-peaked domain. *Else* in Table 8.2 denotes all conditions except the ones noted.

**Table 8.2** Summary: Chapter 5

| Fuzzy Social Choice Function (FSCF) | Assumptions of the FSCF | Properties of the FSCF | Social Choice Results | Reference |
|---|---|---|---|---|
| Fuzzy Augmented Median Rule | Choice intensities are separable | Sigma-only Weakly-Paretian Monotonic IIA-3 | Strategy-proof (Dictatorial) Allows for unrestricted choice domain | Definition 6.32 |
| Else | Else | Else | Manipulable | Theorem 6.35 |

In Chapter 6, we showed that when fuzzy simple rules allow the social strict preference relation to be regular, the fuzzy maximal set is not necessarily equal to the set of $f$-medians under a single-peaked profile. As Table 8.3 illustrates, the conditions placed on $\pi$ must be quite restrictive. Else, the maximal set may be empty, or it may contain more alternatives than the set of $f$-medians. In other words, Black's

Median Voter Theorem does not hold under many conceptualizations of the fuzzy framework.

Table 8.3 Summary: Fuzzy Simple Rules vs. Fuzzy Voting Rules

|  | Fuzzy Non-Dominated Set $ND(\pi,\mu)$ | Fuzzy Maximal Set $M(\pi,\mu)$ |
|---|---|---|
| Fuzzy Simple Rules | Non-empty when $\pi$ is weakly single-peaked | Non-empty when $\pi$ is weakly single-peaked |
| Fuzzy Voting Rules | Non-empty when $\pi$ is single-peaked | Non-empty when $\pi$ is weakly single-peaked |

However, Chapter 6 also demonstrates that Black's Median Voter Theorem can be made to hold under both the fuzzy non-dominated and maximal sets in the fuzzy framework if the strict preference relation, $\pi$, is partial. Unfortunately, the non-emptiness of neither the maximal set nor the non-dominated set can be guaranteed under any other type of preference relation, as Table 8.4 illustrates.

Table 8.4 Summary: Results with Black's Median Voter Theorem

|  | Fuzzy Non-Dominated Set $ND(\pi,\mu)$ | Fuzzy Maximal Set $M(\pi,\mu)$ |
|---|---|---|
| Fuzzy Augmented Median Voter Theorem | Holds | Holds |
| Else | Does Not Hold | Does Not Hold |

In the final chapter of this book, we turn our focus to the spatial model. In Section 7.1, we consider several problems that can occur when modeling crisp social preference, indifference and majority rule. A particular area of concern is the inherent difficulty of modeling indifference curves which are thick or irregularly-shaped, as is often the case in empirical reality. Traditionally, mathematicians have only been able to do this with great difficulty and limited accuracy. Additionally, traditional crisp spatial models tend to do a rather poor job of predicting majority rule outcomes. A remarkable example of this is that many Euclidian rational choice models predict empty or cycling maximal sets with a high degree of frequency. However, emptiness or cycling problems are very rarely observed in real world elections.

The application of fuzzy set theory to the traditional spatial model addresses both of these concerns surprisingly well. Fuzzy spatial models can successfully accommodate thick, highly-irregular indifference curves in addition to predicting the existence of non-empty, non-cycling maximal sets. We detail this approach, which relies on a specific isomorphism, in Section 7.2. Section 7.4 proves this homorphism with Proposition 7.7, while Section 7.3 provides an empirical application of our approach.

**Table 8.5** Summary: Chapter 7

| Spatial Model | $f_*$ | Necessary Assumptions on $R$ | Maximal Set | Reference |
|---|---|---|---|---|
| Any $n > 2$ game | Homomorphism as defined in Def 6.5 | $PS_N(\overline{p}) = (\bigcup_{k=1}^n C_k) \cup (\bigcup_{j=1}^m C'_j) \cup N''_1$, where $N''_1 \subseteq N'_1, C_k$ are cycles, $k = 1, \ldots, n, C'_j$ subsets of cycles which are not themselves cycles, $j = 1, \ldots, m$, and <br> (1) $\forall s \in \bigcup_{j=1}^m C'_j$, there exists $c \in (\bigcup_{k=1}^n C_k) \cup (\bigcup_{j=1}^m C'_j)$ such that $cPs$, <br> (2) $\forall s \in N''_1$ there exists $c \in (\bigcup_{k=1}^n C_k) \cup (\bigcup_{j=1}^m C'_j)$ such that $cPs$. | Empty only if PS contains a union of cycles or a subset of a union of cycles | Theorem 7.27 |
| Specific Case of 3-Player Spatial Models | Homomorphism as defined in Def 6.5 | $PS_N(\overline{p}) = (\bigcup_{k=1}^n C_k) \cup (\bigcup_{j=1}^m C'_j) \cup N''_1 \cup I'_1$, where $N''_1 \subseteq N'_1, I'_1 \subseteq I_1, C_k$ are cycles, $k = 1, \ldots, n, C'_j$ are subsets of cycles which are not themselves cycles, $j = 1, \ldots, m$, and <br> (1) for all $\overline{s} \in \bigcup_{j=1}^m C'_j$, there exists $\overline{c} \in (\bigcup_{k=1}^n C_k) \cup (\bigcup_{j=1}^m C'_j)$ such that $\overline{c}P\overline{s}$, <br> (2) for all $\overline{s} \in N''_1$ there exists $\overline{c} \in (\bigcup_{k=1}^n C_k) \cup (\bigcup_{j=1}^m C'_j)$ such that $\overline{c}P\overline{s}$ and <br> (3) for all $\overline{i} \in I'_1$, there exists $\overline{d} \in (\bigcup_{k=1}^n C_k) \cup (\bigcup_{j=1}^m C'_j) \cup N''_1$ such that $\overline{d}P\overline{i}$. | Can be empty even if its elements are not part of a cycle or subset of a cycle, and they are defeated by 1 or more elements in the Pareto Set. | Theorem 7.35 |

The chapter's main theorem is presented in Section 7.5. Specifically, Theorem 7.27 proves the possibility of a non-empty majority rule maximal set, while the rest of the section describes the necessary conditions for its existence. As illustrated in Table 8.5, we found that in nearly all cases, a spatial model with thick indifference curves results in an empty maximal set if and only if its Pareto set contains a union of cycles a subset of a union of cycles. The certain exceptions which exist in the case of specific 3-player games are listed in their entirety in Subsection 7.5.2 and explained in further detail therein. Section 7.6 concludes our final chapter with a summary of our findings and a discussion of applications relating to the theoretical approach utilized in Theorem 7.27.

We hope that our work lends itself in the future to further empirical studies of how these theorems behave. We are confident that the benefits of this novel approach to social choice will justify the extensive theoretical work which has been done, and much of which remains to be completed. In particular, we hope future research expounds upon the applications of fuzzy topology and fuzzy calculus to the spatial model, as well as the relevance of fuzzy propositional logic to social choice theory.

# References

Banerjee, A.: Fuzzy preferences and Arrow-type problems. Social Choice and Welfare 11, 121–130 (1994)

Billot, A.: Economic theory of fuzzy equilibria: an axiomatic analysis. Lecture notes in economics and mathematical systems. Springer (1992),
http://books.google.com/books?id=ml-7AAAAIAAJ

Dasgupta, M., Deb, R.: An impossibility theorem with fuzzy preferences. In: Logic, Game Theory and Social Choice: Proceedings of the International Conference, LGS, vol. 99, pp. 13–16 (1999)

Duddy, C., Perote-Peña, J., Piggins, A.: Arrow's theorem and max-star transitivity. Social Choice and Welfare 36(1), 25–34 (2011),
http://dx.doi.org/10.1007/s00355-010-0461-x

Dutta, B.: Fuzzy preferences and social choice. Mathematical Social Sciences 13(3), 215–229 (1987)

Fono, L.A., Andjiga, N.G.: Fuzzy strict preference and social choice. Fuzzy Sets Syst. 155, 372–389 (2005), http://dx.doi.org/10.1016/j.fss.2005.05.001

Fono, L.A., Donfack-Kommogne, V., Andjiga, N.G.: Fuzzy Arrow-type results without the pareto principle based on fuzzy pre-orders. Fuzzy Sets and Systems 160(18), 2658–2672 (2009)

Mordeson, J.N., Clark, T.D.: Fuzzy Arrow's theorem. New Mathematics and Natural Computation 5(2), 371–383 (2009),
http://www.worldscientific.com/doi/abs/10.1142/S1793005709001362

Richardson, G.: The structure of fuzzy preferences: Social choice implications. Social Choice and Welfare 15, 359–369 (1998)

# Index

Printed in the United States
By Bookmasters

Printed in the United States
By Bookmasters